LIPID BINDING PROTEINS WITHIN MOLECULAR AND CELLULAR BIOCHEMISTRY

Lipid Binding Proteins within Molecular and Cellular Biochemistry

Edited by

LEONARD BANASZAK

Department of Biochemistry
University of Minnesota
Minneapolis, MN, USA

DAVID A. BERNLOHR

Department of Biochemistry
University of Minnesota
St Paul, MN, USA

Reprinted from *Molecular and Cellular Biochemistry*, Volume 192 (1999)

Kluwer Academic Publishers
Dordrecht / Boston / London

Library of Congress Cataloging-in-Publication Data

Lipid binding proteins within molecular and celullar biochemistry /
edited by David A. Bernlohr, Leonard Banaszak
p. cm. -- (Developments in molecular and cellular
biochemistry)
Proceedings of the Third International Conference on Lipid-Binding
Proteins, held at the University of Minnesota in May 1997.
ISBN 0-7923-8223-4
1. Lipids--Physiological transport--Congresses. 2. Molecular
chaperones--Congresses. I. Bernlohr, David A. II. Banaszak
Leonard. III. International Conference on Lipid-Binding Proteins
(3rd : 1997 : International Conference on Lipid-Binding Proteins
QP751.L5455 1998
572'.57--dc21 98-28801
CIP

ISBN 0-7923-8223-4

Published by Kluwer Academic Publishers,
P.O. Box 17, 3300 AA Dordrecht, The Netherlands.

Sold and distributed in North, Central and Latin America
by Kluwer Academic Publishers,
101 Philip Drive, Norwell, MA 02061, U.S.A.

In all other countries, sold and distributed by
Kluwer Academic Publishers,
P.O. Box 322, 3300 AH Dordrecht, The Netherlands.

Printed on acid-free paper

Printed in the Netherlands

Molecular and Cellular Biochemistry:

An International Journal for Chemical Biology in Health and Disease

CONTENTS VOLUME 192, Nos. 1 & 2, February 1999

LIPID BINDING PROTEINS WITHIN MOLECULAR AND CELLULAR BIOCHEMISTRY
Drs. David A. Bernlohr and Leonard Banaszak

Molecular and Cellular Biochemistry **192:** 1, 1999.

Preface

The breadth of modern biology is characterized by a comprehension of phenomena at many levels of organization. Such levels of understanding range from the organismal to the molecular. It is when all these levels can be discussed together that a sense of true achievement begins to be felt. The tropical area of fatty acid transport and metabolism was the focus of the Third International Conference on Lipid-Binding Proteins held at the University of Minnesota in May of 1997. The articles which are contained in this issue of 'Molecular and Cellular Biochemistry' are a sampling of the proceedings from this meeting.

A brief summary of the meeting is as follows. Centralmost to lipid dynamics are the intracellular lipid binding proteins or iLBPs. They are small, approximately 130 residue, proteins that are believed to be present to buffer and/or chaperone hydrophobic compounds within the cell. In addition to the iLBPs, a number of other proteins also have an important role in the uptake and metabolism of fatty acids including membrane transporters and nuclear lipid-binding transcription factors. In the period since the Second International Workshop held at the University of Limburg in Maastrict, The Netherlands, several new genotypes of the iLBP family as well as other lipid binding proteins have been discovered. The overwhelming conclusion form the analysis of such proteins is that nature has evolved a complex set of proteins which bind a variety of hydrophobic ligands, from retinoids to fatty acids. By virtue of their binding properties, it is likely that iLBPs facilitate the signal transduction activities of such lipophilic molecules. In addition to extensive knowledge concerning the distribution of lipid-binding proteins, a variety of new molecular structures have been determined - some by multi-dimensional NMR others by X-ray crystallographic methods. Many of the structural results have been correlated with thermodynamic and kinetic studies of the binding process. In several cases, the unusual cavity nature of the lipid binding site has let to studies focusing on the binding mechanism and the molecular dynamics of segments of the binding proteins.

The workshop included talks from nearly 20 speakers and over 70 poster presentations from amongst the 130 conferees. Adding to the conference was the general cordiality and social environment which was particularly conductive to having nearly everyone participate in the scientific discussions. The city of Minneapolis and the University of Minnesota provided an excellent venue to partake of both the scientific and cultural richness of the region. All taken into account, there was a great deal of new and exciting results - enough so that we look forward to a Fourth International Workshop on Lipid Binding Proteins!

Leonard Banaszak, Twin Cities, Minnesota, USA
David Bernlohr, Twin Cities, Minnesota, USA

Molecular and Cellular Biochemistry **192:** 3–8, 1999.

CD36 antisense expression in 3T3-F442A preadipocytes

Zeina Sfeir,[1] Azeddine Ibrahimi,[1] Ez-zoubir Amri,[2] Paul Grimaldi[2] and Nada Abumrad[1]

[1]*Department of Physiology and Biophysics, State University of New York at Stony Brook, NY, USA;* [2]*Department of Biochemistry, University of Nice, France*

Abstract

An adipocyte membrane glycoprotein, FAT, homologous to CD36, has been implicated in the binding/transport of long-chain fatty acids. FAT/CD36 was identified by reaction with reactive long chain fatty acids derivatives under conditions where they inhibited FA uptake. Expression of CD36 in fibroblasts lacking the protein led to induction of a saturable high affinity, phloretin-sensitive component of oleate uptake. In this report, we have examined the effects of FAT/CD36 antisense expression in 3T3-F442A preadipocyte cells, on FA uptake and cell differentiation. Cells were transfected with pSG5-TAF vector obtained by insertion of antisense coding sequence of FAT/CD36 into the BamH 1 site of pSG5. Four clones were selected based on expression of antisense CD36 mRNA. Levels of CD36 protein were determined by flow cytometry and correlated with rates of oleate uptake. Three clones, TAF13, TAF25, and TAF38 exhibited low CD36 expression and one clone TAF 18 had expression comparable to that of F442A control cells. FA uptake rates in clones TAF13, TAF25 and TAF3 8 were lower than those observed in TAF18. At confluence, adipocyte differentiation could be promoted by addition of insulin and triiodothyronine only in TAF18 cells but not in TAF13, TAF25 or TAF38. Addition of fatty acids to clones TAF13, TAF25 and TAF38 lead to an induction of CD36 expression, an enhancement of FA uptake and better cell differentiation. The data support a role of CD36 in the membrane uptake of long chain FA. CD36 expression and FA uptake appear to be closely linked to preadipocyte differentiation. (Mol Cell Biochem **192**: 3–8, 1999)

Key words: fatty acid transport, CD36-antisense RNA, transfection, adipocyte differentiation

Abbreviations: FA – fatty acids; GAPDH – glyceraldehyde phosphate dehydrogenase; BSA – bovine serum albumin; ALBP – Adipose Lipid Binding Protein; LPL – Lipoprotein lipase

Introduction

Kinetic characterization of membrane permeation of long-chain fatty acid (FA) by rat adipocytes identified a saturable component and suggested the involvement of a high affinity carrier system [1–3]. Later studies, using reactive sulfo-N-succinimidyl derivatives of oleate (SSO), implicated a membrane protein with an apparent molecular weight of 88 kDa in the uptake process [4, 5]. The protein was specifically labeled under conditions where FA uptake into the cell was inhibited by about 75%. The isolated protein (FAT) had an amino terminal sequence similar to that of human glycoprotein IV or CD36 [6, 7]. The complete protein sequence, deduced from the cDNA, was 85% similar to that of CD36 [8].

Multiple lines of evidence were consistent with a role for the protein in FA uptake. Purified CD36, *in vitro*, bound FA at nM concentrations and low ligand to protein ratio [9]. Distribution of FAT/CD36 mRNA favored tissues with a high metabolic capacity for FA [8, 10]. It was high in the heart and in muscle tissues with a predominance of red oxidative fibers and was upregulated during heart development when FA utilization increases [10]. In culture, CD36 mRNA was a

Address for offprints: N.A. Abumrad, Department of Physiology and Biophysics, State University of New York at Stony Brook, New York 11794-8661, USA

strong marker of preadipocyte differentiation and was modulated by the same factors effective on mRNAs coding for other proteins involved in FA metabolism (FA, dexamethasone, cAMP) [11].

Expression of CD36 in Ob17PY fibroblasts lacking the protein increased both binding and membrane permeation of FA by these cells [12] by addition of a saturable, phloretin-sensitive system to the unsaturable phloretin-insensitive component already present in the cell [12]. The Km of the saturable component was in the low nanomolar and within the range of physiologic concentrations of unbound FA (FA:albumin ratios between 0.2 and 1.0) [12].

In this report, we have examined the effect of CD36 antisense mRNA expression in 3T3-442A preadipocytes on FA uptake and on differentiation into adipocytes.

Materials and methods

Materials

Vectors were from Stratagene (pSG5) and Clontech (pMAM neo). The [α-^{32}P]dCTP was from ICN. The nylon membranes (Hybond N+) were from Amersham. Enzymes for RNA and DNA manipulation, the random-primed DNA labeling kit and the transfection reagent (DOTAP) were from Boehringer Manheim. RNA STAT-60 for preparation of total RNA was from Tel-Test Inc. Cell culture media and Geneticin were from Gibco. Fetal calf serum was from Atlanta biologicals. All labeled fatty acids were from Du Pont-New England Nuclear. Fluorescein- conjugated goat IgG fraction to rabbit immunoglobulin(lgG, lgA, IgM) was obtained from Cappel. Kodac films and all others chemical products were obtained from Sigma.

Cell culture and treatment

3T3-F442A preadipocytes were plated in 35-mm dishes at a density of $2 \times 10^3/cm^2$ and were maintained in Dulbecco's modified Eagle medium (DMEM) supplemented with Hepes (15 mM) biotin (33 μM), pantothenate (17 μM), penicillin (200 U/ml), streptomycin (50 μg/ml) and 8% fetal bovine serum. To promote differentiation, insulin (17 nM) and tdiodothyronine (2 nM) were added at confluence. For cell transfection, an expression vector was constructed by inserting the coding region of CD36 cDNA in the antisense orientation in the BamH1 site of the pSG5 expression vector. Because of the specificity of the nucleotide sequence of this fragment, expression of the antisense RNA would be expected to interfere selectively with transcription of the CD36 gene and with the translation or stability of its mRNA. Clones were isolated based on expression of antisense CD36 mRNA and on the levels of CD36 protein as determined by flow cytometry. Where indicated, transfected cells were treated with bromopalmitate one day before confluence, day (A), for 60 h.

FA transport assays

Before assays, all dishes were washed and incubated at 37°C with KRH buffer containing 0.5% BSA for 20 min. They were then washed with buffer minus BSA.Transport buffer (0.7 ml) with 2.1×10^6 cpm ^{3}H-oleate (80 μM complexed to 80 μM BSA) was added, to washed cells followed by immediate swirling of the dish. At the desired times, ice-cold KRH (3.5 ml) was added and the medium was swirled and aspirated. The cells were washed twice with cold KRH and lysed in 1 ml of 0.1 N NaOH. Aliquots of the lysate were taken for scintillation counting. Uptake of FA is expressed per 10^6 cells. Cell count was determined using a hemocytometer following recovery of cells in suspension by trypsin-EDTA treatment.

RNA analysis

RNA was prepared as described by Chomezynski and Sacchi [13] or using the RNA STAT-60 kit [8]. RNA was electrophoresed on denaturing agarose gel, transferred to hybond-N+ membranes. RNA was hybridized with approximatively 10^6 cpm/ml of randomly primed ^{32}P-labeled DNA probes as previously described [8]. After washing, the membranes were exposed to Hyperfilm at –75°C. mRNA for Glyceraldehyde-3-phosphate dehydrogenase (GAPDH) was used as an internal standard.

Determination of CD36 protein expression

Transfected 3T3 -F442A cells (from 2, 35 mm dishes) were treated with trypsin/EDTA, centrifuged and incubated with a polyclonal anti-CD36 antibody (1:100 dilution) in PBS with 1% BSA and 0 .01% sodium azide buffer. After washing 2 times with 0.5 ml PBS with 1% BSA buffer, cells were incubated with fluorescein-conjugated secondary antibody (1:75 dilution), washed 3 times with 0.5 ml then fixed with 1% formaldehyde in PBS, and resuspended in 100 μl of PBS. Labeling was performed at 4°C. As a control, cells were incubated with secondary antibody only. Cells were analyzed using an EPICS Profile II Flow Cytometer (Coulter Corp.) equipped with an air-cooled 488-nm argon-ion laser and 635-mn band-pass filter in the emission path allowing approximately 90% transmission of light in the 615–655-nm range. The excitation maximum for these cells is at 522-nm. The emission maximum is at 628-mn. Analysis involved measuring forward-angle light scatter and log FITC fluorescence (FL1).

Results

Transfection of 3T3-F442A cells with expression vectors encoding CD36-antisense RNA

The antisense CD36 expression vector, TAF, was co-transfected into 3T3-F442A predipocytes with pMAMneo vector for initial clones selection based on neomycin-resistance. Four clonal cell lines (TAF13, TAF18, TAF25 and TAF38) harboring the TAF vector were selected for further analysis based on expression of the antisense CD36 RNA.

As shown in Fig. 1, the antisense CD36 mRNA expressed in the four lines (TAF13, TAF18, TAF25 and TAF38) had a size of ~1.2 kb, as predicted from the inserted sequence and could be easily distinguished from the endogenous CD36 mRNA (~2.8 kb). The highest level of antisense CD36 expression was observed in TAF38 and TAF13 cells followed by TAF25 and TAF18. Expression of CD36 antisense resulted in reductions of the cellular levels of the encoded protein (Fig. 2). Flow cytometry of cells revealed that three of the clones expressing the antisense RNA had reduced levels of CD36 protein. TAF13, TAF25 and TAF38 cells which expressed high levels of antisense CD36 RNA, exhibited a big suppression of CD36 protein. In contrast TAF 18 cells exhibited CD36 protein expression that was comparable to that of F442A controls despite good expression of antisense mRNA.

Oleate transport (Fig. 3) measured at 23°C was significantly lower in clones TAF13, TAF25 and TAF38 as compared to TAF18 based on assays conducted with six separate cell series from each clone. Similar data were also obtained when ^{3}H-palmitate was used instead of oleate (data not shown).

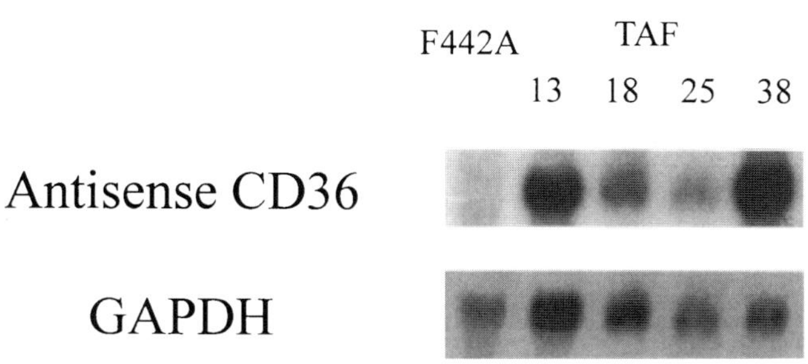

Fig. 1. Expression of CD36 antisense in F442A. 3T3-F442A cells were co-transfected with pMAMneo and pSG5 driving CD36 antisense sequence. CD36 antisense RNA expression was analyzed by northern blot (20 μg/lane) in 3T3-F442A cells and in four clones transfected with the CD36 antisense (TAF13, TAF18, TAF25 and TAF38). mRNA for GAPDH is shown as an internal control.

In contrast to FA transport, transport of 2-deoxyglucose was not significantly altered in CD36 antisense-expressing clones (not shown). This indicated that the decrease in FA transport observed in clones TAF13, TAF25 and TAF38 was specific.

Effect of FA treatment

To investigate whether FA can rescue CD36 expression in antisense-expressing clones, we conducted studies with TAF38 cells. As demonstrated in Fig. 4, treatment of cells at confluence for 60 h with bromopalmitate, a non metabolizable FA, moderately induced CD36 protein expression in TAF38 cells as measured by flow cytometry. This correlated with an increase in FA uptake. Oleate transport, measured at 23°C was higher in TAF38 cells treated with FA. Parallel changes in CD36 protein expression and in oleate uptake following FA treatment were observed in TAF13 and TAF25 cells (data not shown).

Effect of antisense CD36 RNA expression on adipocyte differentiation

When different TAF clones and the 3T3-F442A control cells were treated with insulin and triiodothyronine to promote differentiation, control cells and TAF18 cells differentiated (about 80% of cells) into mature adipocytes with multiple cytoplasmic lipid droplets. In contrast, TAF13, TAF25 and TAF38 cells showed minimal to undetectable differentiation. Figure 5 compares mRNAs levels for two differentiation marker genes in F442A and the TAF clones. At 10 days following confluence, TAF13, TAF25 and TAF38 cells harboring the antisense CD36 vector exhibited drastically reduced levels of ALBP [Adipose lipid binding protein] and LPL [Lipoprotein lipase] mRNAs compared with control F442A cells. On the other hand, TAF18 cells which exhibited levels of CD36 protein comparable to F442A control cells, also had comparable level of ALBP and LPL mRNAs.

Discussion

CD36 antisense RNA expression in F442A preadipocytes was associated with a decrease in the uptake of long-chain fatty acids and the magnitude of the decrease generally correlated with the level of protein expression.

These data lend further support to the role of CD36 in FA uptake and would imply that CD36 deficiency or defects in its expression should lead to impaired FA uptake. Some recent data in humans would suggest that this may be the case.

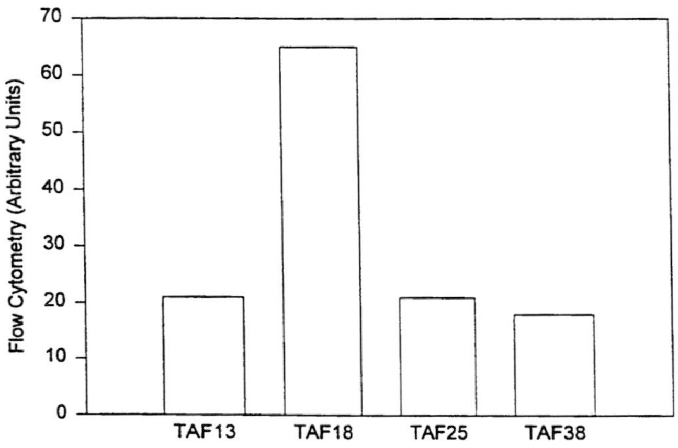

Fig. 2. Expression of CD36 protein in clones expressing CD36 antisense. Cells were incubated at 4°C with fluorescein conjugated secondary antibody (1:75 dilution) in PBS + 1% BSA buffer, fixed with 1% formaldehyde in PBS, and resuspended in 100 μl of PBS. As a control, cells were incubated with secondary antibody only. Cells were analyzed by flow cytometry within 24 h. Analysis involved measuring forward-angle light scatter and log FITC fluorescence (FL1).

Platelet CD36 deficiency is present in about 3% of the Japanese population and in 0.3 % of the US population [14] and can be divided into type I or type II according to wether CD36 is (type II) or is not (type I) expressed on monocytes [14]. Type I subjects have different CD36 gene abnormalities (deletion, frame-shift) resulting in the absence of CD36 protein [14–16].

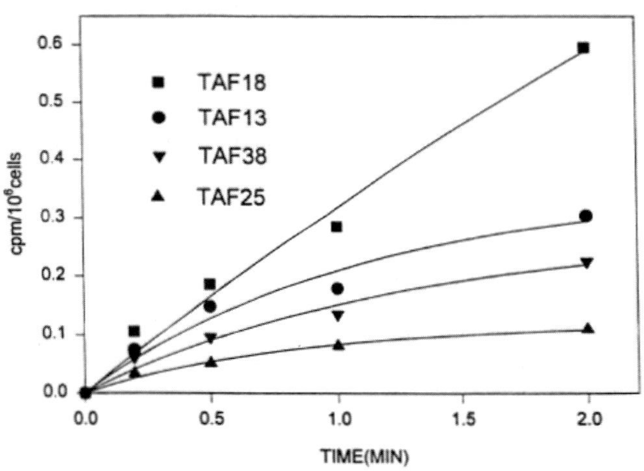

Fig. 3. Correlation between CD36 expression and oleate uptake. Transport of oleate (80 μM FA, 3000 cpm/μl) complexed to BSA at a 1:1 ratio, was determined at 23°C. Cells were equilibrated for 5 min at room temperature and were washed with KRH before addition of the isotopic solution. Data shown for each clone are typical of at least six experiments. Similar results were obtained with palmitic acid.

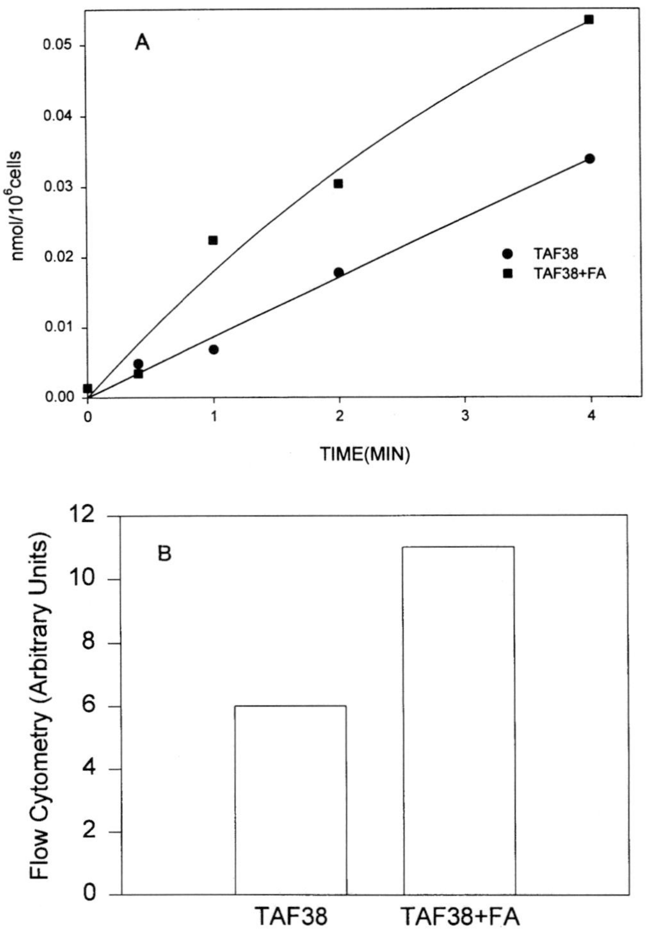

Fig. 4. Effect of FA treatment in TAF38 cells. CD36 flow cytometry (Panel A) and oleate transport (Panel B) in clone TAF38 without and with treatment with bromopalmitate (100 μM) were determined as detailed in the text and under the legends to Figs 2 and 3.

Tanaka *et al.* [17, 18] studied CD36 expression in 55 patients with hypertrophic cardiomyopathy (HCM). In 23 out of the 55 HCM patients (41.8%) they measured negligible, reduced, or mutated CD36 expression. In 95% of these patients, scintigraphy experiments showed impaired long chain FA accumulation in the heart. In contrast, patients with normal CD36 expression generally (14 out of 20) showed no severe impairment of myocardial long chain FA uptake [17–19]. CD36 has been proposed to function as a receptor for oxidized lipoproteins (ox-LDL) [20]. We had previously speculated that CD36 might recognize FA oxidation products on apo-proteins. It would be expected that CD36 deficiency could also result in alterations in ox-LDL binding and possibly uptake. Although, this could not be determined with TAF cells in this study, other data with macrophages would support such an interpretation. Monocytes from subjects with CD36 deficiency exhibited 40% reduced ox-LDL binding [21]. Down regulation of CD36 protein by cAMP was associated with reduction of

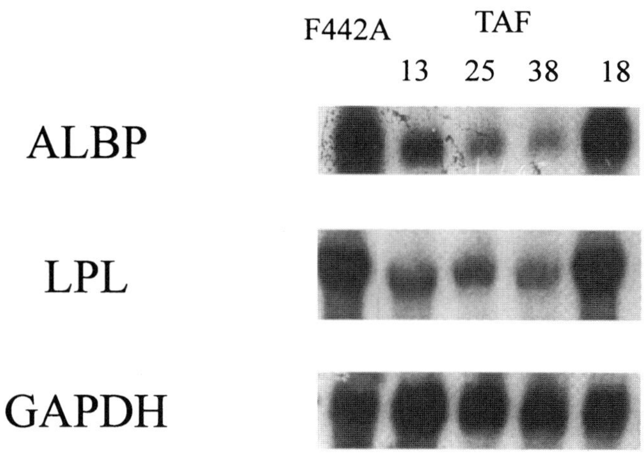

Fig. 5. Effect of antisense CD36 RNA expression on expression of various adipose gene markers. At confluence, control 3T3-F442A cells and transfected clones were treated with insulin and tdiodothyronine to promote differentiation. Ten days after confluence, RNA was prepared and analyzed for ALBP and LPL mRNAs. mRNA for GAPDH is shown as an internal control.

ox-LDL uptake by human macrophages in culture (W.L. Chang and N.A. Aburnrad, unpublished results).

In addition to FAT/CD36, several proteins have been identified as potential membrane FA receptors and/or transporters. These proteins differ in structure and in their tissue distribution. $FABP_{pm}$ (43 kDa) or membrane-bound aspartate aminotransferase, isolated by oleate-agarose chromatography of solubilized plasma membranes, is thought to adhere to the membrane by a specific N-terminal peptide [22]. It is ubiquitously distributed but only associated with the membrane in specific tissues such as liver [22]. FATP (63 kDa), isolated by expression cloning using fluorescent FA, was shown to have six predicted membrane-spanning domains [23]. In contrast to FAT/CD36, FATP expression was detected in all tissues and cells studied with a relatively high expression in the brain. In a recent study Berk *et al.* examined regulation of the three proteins FAT/CD36, FATP and $FABP_{pm}$, [24] in fatty and diabetic rats. Expression of these proteins was comparable in fatty rats and correlated with FA uptake. However, in diabetic rats differences in expression were observed and possibly reflected differential regulation of these proteins by metabolic factors. FATP mRNA is down regulated by insulin [25] while FAT/CD36 mRNA is not. In contrast, FAT/CD36 is closely regulated by FA, PPARs and C/EBP.

Acknowledgments

This work was supported by NIH grant DK33301 (to N.A.A.) and by grant CRE 94G0177 from Le Ministere de L'Enseignement Superieur et de la Recherche (to P.A.G.)

References

1. Aburnrad NA, Perkins RC, Park M, Park CR: Mechanism of long chain fatty acid permeation in the isolated adipocyte. J Biol Chem 256: 9183–9191, 1981
2. Abumrad NA, Park JH, Park CR: Permeation of long chain fatty acids into adipocytes. J Biol Chem 259: 8945–8953, 1984
3. Schwieterman W, Sorrentino D, Potter BJ, Rand J, Kiang CL, Stump D, Berk PD: Uptake oleate by isolated rat adipocytes is mediated by a 40-kDa plasma membrane fatty acid binding protein closely related to that in liver and gut. Proc Natl Acad Sci USA 85: 359–363, 1988
4. Harmon CM, Luce P, Beth A, Abumrad NA: Labeling of adipocyte membranes by, Sulfo-N-Succinixndyl derivatives of long chain fatty acids: Inhibition of fatty acid transport. J Mem Biol 124: 261–268, 1991
5. Harmon CM, Aburnrad NA: Binding of sulfosuccinimidyl fatty acids to adipocyte membrane proteins: Isolation and amino-terminal sequence of an 88 kDa protein implicated in transport of fatty acids. J Mem Biol 131: 431–449, 1993
6. Oquendo P, Hundt E, Lawler J, Seed B: CD36 directly mediates cytoadherence of Plasmodium falcipamm parasitized erythrocytes. Cell 58(1): 95–101, 1989
7. Greenwalt DE, Lipsky RH, Ockenhouse CF, Ikeda H, Tandon NN, Jamieson GA: Membrane glycoprotein CD36: A review of its role in adherence, signal transduction, and transfusion medicine. Blood 80: 1105–1115, 1992
8. Aburnrad NA, El Maghrabi R, Amri EZ, Lopez E, Grimaldi P: Cloning of a rat adipocyte membrane protein implicated in binding or transport of long-chain fatty acids that is induced during adipocyte differentiation. Homology with human CD36. J Biol Chem 268: 17665–17668, 1993
9. Baillic AGS, Cobum CT, Aburnrad NA: Reversible binding of long-chain fatty acids to purified FAT, the adipose CD356 homolog. J Mem Biol 153: 75–81, 1996
10. Van Nieuwenhoven FA, Verstijnen CPW, Aburnrad NA, Willemsen PHM, Van Eys GJJM, Van Der Vusse GJ, Glatz MC: Putative membrane fatty acid translocase and cytoplasmic fatty acid binding protein are co-expressed in rat heail and skeletal muscles. Biochem Biophys Res Commun 207: 747–752, 1995
11. Sfeir Z, Ibrahimi A, Amri E, Grimaldi P, Abumrad NA: Regulation of FAT/CD36 gene expression: Further evidence in support of a role of the protein in fatty acid binding/transport. Prostaglandins, Leukotrienes and essential Fatty Acids 57: 17–21, 1997
12. Ibrahimi A, Sfeir Z, Magharaic H, Amri EZ, Grimaldi P, Abumrad NA: Expression of the CD36 homolog (FAT) in fibroblast cells: Effects on fatty acid transport. Proc Natl Acad Sci USA 93: 2646–2651, 1996
13. Chomezynski P, Sacchi N: Single-step method of RNA isolation by acid guanidinium thiocyanate-phenol-chloroform extraction. Anal Biochem 162(1): 156–159, 1987
14. Kashiwagi H, Tomiyama Y, Kosugi S, Shiraga M, Lipsky RH, Kanayarna Y, Kurata Y, Matsuzawa Y: Identification of molecular defects in a subject with type I CD36 deficiency. Blood 83(12): 3545–3552, 1994
15. Kashiwagi H, Tomlyama Y, Nozaki S, Honda S, Kosugi S, Shiraga M, Nak-agawa T, Nagao N, Kanakura Y, Kurata Y, Matsuzawa Y: A single nucleotide insertion in codon 317 of the CD36 gene leads to CD36 deficiency. Arterioscler Throm Vasc Biol 16(8): 1026–1032, 1996
16. Kashiwagi H, Tomiyama Y, Honda S, Kosuel S, Shiraga M, Nagao N, Sekiguchi S, Kanayarna Y, Kurata Y, Matsuzawa Y: Molecular basis of CD36 deficiency. Evidence that a 47SC--->T substitution (proline90-->serine) in CD36 cDNA accounts for CD36 deficiency. J Clin Inv 95(3): 1040–1046, 1995

17. Tanaka T, Sohmiya K, Kawamura K: Is CD36 deficiency an etiology of hereditary hypertrophic cardiomyopathy? J Mol Cell Cardiol 29: 121–127, 1997
18. Tanaka T, Okamoto F, Sohmiya K, Kawamura K: Lack of myocardial iodine-123 15-(p-iodiphenyl)-3-R,S-methylpentadecanoic acid (BMIPP) uptake and CD36 abnormality--CD36 deficiency and hypertrophic cardiomyopathy. Japanese Circ J 61(8): 724–725, 1997
19. Inoue F, Hashimoto T, Nishida Y, Dohi K, Matsushima A, Sakakibara H, Ishida Y: Absence of myocardial 123I-BMIPP uptake in the presence of a normal coronary angiogram and normokinetics on a left ventriculogram. Japanese Circ J 61(3): 263–267, 1997
20. Endemann G, Stanton LW, Madden KS, Bryant CM, White RT, Protter AA: CD36 is receptor for oxidized low density lipoprotein. J Biol Chem 268: 11811–11816, 1993
21. Nozaki S, Kashiwagi H, Yamashita S, Nakagawa T, Kostner B, Tomiyama Y, Nakata A, Ishigarni M, Miyagawa J, Kameda-Takemura K, Kurata Y, Matsuzawa Y: Reduced uptake of oxidized low density lipoproteins in monocyte-derived macrophages from CD36-deficient subjects. J Clin Inv 96(4): 1859–1865, 1995
22. Stump DD, Zliou SL, Berk PD: Comparison of plasma membrane FABP and mitochondrial isoform of aspartate aminotransferase from rat liver. Am J Physiol 265(5 Pt 1): G894–G902, 1993
23. Shaffer JE, Lodish HF: Expression cloning and characterization of a novel adipocyte long chain fatty acid transport protein. Cell 79: 427–436, 1994
24. Berk PD, Zhou SL, Kiang CL, Stump D, Bradbury M, Isola LM: Uptake of long-chain fatty acid is selectively up-regulated in adipocytes of Zucker rats with genetic adiposity and noninsulin-dependent diabetes mellitus. J Biol Chem 2772: 8830–8835, 1997
25. Man MZ, Hui TY, Schaffer JE, Lodish HF, Bernlohr DA: Regulation of the murine adipocyte fatty acid transporter gene by insulin. Mol Endocrinol 10(8): 1021–1028, 1996

Molecular and Cellular Biochemistry **192**: 9–16, 1999.

The liver fatty acid binding protein - comparison of cavity properties of intracellular lipid-binding proteins

James Thompson, Jeramia Ory, Amy Reese-Wagoner and Leonard Banaszak

Department of Biochemistry, University of Minnesota, Minneapolis, MN, USA

Abstract

The crystal and solution structures of all of the intracellular lipid binding proteins (iLBPs) reveal a common β-barrel framework with only small local perturbations. All existing evidence points to the binding cavity and a poorly delimited 'portal' region as defining the function of each family member. The importance of local structure within the cavity appears to be its influence on binding affinity and specificity for the lipid. The portal region appears to be involved in the regulation of ligand exchange. Within the iLBP family, liver fatty acid binding protein or LFABP, has the unique property of binding two fatty acids within its internalized binding cavity rather than the commonly observed stoichiometry of one. Furthermore, LFABP will bind hydrophobic molecules larger than the ligands which will associate with other iLBPs. The crystal structure of LFABP contains two bound oleate molecules and provides the explanation for its unusual stoichiometry. One of the bound fatty acids is completely internalized and has its carboxylate interacting with an arginine and two serines. The second oleate represents an entirely new binding mode with the carboxylate on the surface of LFABP. The two oleates also interact with each other. Because of this interaction and its inner location, it appears the first oleate must be present before the second more external molecule is bound. (Mol Cell Biochem **192**: 9–16, 1999)

Key words: lipid binding proteins, fatty acid binding protein, intracellular lipids and proteins

The β-barrel of intracellular lipid binding proteins

The structural similarities and differences between the intracellular lipid binding proteins (iLBPs) have been documented through numerous x-ray and NMR studies of family members. Citations and summary data are given in Table 1 and we are grateful for the coordinates which have been used in preparing the discussion which follows. Members of the iLBP family bind primarily either fatty acids in liver (LFABP)[1], myelin (P2)[2], adipocytes (ALBP)[3], intestine (IFABP)[4], heart muscle (HFABP)[5], locust muscle (L-MFABP)[6], and hornworm midgut (MFB2)[7], or retinol in liver (CRBPI)[2] and intestine (CRBPII)[8], or retinoic acid in testis (CRABPI)[9] and skin (CRABPII)[9]. Though the iLBPs are generally named after the tissue in which they were first found, their gene expression patterns vary in differing cell types involved with lipid metabolism[10]. Some cells also co-express more than one of these proteins found in Table 1.

Although the identity of the primary structure throughout the iLBP family varies from 20–70% (see Fig. 1a), the fundamental tertiary motif is a slightly flattened β-barrel surrounding an interior cavity which is the lipid binding site[11]. Comparison of the crystal coordinates of all family members fitted to the LFABP model reveals the same basic skeleton. The ten strands comprising the β-barrel are found in an antiparallel arrangement, with the intrastrand hydrogen bonding pattern shown in Fig. lb. The ten strands are labeled βA through βJ beginning from the N-terminal. A bulge

Address for offprints: L. Banaszak, Department of Biochemistry, University of Minnesota, Minneapolis, MN 55455, USA

Table 1. LFABP and iLBP Comparison

	PDB code	Cavity (Å^2)	(Å^3)	rmsd (Å)	Ref.
Liver Fatty Acid Binding Protein (LFABP)	1lfo	610	440	0.0	1
Locust Muscle Fatty Acid Binding Protein (L-MFABP)	1ftp	510	350	1.49	6
Myelin P2 Lipid Binding Protein (P2)	1pmp	420	330	1.34	2
Manduca sexta Fatty Acid Binding Protein (MFB2) ¥	1mdc	388	324	1.89	7
Heart Muscle Fatty Acid Binding Protein (HFABP) ¥	1hmr	391	323	1.46	5
Adipocyte Lipid Binding Protein (ALBP)	1adl	450	310	1.45	3
Cellular Retinol Binding Protein II (CRBPII)¥	1opb	372	261	1.46	8
Cellular Retinoic Acid Binding Protein II (CRABPII)	1cbq	470	260	1.62	9
Cellular Retinoic Acid Binding Protein I (CRABPI)	1cbr	410	210	1.67	9
Cellular Retinol Binding Protein 1 (CRBPI)¥	1crb	334	235	1.70	2
Intestinal Fatty Acid Binding Protein (IFABP)¥	1ifc	353	234	1.81	4

Rmsd's were calculated by comparison with the crystallographic coordinates of LFABP. The assigned accession code from the Brookhaven Protein Databank is indicated in the second column of the table. Also a reference is provided for each protein. These structures are commented on throughout this work. The symbol, ¥, indicates no artificial cutoff was needed to delimit the binding cavity boundary as it is entirely enclosed by protein atoms (see text). Thus, the measurements are more precise than about the ± 30 Å^2 or ± 20 Å^3 found for cavities partially open to the exterior. These are solvent accessible volumes using a 1.4 Å probe.

present in strand βA allows interactions with both βB and βJ. Some variation exists in the interstrand hydrogen bonds. Generally there are 7–9 such bridges except for the pairs βA-βB, βA-βJ and βD-βE. For the latter strand pair, the separation is too large to produce meaningful mainchain interactions and in the overall structure βD-βE is referred to as the 'gap'. In the static structures, the gap is filled with sidechains and ordered water.

While the general fold in Fig. 1b is that of all iLBPs, the hydrogen bonding pattern of LFABP that is shown differs in the region from βF to βG. At least four interstrand interactions are missing from between these two β-strands. Furthermore, the neighboring β-turn joining βG and βH is two residues shorter than the average for the iLBP family. Therefore, the β-barrel framework can have two gaps but presently, the second gap is unique to LFABP.

The striking number of interchain interactions may have some interesting implications. In the β-barrel framework, each strand interacts with its two neighbors except as noted above. Therefore, the formation of the β-barrel involves the formation of 10–17 hydrogen bonds per strand. As is visible in Fig. 1b, this checkerboard of interactions gives the appearance of an extremely stable supersecondary structure. One assumption is that the hydrogen bonding framework found in the mainchain is cooperative. That is, breaking one or more of the hydrogen bonds will weaken the barrel. This being the case, the network of buried interstrand hydrogen bonds are probably maintained during any conformational change. The conformational degrees of freedom would then be either limited to cooperative changes in the angles and distances of these mainchain hydrogen bonds or largely confined to residues lacking these restraining interactions.

Other than the mentioned gaps, the iLBP conformation does have other elements which are not part of the hydrogen bonding network of the β-barrel. A good example is the connection between βA and βB that covers one end of the barrel. The connection, also visible in Fig. 1b, consists of a helix-turn-helix usually containing about 20 amino acids. The two α-helices are linked to the main framework by the βA-αI and αII-βB polypeptide segments. Because of their limited interaction with the β-barrel framework, the two α-helices could be considered as rigid segments distinct from the remaining barrel. Some conformational variation also is likely on the ends of the tight β-turns linking the strands. At the tips of the turns connecting the β-strands, interstrand hydrogen bonds are missing and conformational variation in the mainchain torsional angles ϕ and φ seem feasible.

Cavity and portal properties of the liver fatty acid binding protein

The ligand specificity of LFABP in terms of size and lipid-type is broad and its binding stoichiometry is unique to the family. While LFABP binds fatty acids with high affinity, it can also bind heme, bilirubin, lysophospholipids, bile salts, acyl-CoA esters, bromosulphothalein, hexachlorophene, cyclopentenone prostaglandins, and a host of other hydrophobic compounds (For a review, see [12]). Moreover, one mole of LFABP is capable of binding two moles of long-chain fatty acids [13]. For the larger ligands, such as heme, bile salts, and lysophospholipid, the ratio is one to one [12, 14]. Although no direct structural information is available, competition binding assays show that both fatty acid sites are required to bind the larger ligands such as heme [15].

A

```
CRABPI     1  ----PNFAGTWKMRSSENFDELLKALGVNAMLRKVAVAAASKPHVEIRQD
CRABPII    1  ----PNFSGNWKIIRSENFEELLKVLGVNVMLRKIAVAAASKPAVEIKQE
CRBPI      1  ---PVDFTGYWKMLVNENFEEYLRALDVNVALRKIA--NLLKPDKEIVQD
CRBPII     1  ---TRDQNGTWEMESNENFEGYMKALDIDFATPKIA--VRLTQTKVIDQD
P2         1  ---SNKFLGTWKLVSSENFDDYMKALGVGLATRKLG--NLAKPTVIISKK
ALBP       1  ---CDAFVGTWKLVSSENFDDYMKEVGVGFATRKVA--GMAKPNMIISVN
MFABP      1  ---VDAFLGTWKLVDSKNFDDYMKSLGVGFATRQVA--SMTKPTTIIEKN
KLBP       1  MATVQQLEGRWRLVDSKGFDEYMKELGVGIALRKMG--AMAKPDCIITCD
IFABP      1  -----AFDSTWKVDRSENYDKFMEKMGVNIVKRKLA--AHDNLKLTITQE
LFABP      1  ----MSFSGKYQLQSQENFEAFMKAIGLPEELIQKG--KDIKGVSEIVQN
ILBP       1  -----AFTGKFEMESEKNYDEFMKLLGISSDVIEKA--RNFKIVTEVQQD
                     βA            α1          α2          βB

CRABPI    47  GDQFYIKTSTTVRTTEINFKVGEGFEEET--VDGRKCRSLATWENENKIH
CRABPII   47  GDTFYIKTSTTVRTTEINFKVGEEFEEQT--VDGRPCKSLVKWESENKMV
CRBPI     46  GDHMIIRTLSTFRNYIMDFQVGKEFEEDLTGIDDRKCMTTVSWDG-DKLQ
CRBPII    46  GDNFKTKTTSTFRNYDVDFTVGVEFDEYTKSLDNRHVKALVTWEG-DVLV
P2        46  GDIITIRTESTFKNTEISFKLGQEFEETT--ADNRKTKSIVTLQR-GSLN
ALBP      46  GDVITIKSESTFKNTEISFILGQEFDEVT--ADDRKVKSTITLDG-GVLV
MFABP     46  GDILTLKTHSTFKNTEISFKLGVEFDETT--ADDRKVKSIVTLDG-GKLV
KLBP      49  GKNLTIKTESTLKTTQFSCTLGEKFEETT--ADGRKTQTVCNFTD-GALV
IFABP     44  GNKFTVKESSAFRNIEVVFELGVTFNYNL--ADGTELRGTWSLEG-NKLI
LFABP     45  GKHFKFTITAGSKVIQNEFTVGEECELET--MTGEKVKTVVQLEGDNKLV
ILBP      44  GQDFTWSQHYSGGHTMTNKFTVGKESNIQT-MGGKTFKATVQMEG-GKLV
                βC        βD       βE             βF

                              *                   * *
CRABPI    95  CTQTLLEGDGPKTYWTRELAN-DELILTFGADDVVCTRIYVRE
CRABPII   95  CEQKLLKGEGPKTSWTRELTNDGELILTMTADDVVCTRVYVRE
CRBPI     95  CVQKGE---KEGRGWTQWIEG-DELHLEMRVEGVVCKQVFKKVQ
CRBPII    95  CVQKGE---KENRGWKQWIEG-DKLYLELTCGDQVCRQVFKKK
P2        93  QVQRWD---GKETTIKRKLVN-GKMVAECKMKGVVCTRIYEKV
ALBP      93  HVQKWD---GKSTTIKRKRED-DKLVVECVMKGVTSTRVYERA
MFABP     93  HLQKWD---GQETTLVRELID-GKLILTLTHGTAVCTRTYEKEA
KLBP      96  QHQEWD---GKESTITRKLKD-GKLVVECVMNNVTCTRIYEKVE
IFABP     91  GKFKRT-DNGNELNTVREIIG-DELVQTYVYEGVEAKRIFKKD
LFABP     93  TTF---K-N---IKSVTELNG-DIITNTMTLGDIVFKRISKRI
ILBP      92  VNF---P-N---YHQTSEIVG-DKLVEVSTIGGVTYERVSKRLA
              βG          βH        βI       βJ
```

B

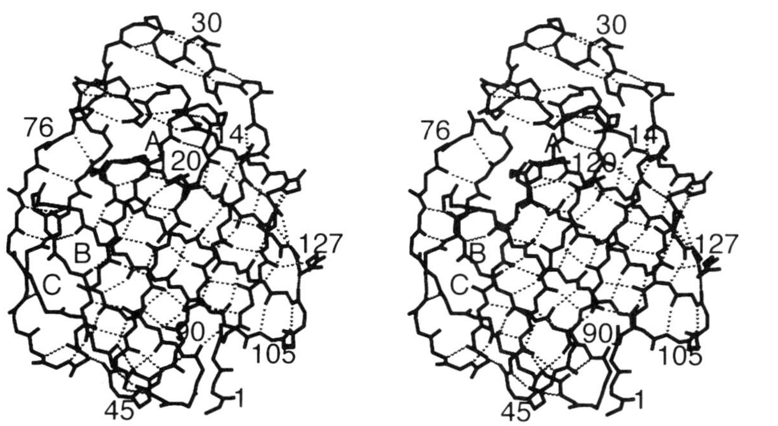

Fig. 1. The iLBP family. (A) The human amino acid sequences of several members of the iLBP family are aligned. The alignment is based on their respective crystal structures or models created from similar iLBP structures of other species. The corresponding secondary structure is shown. The symbol '*' indicates important cavity residues that partly define binding specificity and affinity in the family. KLBP or keratinocyte lipid binding protein is equivalent to epidermal fatty acid binding protein (EFABP). (B) The stereodiagram shows the β-barrel framework with mainchain hydrogen bonds as observed in the crystal structure of LFABP. The hydrogens bonds are represented by the dotted connections and the stick model contains only mainchain atoms. The letter 'A' is placed in the stereodiagram in the approximate region of the portal. The letter 'B' is the gap found in all of the iLBP family members between βD and βE. 'C' marks a smaller gap present only in LFABP.

In terms of the two fatty acid sites found in LFABP, the two ligands have nearly the same affinity. For example, in the dissociation of oleate from rat and bovine LFABP, one site has a Kd ranging from 0.009–0.2 μM and the other, a Kd of from 0.06–4.0 μM [16–18]. The weaker binding data were determined by titration calorimetry while the smaller Kd's were measured with a fluorescent reagent, itself a fatty acid binding protein [18].

Of the long list of ligands given above, only oleate binding to LFABP has been studied by x-ray crystallography[1]. The result of this study provided some structural insight into the broader specificity and the unusual stoichiometry observed with hydrophobic ligands binding to LFABP. Using the crystallographic coordinates, the solvent accessible volume of the ligand binding cavity in LFABP is estimated to be 610 $Å^3$, as compared to 450 $Å^3$ in ALBP and 353 $Å^3$ in IFABP. In fact, the LFABP cavity is the largest of 10 family members whose structures are known; note the listing in Table 1. The cavity is essentially a box with rough dimensions of 13 Å by 9 Å by 4 Å. The portal region is linked to a corner of the box by a narrow but contiguous channel of about 10 Å in length and 3–4 Å width. Much like the other iLBP family members, 48% of this cavity surface is identified with hydrophobic sidechains while 52% is associated with the sidechains of charged or polar residues.

Oleates bound to crystalline LFABP and ALBP are shown in Fig. 2a and b respectively. In these stereodiagrams, the two crystal structures are in the same orientation to simplify comparisons. To make the drawings, the crystallographic coordinates of the oleate complex with ALBP were fit to those of LFABP using the least squares method. For Cα, atoms only, the RMS difference is 1.2 Å. Two bound oleate molecules should be immediately visible in LFABP while only one is present in ALBP. The additional space within the binding cavity in LFABP and an alteration in the overall cavity interactions from that found in other iLBPs are the two factors which appear to account for the unusual binding stoichiometry.

In the orientation of LFABP that is shown at the top of Fig. 2, one fatty acid molecule is in a roughly U-shaped conformation and is clearly at a lower point within the binding cavity. The carboxylate of this oleate is not accessible to solvent although it is involved in a hydrogen bond network with bound waters. The cavity space occupied by the second bound oleate more closely resembles that found for ALBP as shown at the bottom of Fig. 2b. The exception of course, is that the second molecule bound to LFABP is both more extended and in an inverted conformation compared to ALBP; it has the carboxylate moiety on the surface of the LFABP molecule.

The buried carboxyl group of the lower U-shaped oleic acid is involved in a more extensive hydrogen bonding network than is found for the other oleate. The cavity residues involved in this network are R122, which was previously identified by site directed mutagenesis[19] and chemical modification studies[20, 21], but also S39 and S124. Three of the six bound waters found inside the cavity are members of this system of interactions. Additional amino acids with atoms in contact with this oleate are: I41, F63, E72, T73, T93, and T102. A fourth solvent molecule is within van der Waals distance of the hydrophobic acyl chain.

The bound oleate with its carboxylate group away from the cavity is near residues forming the portal. The hydrogen bonding network involves K31, Y54, S56, D88′, and a bound water. The interaction of the amino acid D88′ is due to crystal packing from a symmetry related molecule and therefore is of lesser importance. Being more solvent exposed, the temperature factors for ligand atoms near the surface are higher than for atoms of the carboxyl moiety belonging to the inner most oleate. The hydrophobic chain of the oleate toward the surface continues down into the cavity residing in contact with the innermost oleate and residues: L28, G32, I35, I52, Y54, G55, K57, M113, and R122. Because of the interactions with the innermost oleate, it appears as if an ordered binding mechanism must be present with the innermost site being filled first. In fact from the crystallographic data, it appears that the second site may not exist until the interior primary site is filled.

An assessment of the cavity volume and area of holo-LFABP compared to that of other family members is contained in Table 1. The LFABP binding cavity is clearly the largest of the iLBPs, exceeding by an additional 26% in volume and 20% in area that of L-MFABP. From the comparison of the LFABP-oleate cavity to that of ALBP, and even other iLBPs, an importance can be assigned to several residues lining the surface that are different but occupy the same location within the proteins. These amino acids are also shown in Fig. 2. The sidechains change from 'small' to 'bulky' as one goes from the large (LFABP) to the smaller (ALBP) cavity. They include S39 to M40, T93 to Q93, S100 to I104, T102 to R106, and S124 to Y128. The majority of this increase in volume is located around the binding site of the more deeply buried oleic acid. These five amino acid differences alone account for the space equivalent of 16 atoms of C,N,O or S. The largest differences are caused by the T102 to R106 and S124 to Y128 replacements.

The T102 to R106 and S124 to Y128 replacements are significant for another reason. In ALBP, both R106 and Y128 are part of the amino acid triad that have ionic interactions with the carboxylate of the bound oleate[22] (see below). Hence two of the major cavity reducing features in the LFABP to ALBP comparison also stabilize the position of the fatty acid headgroup when bound to ALBP.

The affinity of ferriheme for the LFABP cavity was reported to be an order of magnitude greater than that for oleate and the reduced, physiological form, ferroheme,

Fig. 2. The ligand binding cavities in LFABP and ALBP. The topmost stereodrawing shows a Cα model of LFABP with two oleic acids as was found in the crystal structure. The lower drawing is a representation of the crystal structure of ALBP with a single molecule of oleic acid. The two drawings are in exactly the same orientation for comparative purposes. To answer the question as to why LFABP can bind two fatty acids, the figures also contain stick representations of the amino acid changes which are believed to account for the different stoichiometry as described in the accompanying text.

competes with oleic acid three-fold better than ferriheme[15]. With knowledge of the conformation of the binding cavity of LFABP from the oleate complex, we have been able to suggest a model for binding heme. The energy minimized model of LFABP-heme is shown in Fig. 3. The structure as shown was the best of the seven different heme orientations tried. It is also supported by the demonstration of an electrostatic interaction between the heme and R122 that is reproduced despite the fact that electrostatic energy terms were not included in the energy refinement of the hypothetical model [19].

The cavity and portal are key to intracellular lipid binding protein function

Together as a family, the iLBPs reversibly and non-covalently bind a diverse group of ligands. The function of these proteins is to enhance the aqueous solubility of the lipids and thus facilitate their intracellular transport, compartmentalization, and metabolism. Individually, the variations in the family member's ligand specificity and affinity are widespread. For example, LFABP has a long list of possible ligands while family members such as CRBP and CRABP bind retinoids. ALBP, HFABP, IFABP, and epidermal fatty acid binding protein (EFABP) primarily bind long-chain fatty acids with 16–22 carbon atoms. Ileal fatty acid binding protein (ILBP), the most identical in amino acid sequence to LFABP, binds the bile salt chenodeoxycholate with greater affinity than fatty acids[23].

The observation that all of the iLBPs are of the same general supersecondary structure emphasizes the idea that the varied binding properties of these proteins are attributable to subtle differences in their molecular structures. Based on a comparison of the crystal and NMR structures and our prior arguments as to the β-barrel stability, the elements which most likely are assignable to this conformational variation include cavity sidechains interacting with the ligand, the helix turn helix segment of the portal region, turns between the antiparallel β-strands - mainly

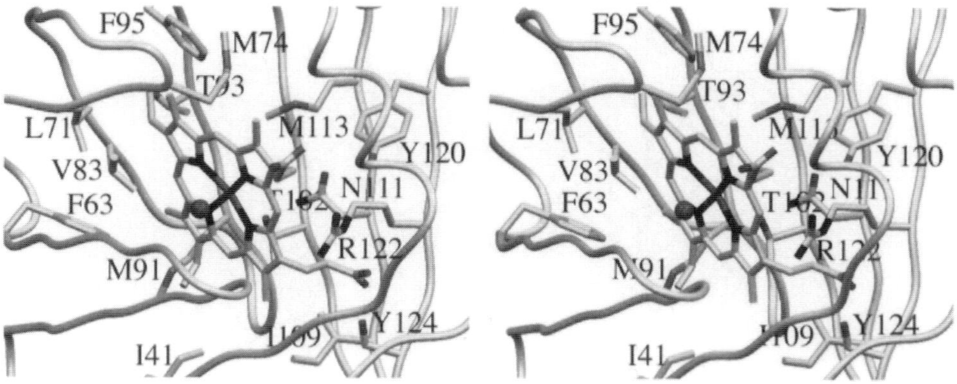

Fig. 3. Modeled interactions in the cavity of the LFABP-heme complex. The heme molecule was positioned into the LFABP binding cavity in seven different orientations, based mainly on packing restrictions. These models were refined without any electrostatic energy terms or water molecules included; the best is shown here. There is a predominance of hydrophobic contacts, though a hydrogen bond is evident between one propionic acid moiety and R122. The other propionic acid continues up the passage formed by the portal region of LFABP. The sphere, located where an oxygen is found in cytochromes and hemoglobin, is present merely to show additional cavity volume next to the iron atom large enough for 2–4 water molecules. The coordinates of the LFABP-heme model are available on request from the authors.

βC-βD, βE-βF, βG-βH and the gap between strands βD and βE. In simple terms, species or genotype variation in iLBP function will likely be attributable to the structure of the cavity and portal region.

For example, bound fatty acid in IFABP occupies cavity volume not found in other iLBPs and follows a comparatively elongated cavity shape[24]. LFABP also has unique and differently shaped cavity. The significance of the turns is in part suggested by the fact that in most iLBP crystal structures, the temperature factors for residues in the segments connecting β-strands and near the helices are higher than the average. As noted in nearly all of the structural reports on holo-iLBP, a few homologous cavity residues interact with the polar headgroup of the hydrophobic ligand. The amino acids found in these important cavity positions are emphasized by the symbol * in Fig. 1a and are reviewed elsewhere [11]. Many mutational studies support their significance and these so called recognition sites have a large effect on the individual iLBP's binding affinity and specificity [25–28].

Aside from the affinity and specificity arguments associated with cavity residues, there is a further question regarding the diversity of iLBP expression. Distinct iLBP family members are both associated with diverse cell types and frequently co-expressed in the same cells. Although it is possible there is some cellular redundancy in iLBP function, the portal region may be key to defining the different functions. If one can assume that differing specialized interactions with other cellular components are possible, it is plausible that other enzymes, proteins, or membranes may influence the entry or exit of lipid from the iLBP. A potential mechanism could involve stabilizing a normally transient open state of the ligand portal.

The positioning of a special ligand entrance site or portal near the helical lid is supported by holo-crystal structures where either the β-ionone ring of retinoic acid[9] or the fatty acid's hydrocarbon tail[5, 22] spans the length of this passage into or nearly into the solvent. In fact, small conformational changes induced by chemical modification[29], mutation[30, 31] or deletion[32] in this region play a role in the protein's binding affinity and kinetics. Apo-iLBPs are also measured as slightly less resistant to proteolysis[33–35] occurring just to carboxy-terminal side of the second helix and less thermodynamically stable in solution[25, 36].

Several examples of conformational variation at the portal have been studied. One example comes from the crystallographic studies of the CRABPs, another from NMR studies of IFABP (See Cistola *et al.* in this volume). In the case of the CRABPs, the mainchain trace of the crystal structures of apo-CRABPI and holo-CRABPI are the least similar to the rest of the iLBP family[9, 37]. In the x-ray structure of apo-CRABPI, a major conformational change of up to 8 Å was found between the two molecules found in the crystal asymmetric unit[37]. The difference involves one turn between the βC- and βD-strands. Although the conformational difference is attributable to crystal packing, an intermolecular β-sheet forms across the gap region of two CRABP molecules forming a homodimer. A comparable intermolecular interaction may help stabilize a normally transient open state of the portal as mentioned above. Examination of the differences in the crystal structures shows that a single mainchain hydrogen bond is lost when the βC-βI) turn moves away from the cavity. The apo-CRABPI structure differs in the positioning of the βE-βF loop and at the C-terminal end of the second helix - all again in the portal region[38]. These differences are found in one of the two molecules of CRABPII present in the asymmetric unit and are not assignable to crystal packing.

In summary, the β-barrel framework of LFABP is very similar to the other iLBP family members. A significant increase in cavity volume results from a few amino acid changes that occur on residues lining the binding site. Because the cavity is larger, two oleic acids can bind rather than one. The larger binding site also can accommodate a single larger ligand and it has been possible using the crystallographic coordinates to devise a heme orientation within the cavity of LFABP. Moreover, as a growing number of crystallographic and NMR structures are overlaid, the framework of the β-barrel along with its many mainchain hydrogen bonds suggests that conformational changes must accompany the binding of ligands. These changes are likely to be found in the portal region, in which the gap(s) may also be included. The β-turns in this region of the barrel interact with the helix-turn-helix lid. It is possible to envision transient conformational changes with few or no alteration in the number of β-barrel hydrogen bonds. It is also the portal region which might distinguish the functional properties associated with the different genotypes, particularily those with similar ligand specificities.

Acknowledgements

The authors are grateful for research support from the NIH, grant number GM13925. Two of the authors, J.O. and A. R.-W. acknowledge predoctoral support from the Molecular Biophysics Training Grant, NIHGM08277. We also appreciate continued grant support from the Minnesota Supercomputer Institute for funds to carry out some of the calculations. Coordinates for the LFABP-heme complex can be obtained from the authors and that for the LFABP-oleate complex from the Protein Data Bank (identification code lIfo).

References

1. Thompson J, Winter N, Terwey D, Bratt J, Banaszak L: The crystal structure of the liver fatty acid binding protein. A complex with two bound oleates. J Biol Chem 272: 7140–7150, 1997
2. Cowan SW, Newcomer ME, Jones AT: Crystallographic studies on a family of cellular lipophilic transport proteins - refinement of P2 myelin protein and the structure determination and refinement of cellular retinol-binding protein in complex with all-*trans*-retinol. J Mol Biol 230: 1225–1246, 1993
3. LaLonde JM, Levenson M, Roe JJ, Bernlohr DA, Banaszak L: Adipocyte lipid-binding protein complexed with arachidonic acid. Titration calorimetry and x-ray crystallographic studies. J Biol Chem 269: 25339–25347, 1994
4. Scapin G, Gordon R, Sacchettini JC: Refinement of the structure of recombinant rat intestinal fatty acid-binding apoprotein at 1.2 Å resolution. J Biol Chem 267: 4253–4269, 1992
5. Young ACM, Scapin G, Kromminga A, Patel SB, Veerkamp JH, Sacchettini JC: Structural studies on human muscle fatty acid binding protein at 1.4 Å resolution: Binding interactions with three C18 fatty acids. Structure 2: 523–534, 1994
6. Haunerland NH, Jacobson BL, Wesenberg G, Rayment I, Holden HM: Three-dimensional structure of the muscle fatty-acid- binding protein isolated from the desert locust *Schistocerca gregaria.* Biochemistry 33: 12378–12385, 1994
7. Benning MM, Smith AF, Wells MA, Holden HM: Crystallization, structure determination and least-squares refinement to 1.75-Angstrom resolution of the fattyacid binding protein isolated from *Manduca Sexta* L. J Mol Biol 228: 208–219, 1992
8. Winter NS, Bratt JM, Banaszak L: Crystal structures of holo and apo-cellular retinol-binding protein II. J Mol Biol 230: 1247–1259, 1993
9. Kleywegt GJ, Bergfors T, Senn H, Le Motte P, Gsell B, Shudo K, Jones AT: Crystal structures of cellular retinoic acid binding proteins I and II in complex with all-*trans*-retinoic acid and a synthetic retinoid. Structure 2: 1241–1258, 1994
10. Bernlohr DA, Simpson MA, Hertzel AV, Banaszak L: Intracellular lipid-binding proteins and their genes. Annu Rev Nutr 17: 277–303, 1997
11. Banaszak L, Winter N, Xu Z, Bernlohr DA, Cowan S, Jones TA: Lipid binding proteins: A family of fatty acid and retinoid binding proteins. Advan Protein Chem 45: 89–151, 1994
12. Glatz, JFC, van der Vusse GJ: Cellular fatty acid-binding proteins: Their function and physiological significance. Prog Lipid Res 35: 242–282, 1996
13. Cistola DP, Sacchettini J, Banaszak L, Walsh M, Gordon J: Fatty acid interactions with rat intestinal and liver fatty acid-binding proteins expressed in *Escherichia coli.* J Biol Chem 264: 2700–2710, 1989
14. Thumser AE, Wilton DC: The binding of natural and fluorescent lysophospholipids to wild-type and mutant rat liver fatty acid-binding protein and albumin. Biochem J 307: 305–311, 1995
15. Stewart JM, Slysz GW, Pritting MA, Muller-Eberhard U: Ferriheme and ferroheme are isosteric inhibitors of fatty acid binding to rat liver fatty acid binding protein. Biochem Cell Biol 74: 249–255, 1996
16. Rolf B, Oudenampsen-Kruger E, Borchers T, Faergeman NJ, Knudsen J, Lezius A, Spener F: Analysis of the ligand binding properties of recombinant bovine liver type fatty acid binding protein. Biochim Biophys Acta 1259: 245–253, 1995
17. Miller KR, Cistola DP: Titration calorimetry as a binding assay for lipid-binding proteins. Mol Cell Biochem 123: 29–37, 1993
18. Richieri GV, Ogata RT, Kleinfeld AM: Equilibrium constants for the binding of, fatty acids with fatty-binding proteins from adipocyte, intestine, heart, and liver measured with the fluorescent probe ADIFAB. J Biol Chem 269: 23918–23930, 1994
19. Borchers T, Spener F: Involvement of arginine in the binding of heme and fatty acids to fatty acid-binding protein from bovine liver. Mol Cell Biochem 123: 23–27, 1993
20. Thumser AE, Evans C, Worrall AF, Wilton DC; Effect on ligand binding of arginine mutations in recombinant rat liver fatty acid-binding protein. Biochem J 297: 103–107, 1994
21. Thumser AE, Voysey J, Wilton DC: Mutations of recombinant rat liver fatty acid-binding protein at residues 102 and 122 alter its structural integrity and affinity for physiological ligands. Biochem J 314: 943–949, 1996
22. Xu Z, Bernlohr DA, Banaszak L: The adipocyte lipid-binding protein at 1.6 Å. Crystal structures of the apoprotein and with bound saturated and unsaturated fatty acids. J Biol Chem 268: 7874–7884, 1993

23. Sacchettini JC, Hauft SM, Van Camp SL, Cistola DP, Gordon JI: Developmental and structural studies of an intracellular lipid binding protein expressed in the ileal epithelium. J Biol Chem 265: 19199–19207, 1990
24. Sacchettini JC, Gordon R, Banaszak U: Crystal structure of rat intestinal fattyacid-binding protein. Refinement and analysis of the *Escherichia coli*-derived protein with bound palmitate. J Mol Biol 208: 327–339, 1989
25. Prinden CFM, Veerkamp JH: Fatty acid binding and conformational stability of mutants of human muscle fatty acid-binding protein. Biochem J 314: 253–260, 1996
26. Jakoby MG, Miler KR, Toner JJ, Bauman A, Cheng L, Li E, Cistola DP: Ligand-protein electrostatic interactions govern the specificity of retinol- and fatty acid-binding proteins. Biochemistry 32: 872–878, 1993
27. Cheng L, Qian S, Rothschild C, Avignon A, Lefkowith J, Gordon J, Li E: Alteration of the binding specificity of cellular retinol-binding protein II by site-directed mutagenesis. J Biol Chem 267: 24404–24412, 1992
28. Richieri GV, Low PJ, Ogata RT, Kleinfeld AM: Mutants of rat intestinal fatty acid-binding protein illustrate the critical role played by enthalpy-entropy compensation in ligand binding. J Biol Chem 272: 16737–16740, 1997
29. Herr FM, Matarese V, Bernlohr DA, Storch J: Surface lysine residues modulate the collisional transfer of fatty acid from adiptocyte fatty acid binding protein to membranes. Biochemistry 34: 11840–11845, 1995
30. Ory J, Kane CD, Simpson MA, Banaszak U, Bernlohr DA: Biochemical and crystal lographic analyses of a portal mutant of the adipocyte lipid-binding protein. J Biol Chem 272: 9793–9801, 1997
31. Herr FM, Aronson J, Storch J: Role of portal region lysine residues in electrostatic interactions between heart fatty acid binding protein and phospholipid membranes. Biochemistry 35: 1296–1303, 1996
32. Cistola DP, Kim K, Rogl H, Frieden C: Fatty acid interactions with a helix-less variant of intestinal fatty acid-binding protein. Biochemistry 35: 7559–7565, 1996
33. Honma Y, Niimi M, Uchiumi T, Takahashi Y, Odani S: Evidence for conformational change of fatty acid-binding protein accompanying binding of hydrophobic ligands. Biochem J 116: 1025–1029, 1994
34. Rizo J, Liu A, Gierasch LM: ^{1}H and ^{15}N resonance assignments and secondary structure of cellular retinoic acid-binding protein with and without bound ligand. J Biomol NMR 4: 741–760, 1994
35. Jamison RS, Newcomer ME, Ong DE: Cellular retinoic acid binding proteins: Limited proteolysis reveals a conformational change upon ligand binding. Biochemistry 33: 2873–2879, 1994
36. Ropson IJ, Gordon JI, Frieden C: Folding of a predominately beta-structure protein: Rat intestinal fatty acid binding protein. Biochemistry 29: 9591–9599, 1990
37. Thompson JR, Bratt JM, Banaszak U: Crystal structure of cellular retinoic acid binding protein I shows increased access to the binding cavity due to formation of an intermolecular β-sheet. J Mol Biol 252: 433–446, 1995
38. Ji X, Chen X, Tordova M, Gilliland GL, Wang L, Li Y, Yan Honggao: Crystal structure of apo-cellular retinoic acid binding protein type II suggests a mechanism of ligand entry. ACA Annual Meeting Abstracts, St. Louis, MO, 1997

Molecular and Cellular Biochemistry **192:** 17–31, 1999.

Mechanisms of cellular uptake of long chain free fatty acids

P.D. Berk and D.D. Stump
Departments of Medicine and Biochemistry, Mount Sinai School of Medicine, New York, NY, USA

Abstract

Cells take up long chain free fatty acids (FFA) *in vivo* from the non-protein bound ligand pools in extracellular fluid and plasma, which contain ~100 and 600 μM albumin, respectively. The physiologic range of unbound FFA concentrations in such fluids has traditionally been calculated at < 1 μM. Studies of [^{3}H]-oleate uptake by hepatocytes, adipocytes, cardiac myocytes and other cell types demonstrate that FFA uptake within this range is saturable, and exhibits many other kinetic properties indicative of facilitated transport. Within this range, the uptake kinetics of the acidic (pKa = 0.5) FFA analog $\alpha_2,\beta_2,\omega_3$-heptafluorostearate are similar to those of stearate. Thus, uptake of physiologic concentrations of FFA involves facilitated transport of the FFA anion (FA^-). Over a much wider range of unbound FFA concentrations hepatocellular [^{3}H]-oleate uptake exhibits both saturable and non-saturable components. Oleate binding to liver plasma membranes (LPM) also demonstrates such components. Comparing the two components of FFA uptake to the corresponding components of binding permits estimates of trans-membrane transport rates. $T_{1/2}$ for saturable uptake (~ 1 sec) is less than for non-saturable uptake (~ 14 sec). Others have determined the flip-flop rates of protonated FFA (FAH) across small and large unilamellar vesicles (SUV, LUV) and across cellular plasma membranes. These reported flip-flop rates, measured by the decrease in pH resulting from the accompanying proton flux, exhibit a highly significant inverse correlation with cell and vesicle diameter (r = 0.99). Although $T_{1/2}$'s in vesicles are in the msec range, those in cells are > 10 sec, and thus comparable to the rates of non-saturable uptake we determined. Thus, under physiologic conditions, the predominant mechanism of cellular FFA uptake is facilitated transport of FA^-; at much higher, non-physiologic FFA concentrations, passive flip-flop of FAH predominates. Several plasma membrane proteins have been identified as potential mediators of facilitated FFA transport. Studies in animal models of obesity and non-insulin dependent diabetes mellitus demonstrate that tissue-specific regulation of facilitated FFA transport has important pathophysiologic consequences. (Mol Cell Biochem **192**: 17–31, 1999)

Key words: facilitated transport, flip-flop, hepatocytes, adipocytes

Abbreviations: BSA – bovine serum albumin; FFA – long chain free fatty acids; FAH – protonated FFA; FA^- – FFA anion; HFS – $\alpha_2,\beta_2,\omega_3$-heptafluorostearate; LPL – lipoprotein lipase; mAspAT – mitochondrial aspartate aminotransferase; FAT – fatty acid translocase; FATP – fatty acid transporting protein; NIDDM – non-insulin-dependent diabetes mellitus; $[O_u]$ – unbound oleate concentration; TG – triglycerides; *v* – fatty acid:albumin molar ratio.

Introduction

Long chain free fatty acids (FFA) serve a number of important biological functions. They are (1) critical energy substrates, which may be stored as triglycerides (TG) until needed for oxidation as fuel; (2) building blocks for essential glyco- and phospholipid components of cell membranes; and (3) precursors of important biological mediators such as the prostaglandins, leukotrienes and thromboxanes. Recently, it has been recognized that they are also important intracellular mediators of gene expression [1, 2]. These multiple roles suggest that careful regulation of all aspects of FFA disposition, including cellular uptake, would be beneficial to the cell. However, it has long been the conventional view that cellular

Address for offprints: P.D. Berk, Division of Liver Diseases (Box 1633), Mount Sinai School of Medicine, 1 Gustave L. Levy Place, New York, NY 10029, USA

FFA uptake occurs by a passive, unregulated mechanism (reviewed in [3, 4]). Many studies (e.g. [5–7]) report that FFA cross synthetic membranes at rates which greatly exceed rates of cellular uptake or utilization, leading to the argument that there is no need for a facilitated uptake mechanism to meet cellular FFA requirements [3, 4]. However, FFA uptake rates in living cells observed by the same investigators [8, 9] were orders of magnitude slower than those reported with synthetic membranes. These and other data [10] suggest that synthetic membrane vesicles are not good models for cellular plasma membranes, and call into question the biological relevance of FFA transport rates measured in synthetic liposomes.

Other studies suggest that cellular uptake of FFA exhibits the kinetic features of facilitated transport [11]. Indeed, no fewer than five plasma membrane proteins have been identified as possible FFA transporters, of which three have been cloned (reviewed in [12]).

In this review, we will examine the evidence supporting these two, divergent mechanisms for the entry of FFA into cells; briefly summarize data about proposed plasma membrane protein transporters of FFA; and consider the implications of facilitated FFA transport for such disorders as obesity and non-insulin dependent diabetes mellitus (NIDDM).

Mechanisms of FFA uptake

Physiologic conditions for cellular FFA uptake

Mammalian cells sequester FFA *in vivo* from biological fluids containing appreciable concentrations of albumin. In the liver, in which a fenestrated sinusoidal endothelium permits the unhindered passage of plasma albumin into the space of Disse and to the basolateral surface of the hepatocyte (Fig. 1), the concentration of albumin bathing the cell surface approaches 600 μM. In other tissues FFA in plasma, as well as those generated on the luminal surface of the endothelium by the action of lipoprotein lipase (LPL), must first cross a tight capillary endothelial barrier into the interstitial extracellular fluid in order to become available for parenchymal cellular uptake. Even in interstitial fluid, however, albumin concentrations may be 100 μM or greater. How FFA cross the barrier posed by the typical capillary endothelium is unclear, but some evidence suggests that a specific FFA transport mechanism may be involved [13]. At least one of the putative FFA transporters discussed below is expressed on the surface of endothelial cells [14] (Fig. 2).

Although it has been suggested that a cell surface albumin receptor mediates the cellular uptake of albumin-bound ligands, including FFA, from the protein bound ligand pool [15], the bulk of recent evidence suggests, in fact, that FFA and other albumin-bound ligands are taken up almost exclusively from the unbound ligand pool in equilibrium with albumin [16–18]. The presence of multiple binding sites with very high affinities for FFA on each albumin molecule means that the unbound FFA pool available for uptake has a concentration at least 3 orders of magnitude lower than the total FFA concentration in these biological fluids. For nearly a quarter of a century, these unbound FFA concentrations have been computed using stepwise equilibrium constants reported by Spector and colleagues [19]. Based on these constants, cellular FFA uptake was believed to occur from solutions in which the available substrate concentration was ≤ 1 μM. Several more recent studies indicate that the Spector constants overestimate the unbound FFA concentration in equilibrium with BSA by a factor of 20 or greater [20–23], placing the 'true' unbound FFA concentrations in typical physiologic situations at ≤ 50 nM. Accordingly, to study FFA uptake under physiologic conditions requires albumin concentrations of at least 100 μM, with FFA:albumin molar ratios (v's) in the range of 0.1:1 to ~3:1. Studies of FFA uptake in the absence of albumin or in the presence of v's significantly greater than ~4:1 are being conducted under substrate (i.e. unbound FFA) concentrations to which mammalian cells are rarely exposed *in vivo*.

Finally, mammalian cells sequester FFA at a body temperature approximating 37°C. FFA uptake is highly temperature sensitive (Fig. 3), and experiments conducted at temperatures appreciably different from 37°C will result in rates of FFA entry that do not reflect the *in vivo* situation.

Facilitated FFA uptake

The year 1981 marked a turning point in our understanding of cellular FFA transport. In that year Abumrad and associates published the first in a series of now classical studies which documented that the kinetic behavior of FFA uptake by adipocytes exhibited many of the defining features of facilitated transport [24, 25]. In that same year Weisiger, Gollan and Ockner described a series of studies of FFA uptake by the isolated perfused rat liver [15]. With perfusates containing a fixed albumin concentration (150 μM) but various v's up to ~3.3:1, FFA uptake was reported to be a linear function of the total FFA concentration in the perfusate, and to be unrelated to the concentration of unbound FFA. We have virtually duplicated these results in isolated hepatocyte preparations [26] (Fig. 4)*. By contrast, when livers were perfused with increasing concentrations of solutions con-

*Values for $[O_u]$ in the studies described in this review were computed using the 1971 binding constants of Spector *et al.* [19]. Use of the more recently reported sets of binding constants (e.g. [20–23] would each result in different computed values for the various parameters of both uptake and binding, but would not alter the qualitative interpretation of these studies, nor estimates of the rate constants k_s and k_r for saturable and non-saturable trans-membrane FFA movement, respectively.

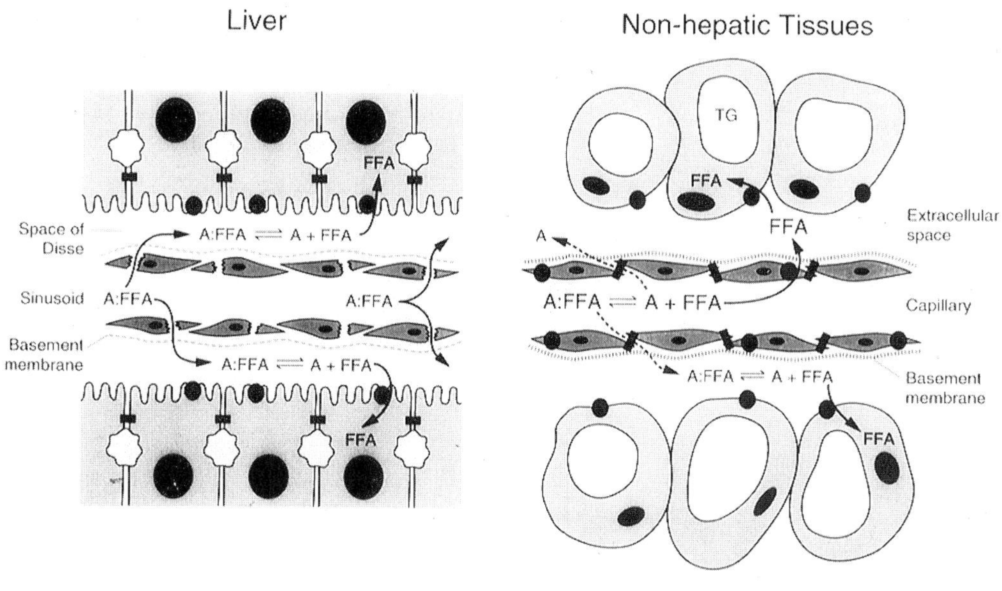

Fig. 1. Influence of microvascular anatomy on cellular uptake of FFA. Cellular FFA uptake occurs from the unbound ligand pool in equilibrium with albumin. The high affinity of albumin for FFA maintains the concentration of unbound FFA at $< 1\ \mu M$ under most physiologic conditions. In the liver, a fenestrated sinusoidal endothelium provides direct access by plasma concencentrations of FFA:albumin complexes to the space of Disse. In other tissues, the capillary endothelium provides a barrier permitting only slow exchange of albumin between plasma and the interstitial space, where albumin concentrations, while still appreciable, are lower than in plasma. Plasma FFA, as well as FFA generated on the luminal surface of the endothelial cell by the action of LPL, cross the endothelial barrier much more rapidly than albumin to become available for cellular uptake. Kinetic evidence suggests that trans-capillary movement of FFA into the interstitial fluid may be facilitated [13].

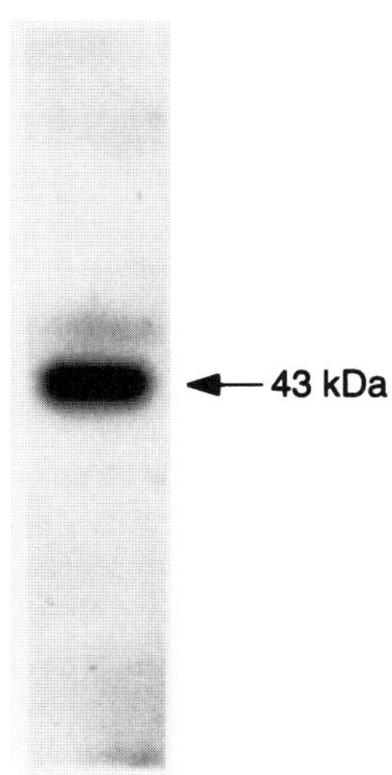

Fig. 2. Western blotting of a putative FFA transporter from endothelial cell plasma membranes. Plasma membranes were prepared from cultured human endothelial cells. Membrane proteins were extracted and separated by SDS-PAGE. After blotting to nitrocellulose filters, the blots were exposed to an antibody to mAspAT. [^{125}I]-protein A was used for detection. Presence of mAspAT on the plasma membranes of these cells was also demonstrated by immunofluorescence. These results confirm observations reported earlier in the rat [14]. As described later in this review, considerable evidence indicates that mAspAT on the plasma membrane serves as an FFA transporter.

taining FFA and albumin at a fixed *v*, saturation of uptake occurred at total FFA concentrations well within the range studied in the earlier set of experiments [15]. The authors concluded from these studies that FFA uptake was limited by the presence of a saturable albumin receptor. However, when we replotted the data from the first set of experiments, or from our own analogous studies with isolated hepatocytes [26], it was clear that uptake was equally well described as a linear function of the total FFA concentration or as a saturable function of the unbound FFA concentration (Fig. 4).

[^{3}H]-oleate uptake by isolated hepatocytes was subsequently examined over a much wider range of *v*'s (≤ 6.7) (Fig. 5), and various possible relationships of uptake to total and unbound FFA concentrations were evaluated by computer [27]. Using conventional criteria for goodness of fit [27], the data were best described by an equation in which uptake was the sum of a saturable plus a non-saturable function of the unbound oleate concentration ($[O_u]$). That is:

$$U([O_u]) = (V_{max} \cdot [O_u])/(K_m + [O_u]) + k_u \cdot [O_u],$$

where $U([O_u])$ = initial oleate uptake velocity at $[O_u]$ (pmol/sec/10^7 cells), V_{max} = maximal initial oleate uptake velocity (pmol/sec/10^7 cells), K_m = $[O_u]$ at half-maximal specific uptake (μM), and k_u = rate constant for non-specific uptake uptake (μl/sec/10^7 cells). Over the total range of ligand

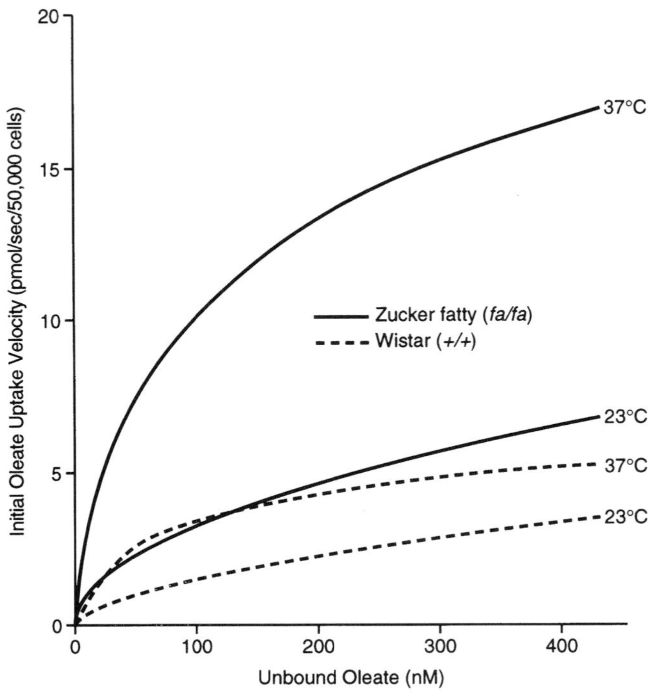

Fig. 3. Effects of temperature on FFA uptake by isolated rat adipocytes. [^{3}H]-oleate uptake kinetics were studied at both 23 and 37°C in epididymal fat pad adipocytes isolated from adult male Zucker fatty and normal Wistar rats. Curves represent computer fits of the cumulative uptake data to the sum of a saturable and a non-saturable function of the unbound oleate concentration. A marked temperature effect, greater with respect to the saturable function, is seen. Values for $[O_u]$ were computed using the binding constants of Spector *et al.* [19].

concentrations studied, it was no longer possible to describe the data by any physiologically meaningful function of the total FFA concentration. Indeed, an extensive body of subsequent work examining the uptake of a wide variety of ligands transported in plasma bound to albumin has confirmed that - consistent with conventional pharmacokinetic theory - uptake of each of these ligands is driven by the unbound ligand concentration in equilibrium with albumin [16–18]. Based on the experimentally determined values for V_{max}, K_m and k_u, more than 90% of hepatocellular FFA uptake occurs by the saturable pathway at physiologic albumin concentrations and v's ($\leq 3{:}1$). At higher but non-physiologic v's, the proportion of cellular uptake via the non-saturable process increases, reaching 81% at $v = 6.7$ [27]. When analyzed in this way, analogous studies of FFA uptake by isolated rat adipocytes [28, 29] lead to comparable conclusions.

When the binding of FFA to purified rat liver plasma membranes was examined [27], it too was seen to consist of a combination of a saturable and a non-saturable component, each of which was a function of the unbound fatty acid concentration in the incubation mixture; i.e.:

$$B([O_u]) = (M_b \cdot [O_u])/(K_d + [O_u]) + k_b \cdot [O_u],$$

where: $B([O_u])$ = total oleate bound at $[O_u]$ (nmol/µmol membrane phospholipid), M_b = maximal specific binding (nmol/µmol membrane phospholipid), K_d = dissociation constant of specific binding (µM), and k_b = non-specific binding constant (ml/µmol membrane phospholipid). We postulated that the component of fatty acid saturably (specifically) bound to the membrane represented the substrate for the saturable process mediating cellular fatty acid uptake and that, similarly, the non-saturably (i.e. non-specifically) bound FFA pool within the membrane represented the substrate for non-saturable uptake. Validation of these hypotheses was obtained both analytically [27] and graphically (Fig. 6). Although cellular FFA uptake and plasma membrane FFA binding were measured in entirely independent experiments, the saturable component of uptake was found to be a linear function of the amount of FFA bound to saturable binding sites on the membrane. Likewise, the non-saturable component of uptake was a linear function of a non-saturable component of binding. Rate constants for the transmembrane movement of saturably bound ($k_s = 0.7$ sec^{-1}) and non-saturably bound ($k_r = 0.05$ sec^{-1}) components of the FFA within the membrane were determined from the slopes of these linear relationships. These rates constants, with $T_{1/2}$'s of ca. 1 and 14 sec, respectively, are similar to rate constants previously reported for the saturable and non-saturable transmembrane movement of various other ions and small molecules [27].

Since FFA exists as an equilibrium between the protonated fatty acid (FAH) and the fatty acid anion (FA^-) it is of interest to consider whether the saturable and non-saturable processes for cellular FFA uptake reflect different mechanisms for the entry of these two different species. Kurz and associates have synthesized a fatty acid derivative, $\alpha_2,\beta_2,\omega_3$-heptafluorostearate (HFS), as well as a photolabile FFA analogue, 11,11-azistearate [30]. Because of the presence of fluoride atoms in the two alpha positions, HFS is a relatively strong acid with a pKa of 0.5. Consequently, at physiologic pH, it is virtually exclusively in the anionic form. Under physiologic conditions of albumin concentration and v, conditions under which > 90% of FFA uptake is via the saturable pathway, the kinetics of hepatocellular HFS uptake are virtually identical to those of natural stearate (Fig. 7). Moreover, photoaffinity labeling of hepatocyte membrane proteins in tissue culture with 11,11-azistearate results in a highly specific inhibition of FFA uptake. The degree of inhibition of HFS uptake is virtually identical to that of stearate. These observations strongly suggest that the saturable component of FFA uptake, observed under physiologic conditions with respect to albumin and FFA concentrations, reflects a protein mediated transport process for fatty acid anions [30].

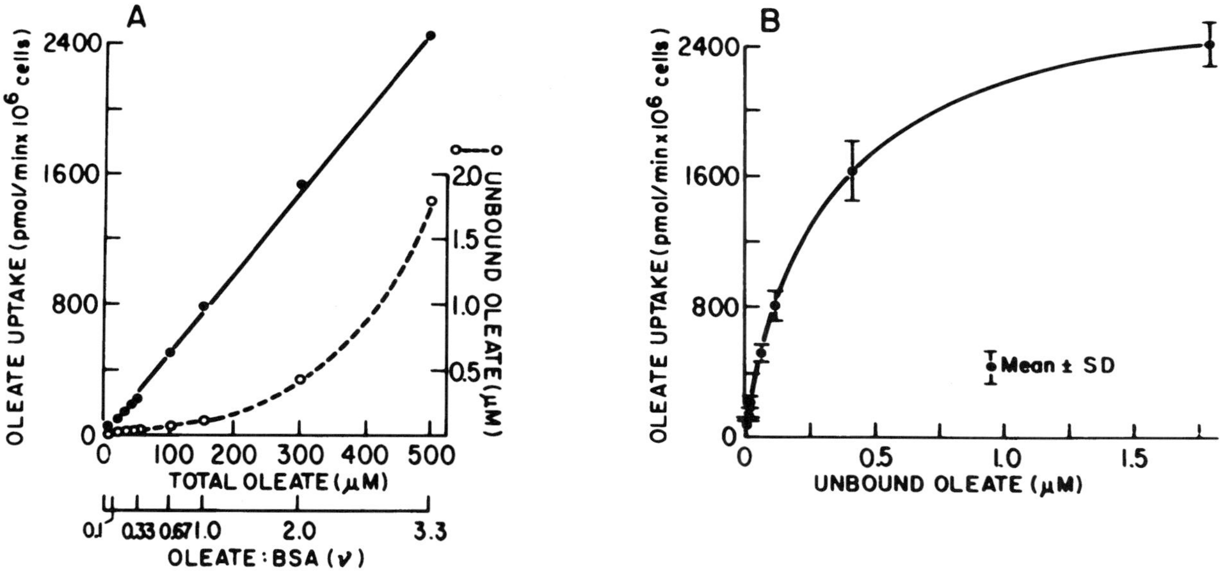

Fig. 4. Oleate uptake by isolated rat hepatocytes. Left hand panel: Initial uptake velocities of [^{3}H]-oleate, determined in the presence of 150 μM BSA at oleate:albumin molar ratios ≤ 4:1, are plotted as a function of the total oleate concentration in the medium. In similar studies with the isolated, perfused rat liver [15], uptake was described as a linear function of the total oleate concentration, unrelated to the equilibrium unbound oleate concentration ([O_u]) in the medium (- - -). Right hand panel: However, as indicated, over the range of concentrations studied, the same results can equally well be expressed as a saturable function of [O_u].

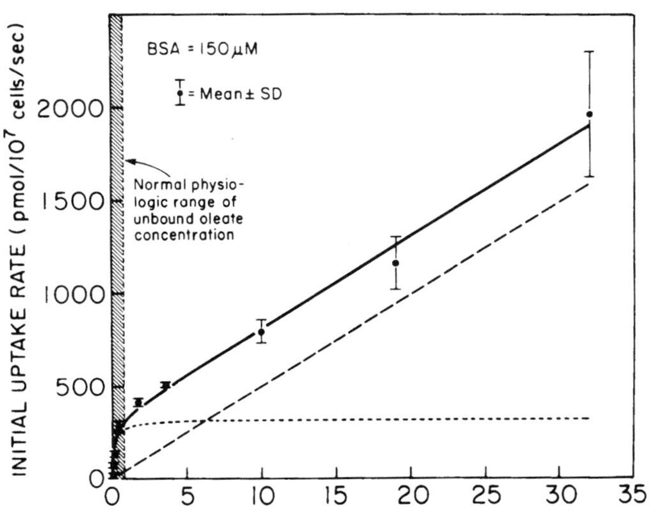

Fig. 5. Oleate uptake by isolated rat hepatocytes. In this study, initial uptake velocities of [^{3}H]-oleate, determined at oleate:albumin molar ratios ≤ 6.7:1, are plotted as a function of the equilibrium unbound oleate concentration in the medium ([O_u]). Data are mean ± S.D. of three separate determinations. Over this range of concentrations, two distinct components are clearly evident. Solid line is the best computer fit to the sum of a saturable plus a non-saturable component. Within the normal physoiologic range of [O_u], > 90% of uptake is via the saturable component. At higher, non-physiologic concentrations of [O_u], the non-saturable component predominates. Reproduced from [27] by permission.

The demonstration of saturability of transport is a necessary but not a sufficient condition to establish that the process is facilitated, i.e. protein mediated. Other kinetic features of specific, e.g. facilitated, transport include cis- (competitive) inhibition, trans-stimulation, counter-transport, and, where applicable, stereospecificity. When studied under appropriate physiologic conditions, the predominant component of FFA uptake in hepatocytes, adipocytes, cardiac and skeletal muscle, jejuneal epithelium, and renal tubules, has been shown to exhibit these defining kinetic features of facilitated transport [24, 25, 27, 28, 31–40]. By contrast, the dominant component of FFA uptake in resting fibroblasts appears to be a non-saturable and presumably passive process [41–43]. These kinetic features have been demonstrated in a variety of systems including whole animals, the isolated perfused liver, isolated cell preparations, and isolated membrane vesicles [40].

Non-saturable FFA uptake processes

Although the studies described above make a compelling argument for the existence of a facilitated FFA uptake process, other investigators have argued strongly that all cellular fatty acid uptake can be explained on the basis of purely passive mechanisms. Several groups of investigators have studied the rates of movement of FFA across synthetic

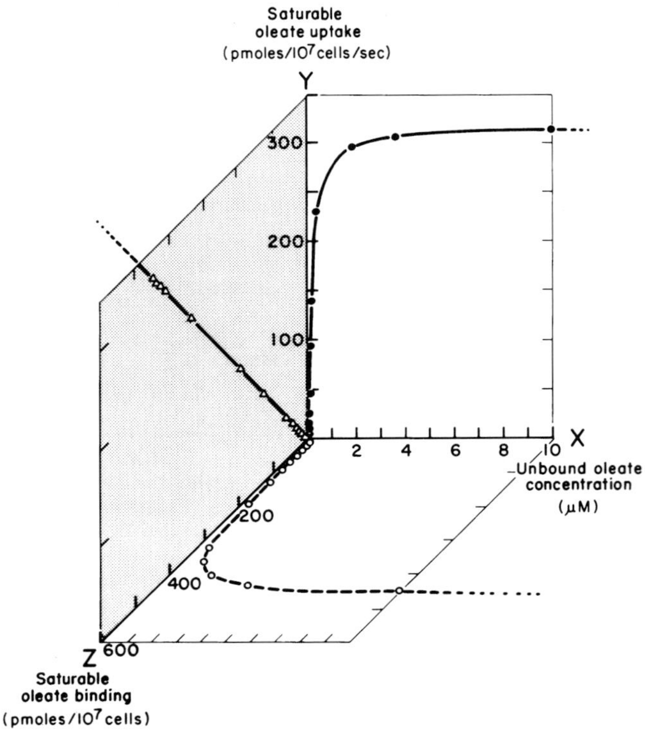

Fig. 6. Relationships between the unbound oleate concentration, saturable oleate uptake, and saturable oleate binding to liver plasma membranes. Unbound oleate concentration ($[O_u]$) is represented on the X axis, saturable oleate uptake (●–●) as Y, and saturable oleate binding to liver plasma membranes (○–○) as Z. Values for Y and Z represent values calculated by computer from fits of the independent experimental measurements of total uptake and total binding to the respective equations described in the text. Data points correspond to $[O_u]$'s at which the actual experimental measurements of uptake were determined. The relationship between saturable oleate binding and saturable uptake is illustrated on the YZ plane Δ–Δ. This indicates that the saturable component of oleate uptake is a linear function of the amount of oleate bound to saturable binding sites in the membrane (r = 0.9997). Reproduced from [27] with permission.

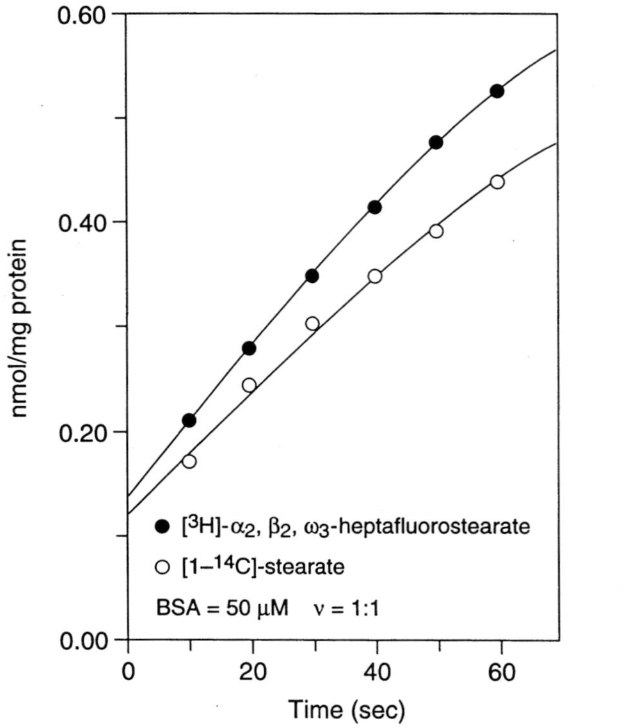

Fig. 7. Uptake of radiolabeled stearate and α_2, β_2, ω_3-hepatofluorostearate (HFS), a highly acidic FFA analogue, by isolated rat hepatocytes. Studies were conducted at a physiologic FFA:albumin molar ratio of 1:1. Kinetic data and the similar inhibitory effects on the uptake of both stearate and HFS of of photoaffinity labeling of cell membrane proteins with 11,11-azistearate suggest that FFA uptake at physiologic concentrations involves the fatty acid anion. Redrawn from [30] with permission.

lipid bilayers. These studies report extremely rapid trans-membrane 'flip-flop' rates, with half-times in the millisecond range (e.g. [5–7, 44]). If FFA traverse lipid bilayers this rapidly, it is argued, cellular needs for FFA would easily be met by passive FFA influx across the plasma membrane, and there would be no need, and indeed no plausible function, for a facilitated FFA transport process [3, 4]. Although reported flip-flop rates across the membranes of unilamellar vesicles are very rapid, FFA uptake rates reported by the same investigators in living cells were orders of magnitudes slower [8, 9]. These and other data suggest that synthetic membrane vesicles are not good models for cellular plasma membranes and call into question the biological relevance of FFA transport rates measured in synthetic liposomes. Extrapolation from liposome studies to cells may be questioned on several grounds. Such studies are typically conducted in the absence of albumin or with trace albumin concentrations and very high, non-physiologic v's. The movement of fatty acids is typically detected by an indirect property, such as the change in intravesicular or intracellular pH resulting from the trans-membrane movement of a protonated fatty acid molecule followed by its subsequent dissociation to FA^- and H^+. In viable cells, however, several different natural buffering systems will dampen the consequent pH changes. Finally, in the case of synthetic liposomes, the studies are conducted with membrane particles having radii of curvature several orders of magnitude smaller than the radius of curvature of cell membranes. There are theoretical reasons to predict that stresses within membranes introduced by these small radii of curvature cause deformations of membrane structure that greatly alter their permeability to FFA [10].

We have examined in detail recent reports in which the entry of FFA into either pancreatic β-cells [8] or adipocytes [9] was assessed by continuous monitoring of the pH changes that reflected the movement of protonated fatty acids into the cell. In a representative study with isolated rat adipocytes, 130 nmol of FFA were added to the 1.3 ml albumin-

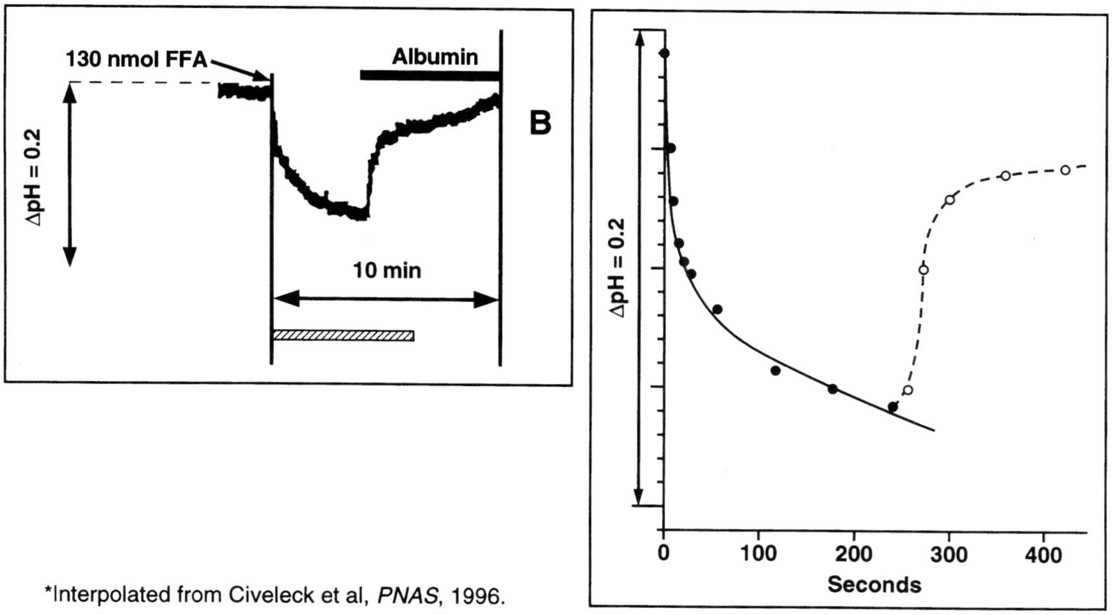

Fig. 8. Time course of pHi following addition of FFA to isolated rat adipocytes. pHi is estimated from fluorescence of the pH indicator BCECF, with which the cells were pre-loaded. Left hand panel: In the illustrated experiment, which is taken from Fig. IB in [9], 130 nmol oleic acid were added to adipocytes suspended in 1.3 ml albumin-free modified Krebs buffer (pH 7.4) at 30°C. Initial pHi was 7.0 ± 0.1. Right hand panel: Individual data points interpolated graphically from the experimental tracing.

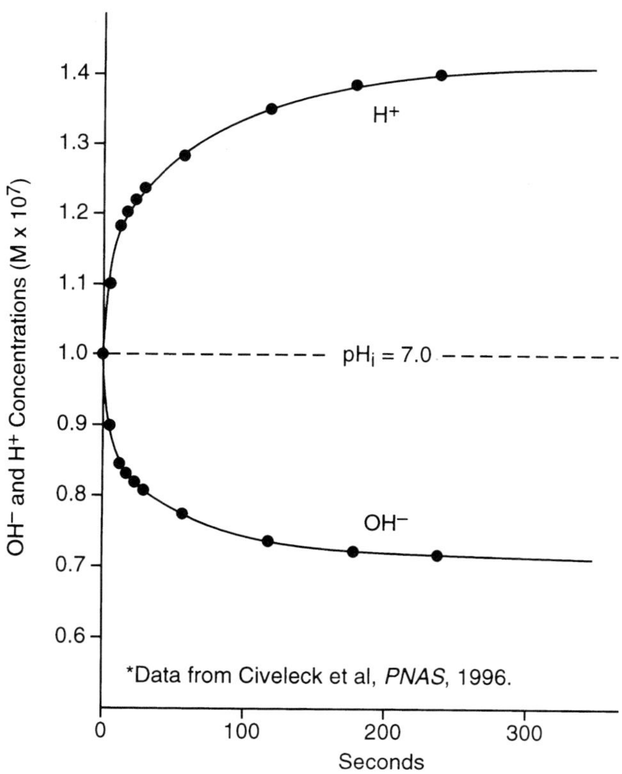

Fig. 9. Effects of FFA uptake by adipocytes on intracellular hydroxyl and hydrogen ion concentrations in isolated rat adipocytes. These data were calculated from the interpolated data points illustrated in Fig. 8.

free incubation mixture (putative $[O_u]$ = 100 μm) and subsequent deviations from the basal intracellular pH (pHi) of 7.0 recorded. The original data [9] from this experiment are shown in the left hand panel of Fig. 8. For subsequent analysis we interpolated individual data points from the original curves at various time intervals. These interpolated data points, illustrated in the right hand panel of Fig. 8, were used to generate curves of the actual intracellular concentrations of H^+ and OH^- ions (Fig. 9). Changes in intracellular pH and H^+/OH^- concentrations over the initial 240 sec were were fitted by computer to several hypothetical models. In all of these analyses, the observed changes (e.g. the fall in intracellular pH or OH-concentration after introduction of FFA) consisted of two exponential components with half-times of 6–10 and ~100 sec, respectively. Thus even the rapid component is nearly an order of magnitude slower than the half-time for facilitated transfer of FA^- across the hepatocyte plasma membrane, as described above. In analogous studies in pancreatic β-cells, only a single exponential component was observed. As reported, it had a half-time of ~ 60 sec [8].

What are the biological phenomena generating these exponential components? Since the end-point used in these studies is a pH change, both components presumably reflect the rate of entry of protonated fatty acids into the cell. Earlier studies by Storch and Kleinfeld [45] suggest that the more rapid component may reflect displacement into the cytoplasm of FAH already present within the inner membrane leaflet, whereas the slower component reflects the trans-membrane 'flip-flop' of FAH from the medium into the cell interior. Based on this assumption, we have plotted the half-times for transmembrane 'flip-flop' of FFA into small and large unilamellar vesicles (SUV, LUV)[7], pancreatic β-cells [8], and adipocytes [9], derived from measurements of intracellular pH changes, as well as the half-time for non-saturable FFA uptake in hepatocytes, derived from studies of [^{3}H]-oleate kinetics [27], against reported measurements of particle or cell diameter (Fig. 10). Although we have studied the uptake kinetics of [^{3}H]-FFA by adipocytes over a wide range of unbound ligand concentrations [28, 29], studies of the binding of FFA to isolated adipocyte plasma membranes have not yet been carried out, and therefore specific rate constants for FFA uptake in adipocytes via the non-saturable and saturable pathways cannot be computed. Similarly, there are as yet no reported data on the 'flip-flop' rates of protonated fatty acids in hepatocytes. Nevertheless, given the differences in methods employed, the general agreement among the measurements in the different cell types is noteworthy. Over a 1,000-fold range of diameters, 'flip-flop' half times are highly correlated with particle size ($r = 0.99$, $p < 0.01$), consistent with theoretical predictions. Even if the rapid component of pH change, observed only in adipocytes, were to more accurately reflect 'flip-flop' rates, the half-times and diameters would still be highly correlated ($r = 0.98$, $p < 0.01$).

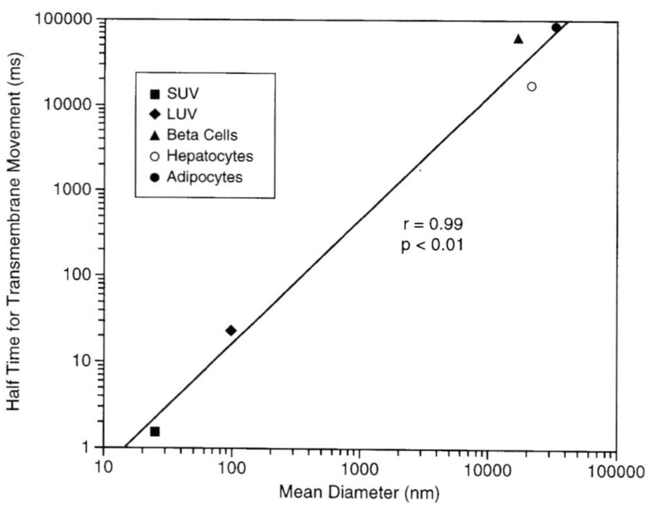

Fig. 10. Rates of 'flip-flop' of FFA into unilamellar vesicles and living cells compared with particle and cell diameters. Flip-flop rates for small and large unilamellar vesicles (SUV, LUV)[7], pancreatic 8-cells [8], and adipocytes [9] were determined from changes of pHi. Only a single component to the change in pHi was seen in pancreatic β-cells. Two components were observed in adipocytes. The data presented here depict the slower component (see text). Data in hepatocytes were determined from analysis of the kinetics of membrane binding and cellular uptake of [^{3}H]-oleate [27].

This observation suggests that extrapolations of trans-membrane 'flip-flop' rates measured in SUV or LUV to events in cell membranes are inappropriate, at least in part because of changes in membrane permeability introduced by large differences in membrane radius of curvature.

Several caveats should be made explicit in considering this analysis. It is based on a synthesis of data derived from experiments with different cell types (e.g. hepatocytes, adipocytes, pancreatic β-cells) performed in different laboratories, using very different methods, and under different conditions (e.g. temperature, particular FFA studied, membrane composition) which may influence the quantitative results. Nevertheless, particle size appears in this analysis to be an independent factor appreciably influencing trans-membrane FFA 'flip-flop' rates.

Based on the analysis just presented, our current hypothesis is that cellular FFA uptake in cells with high trans-membrane FFA fluxes such as hepatocytes and adipocytes consists of two separate components: a facilitated transport process for fatty acid anions which accomplishes trans-membrane FA^- movement with a half-time of less than 1 second, and a slower passive 'flip-flop' which effects the trans-membrane movement of FAH with a half-time of the order of 100 sec. Several lines of evidence suggest that facilitated FA^- transport is, biologically, the dominant process. In hepatocytes, by 2 min after uptake virtually all of [^{3}H]-FFA sequestered under physiologic conditions has been esterified, and is no longer available for efflux from the cell [46]. In adipocytes, esterification is even more rapid; ~90% of sequestered FFA are esterified by 1 min [29]. Thus, FFA taken up at physiologic concentrations of albumin and FFA enter into a metabolically active intracellular compartment and are rapidly metabolized. By contrast, in the studies of FAH 'flip-flop' in adipocytes reported by Civelek *et al.* [9], the effects of FFA addition on intracellular pH were rapidly reversed by addition of a small amount of albumin to the incubation mixture 5 min into the experiment. Thus, under the non-physiologic conditions of these studies, sequestered fatty acids had not entered into the normal, metabolically active pool and had not been esterified. This observation further raises questions about the physiologic significance of this passive 'flip-flop' process, at least as measured under the stated experimental conditions.

The concept that fatty acids can traverse cellular plasma membranes by the dual processes of facilitated transport of FA^- and passive bidirectional 'flip-flop' of FAH should not be surprising. A considerable body of evidence suggests that uncoupling protein in mitochondria effects the facilitated transport of FA^- into the inter-membrane space. After binding a proton generated by oxidative metabolic processes, the resulting protonated fatty acid re-enters the mitochondrial matrix by passive 'flip-flop' across the inner mitochondrial membrane [47, 48].

Table 1. Plasma membrane fatty acid binding proteins

Lab	Year	Size (kDa)	Name	Identification/ Isolation[1]	Sites[2]	Homologies	Proof of Function[3]
Berk[4]	85	43	FABPpm	AC, PhA-L	L, A, My, I, E	mAspAT	Ab-1, LR, GenExp
Fujii	87	56/60	–	AC	K, My	?	N.R.
Trigatti	91	22	–	PhA-L	A	? Caveolin	N.R.
Abumrad[4]	93	88/53	FAT	Coval-L	A, My, M, I, T	CD36	UT-1
Lodish[4]	94	63/71	FATP	Exp-Clon	A, My, M (K,B,L,Lu)	FACS/FACL[5]	GenExpr

[1]AC – affinity chromatography; PhA-L – photoaffinity labeling; Coval-L – covalent labeling; Exp-Clon – expression cloning. [2]A – adipose tissue; B – brain; E – endothelium; I – intestine; K – kidney; L – liver; Lu – lung; M – skeletal muscle; My – cardiac muscle; T – testis. [3]Ab-1 – Selective inhibition of uptake by antibodies to protein; LR – reconstitution of transport by insertion of protein into synthetic liposomes; GenExpr – expression of protein and transport function follows expression of cloned genetic message (cDNA or cRNA); UT-I – Selective inhibition of uptake by covalent labeling of protein; N.R. – Not reported. [4]cDNA cloned. [5]FACS/FACL – fatty acid co-A synthase/ligase.

Fatty acid transporters

Over the past 12 years, five proteins with high affinities for fatty acids have been identified in the plasma membranes of various cell types and proposed to have a function in fatty acid transport [49–54]. Some basic information about these proteins is presented in Table 1. Three of these proteins, plasma membrane fatty acid binding protein (FABPpm)[49], fatty acid translocase (FAT)[53], and fatty acid transporting protein (FATP)[54] have been cloned, and considerable evidence has been developed supporting a role for each of these in FFA transport in various cell types. The roles of the remaining two proteins remain unclear.

Information about the isolation, characteristics, and function of FAT and FATP is presented in detail in other reports from this symposium. We focus, therefore, on FABPpm, the first putative FFA transporter identified. FABPpm was isolated in 1985 from highly purified rat liver plasma membranes by affinity chromatography [49]. It was also shown to label with photoreactive fatty acid analogs [12]. After purification, binding studies indicated a high affinity for long chain fatty acids [49, 55], and polyclonal rabbit antibodies against the isolated rat protein confirmed its location in the plasma membrane (e.g. [49]). A role for FABPpm in FFA transport was strongly suggested by liposome reconstitution [56] and antibody inhibition [28, 32, 34, 36, 41–43, 57] studies. The inhibition of FFA uptake in various cells types by antibodies to FABPpm was highly specific for a variety of saturated and non-saturated long chain fatty acids. The antibodies had no effect on the uptake of medium chain fatty acids, of glucose, or, in the hepatocyte, of bilirubin, sulfobromophthalein or bile acids. During differentiation of mouse 3T3-L1 preadipocytes from a fibroblast to an adipocyte morphology, progressively increased expression of FABPpm on the plasma membrane occurred in parallel with increased expression of saturable FFA uptake [41, 42]. Antibodies to FABPpm had little effect on FFA uptake at the fibroblast stage, but produced significant inhibition of uptake in the differentiated adipocytes which expressed FABPpm on the cell surface.

The role of FABPpm in FFA transport was clouded by the finding that FABPpm was identical to the mitochondrial isoform of the well studied enzyme aspartate aminotransferase (mAspAT). Although this observation initially met with considerable skepticism [58], not the least in our own laboratory, an extensive body of data now documents that the two proteins are identical [55, 59]. They have molecular masses of 43 kDa by SDS-PAGE and 44.5 kDa by electrospray mass spectrometry; identical chromatographic behavior in multiple systems which separate proteins based on differing physical chemical characteristics; a highly basic pI of 9.1 with a characteristic pattern of multiple charge isomers; similar amino acid composition; an identical N-terminal amino acid sequence of at least 35 residues; identical patterns of tryptic digests; similar mAspAT enzymatic specific activities (205–210 mU/mg); similar K_d's for oleate binding; identical labeling with the photo-reactive FFA analogue 11,11-diazarino-HFS; and identical immunologic epitopes. In studies of differentiating 3T3-L1 preadipocytes [41, 42], antibodies to both FABPpm and mAspAT identified an essentially constant quantity of a 43 kDa protein in mitochondria at all stages of differentiation. Only traces of this 43 kDa protein were detected with either antibody in the plasma membrane of 3T3-L1 fibroblasts, whereas a progressive increase in this protein was detected in plasma membranes of differentiating adipocytes using either antibody (Fig. 11). Both antibodies selectively inhibited FFA uptake in 3T3-L1 adipocytes, but not in fibroblasts. Immunoreactivity of either antibody was abolished by pre-incubation with the other antigen. Finally, the increased plasma membrane expression of mAspAT correlated with increased expression of mAspAT cDNA, as indicated initially by slot blotting and subsequently by Northern hybridization [60]. Subsequently, transfection of standard 3T3-L1 fibroblasts with plasmid pMAAT2, con-

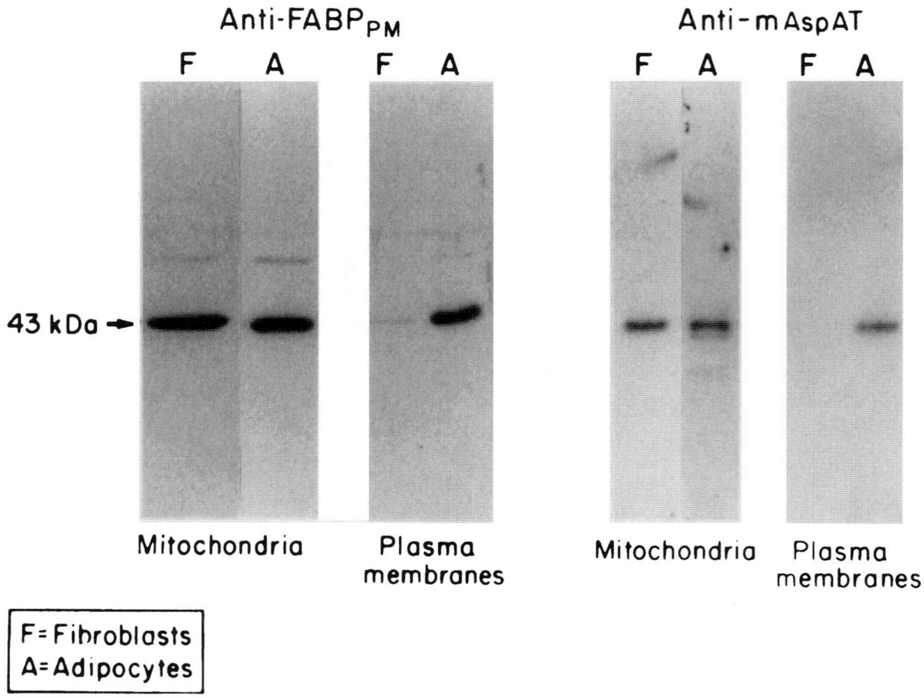

Fig. 11. Immunoblots of mitochondrial and plasma membrane extracts from 3T3-L1 fibroblasts (F) and adipocytes (A). Blots employed either anti-FABPpm (left) or anti-mAspAT (right). Both antisera detect a 43 kDa band in mitochondria isolated from either fibroblasts or adipocytes. Both antisera also detect an identical, abundantly expressed 43 kDa band in adipocyte plasma membranes, which is either absent or only weakly expressed in fibroblast plasma membranes. Immunoreactivity of anti-mAspAT is abolished by pre-incubation with mAspAT, and vice versa. Reproduced from [42] with permission.

taining an mAspAT cDNA under the control of a murine metallothionein promoter, was shown to confer zinc-dependent expression of mAspAT on the cell surface in parallel with development of saturable FFA uptake in this cell line which normally lacks both properties [43]. Microinjection of a capped mAspAT cRNA also stimulates FFA uptake in *Xenopus laevis* oocytes [12].

Although mAspAT thus meets all of the conventional criteria for the characterization of a plasma membrane transporter, important questions remain about its cellular and molecular biology. These include (1) How does mAspAT bind FFA? (2) How is the plasma membrane component of mAspAT transcribed? (3) How is its distribution between mitochondria and plasma membrane regulated? (4) How does mAspAT reach the plasma membrane? and (5) Is mAspAT a peripheral or an integral membrane protein? Each of these questions is being actively pursued. Although definitive answers are not available, there are preliminary observations with regard to each.

(1) Molecular modeling studies of the mAspAT crystal structure have identified a previously unrecognized hydrophilic pocket of approximately 500 Å^3 within the large domain of the protein [61]. This is of sufficient size to easily accommodate the typical long chain free fatty acid. Whether this pocket indeed serves as a fatty acid binding site will be investigated with a combined approach involving peptide mapping of the protein after photoaffinity labeling, and site directed mutagenesis of the putative FFA binding site.

(2) Northern hybridization, primer extension, and RNase protection studies indicate that the mAspAT which reaches the plasma membrane is translated from the same message as that directed to the mitochondrion. Although the 'mitochondrial' leader sequence is encoded by exon 1 of the mAspAT gene, which is spliced to exon 2 precisely within the first codon of the mature protein, we have found no evidence of alternative splicing which might provide a different leader sequence.

(3) We have identified several situations, including the differentiating 3T3-L1 adipocyte [41, 42], pMAAT2 transfected 3T3 cells [43]; and HepG2 hepatoma cells incubated in the presence of ethanol [62], in which a significant upregulation of plasma membrane expression of mAspAT and of saturable FFA transport is observed in the absence of appreciable changes in mitochondrial mAspAT content. Although these studies suggest independent regulation of these two pools of the protein, how this regulation is effected remains to be determined.

(4) Immunoelectron microscopic studies of hepatocytes and HepG2 cells, employing immunogold labeling, confirm the presence of mAspAT on the plasma membrane, in mitochondria, and in several populations of small vesicles, some

of which appear to be fusing with the plasma membrane [63]. Although this suggests that mAspAT reaches the plasma membrane by a vesicular route, it remains to be determined whether this route involves prior passage through the mitochondrion, as has recently been documented for other 'mitochondrial' proteins such as HSP 60 [64].

(5) Although the functional component of mAspAT in the mitochondrion exists in the hydrophilic environment of the mitochondrial matrix, there is also a pool of 'latent' mAspAT activatable only by detergent dissolution of the membrane [65]. Differential extraction studies suggest that a similar partitioning occurs at the plasma membrane: approximately 3/4 of plasma membrane mAspAT is extractable with salt, but the remaining 25% is only accessible after dissolution of the membrane with detergents [55]. Although we believe that the detergent extractable, integral membrane component is the one most likely to function as an FA^- transporter, no firm evidence supports this speculation.

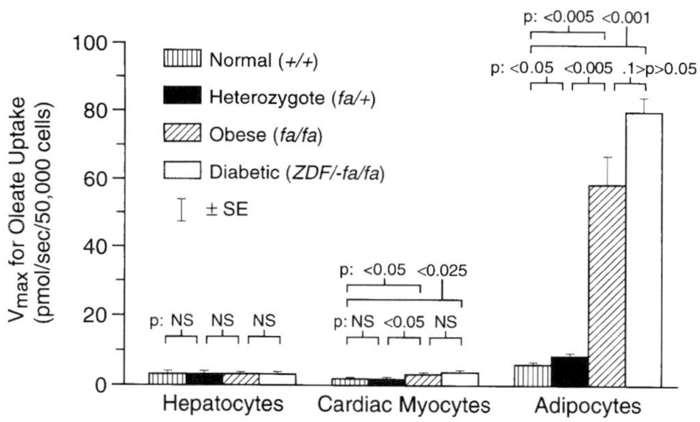

Fig. 12. Uptake of FFA by various cells from Zucker rats with genetic obesity and NIDDM. Figure illustrates computed values for the V_{max} of the saturable component of [^{3}H]-oleate uptake by isolated hepatocytes, cardiac myocytes, and adipocytes from four groups of animals: normal Wistar cntrols (+/+), Zucker lean heterozygotes (*fa*/+), Zucker fatty homozygotes (*fa*/*fa*), and Zucker diabetic fatty *(ZDF)*. Reproduced from [75].

Physiologic implications of facilitated FFA transport

In contrast to passive trans-membrane fluxes, facilitated transport processes are regulatable both physiologically and, potentially, pharmacologically. Therefore, the documentation that cellular FFA uptake had a major facilitated component led us to seek situations in which this process was likely to be altered. We chose for our first example to study FFA uptake in various tissues of the homozygous Zucker fatty rat, a genetically defined model of obesity [66]. The Zucker rat has a mutation in the long form of the leptin receptor [67–70] and is therefore congenitally leptin deficient. Leptin deficiency results both in hyperphagia and in a variety of metabolic abnormalities which have the added effect of increasing the efficiency of caloric utilization, so that Zucker animals become obese even when pair fed with normal controls [71–73]. We conducted studies of FFA uptake kinetics in normal Wistar rats (+/+), in non-obese Zucker heterozygotes (*fa*/+), in fatty Zucker homozygotes (*fa*/*fa*), and in adult males of the Zucker diabetic fatty strain (*ZDF*), which develop NIDDM in addition to obesity [74]. These studies examined [^{3}H]-oleate uptake by isolated adipocytes, hepatocytes, and cardiac myocytes from these strains [29, 75]. As summarized in Fig. 12, FFA uptake was identical in hepatocytes from all 4 strains. Small increases in uptake V_{max} were observed in cardiac myocytes from the *fa*/*fa* and *ZDF* animals. By contrast, there were striking 9–13 fold increases in FFA uptake V_{max} in *fa*/*fa* and *ZDF* adipocytes. Subsequent experiments confirmed that these differences in uptake were followed by comparable differences in the intracellular concentrations of esterified [^{3}H]-derivatives. Likewise, appropriate control studies documented that the observed changes in uptake kinetics were not simply a consequence of increased adipocyte cell size, increased lipolysis or depletion or dilution of the [^{3}H]-FFA substrate present in the incubation medium. These studies documented, therefore: (1) that the increased V_{max} for FFA uptake in *fa*/*fa* and *ZDF* adipocytes reflects upregulation of specific plasma membrane transport mechanisms; and (2) that the regulation of these processes is highly tissue specific. In subsequent studies of oleate uptake kinetics in adipocytes from 20–24 day old weanling +/+, *fa*/+ and *fa*/*fa* rats, among which there were no differences in body weight or plasma glucose, FFA, or insulin levels, there was nevertheless a highly significant increase in the V_{max} for FFA uptake in the *fa*/*fa* animals which could not be explained by the relatively modest increases in adipocyte cell size observed in the sample fat pads. TNFα message levels, which are increased in adult rats of the *fa*/*fa* and *ZDF* strains, were unchanged in weanling animals. Thus, the upregulation of adipocyte FFA transport in the Zucker rat models of obesity and NIDDM is an early event which precedes development of obesity, increased plasma FFA, hyperglycemia, or increased production of TNFα. Although TNFα message levels were not increased in the *fa*/*fa* weanlings, lipoprotein lipase (LPL) and leptin mRNAs were appreciably up-regulated; it is not possible to state at this time whether the increased FFA uptake V_{max} precedes or follows the increased levels of these messages [29]. As illustrated in Fig. 13, the data obtained to date suggest that an increase in adipocyte FFA uptake in Zucker animals is an early step in the cascade of events leading to obesity and NIDDM.

The changes in FFA uptake kinetics just described are independent of the role of particular putative transporters. However, when mAspAT mRNA levels were examined by slot blotting in hepatocytes and adipocytes from +/+, *fa*/*fa*,

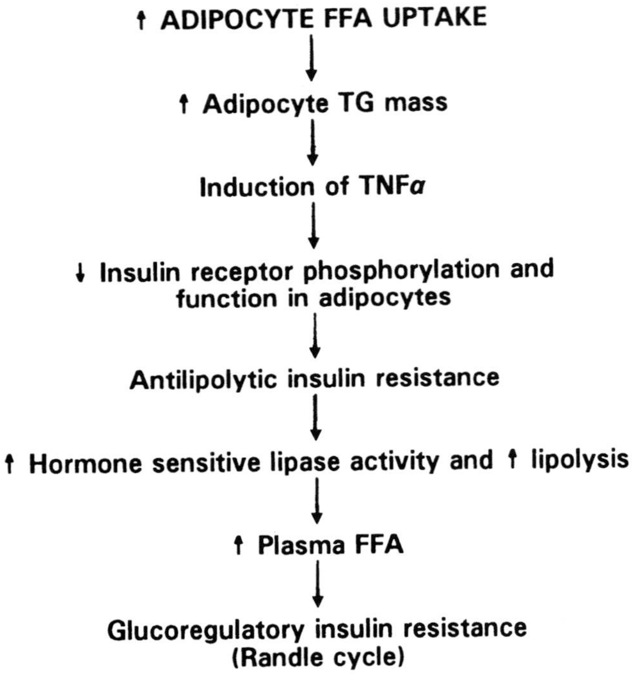

Fig. 13. Proposed cascade of events leading from increased adipocyte FFA uptake to obesity and NIDDM.

and *ZDF* rats, no differences between strains were observed in hepatocytes, consistent with the lack of differences in FFA transport. By contrast, mAspAT mRNA was significantly upregulated in adipocytes from *fa*/*fa* and *ZDF* animals; adipocyte mAspAT mRNA levels were highly correlated with the V_{max} for FFA uptake (Fig. 14). This finding was confirmed by Northern hybridization. Using cDNA clones for FAT and FATP, obtained through the courtesy of Drs. Abumrad and Shaffer, we were able to document by Northern hybridization that message levels for both FAT and FATP were also up-regulated in adipocytes from adult and weanling *fa*/*fa* animals compared to +/+ normal controls [29, 75]. These data are consistent with the hypothesis that upregulation of each of the three cloned putative FFA transporters occurs as part of the evolution of obesity in this particular model. Very similar adipocyte specific changes in FFA transport have been observed (unpublished) in the homozygous obese mouse *(ob*/*ob)* which lacks leptin [76] and in the diabetic *(db*/*db)* mouse which has a mutation in the leptin receptor [77–79].

Most cases of human obesity do not reflect single gene inheritance. Accordingly, we are studying the effects of high fat diets in adult rats of 3 strains; S 5B/P1, Sprague-Dawley, and Osborne-Mendel. S 5B/P1 animals are resistant to the effects of a high fat diet with respect to the development of obesity. Sprague-Dawley and Osborne-Mendel rats are much more prone to developing obesity when fed a high fat diet [80]. This diversity among strains has an obvious human counterpart. Our observations in these animals are qualitatively similar to those observed in the Zucker fatty rat and the *obese* and *diabetic* mouse. As predicted, Sprague-Dawley and Osborne-Mendel rats gained considerably more weight

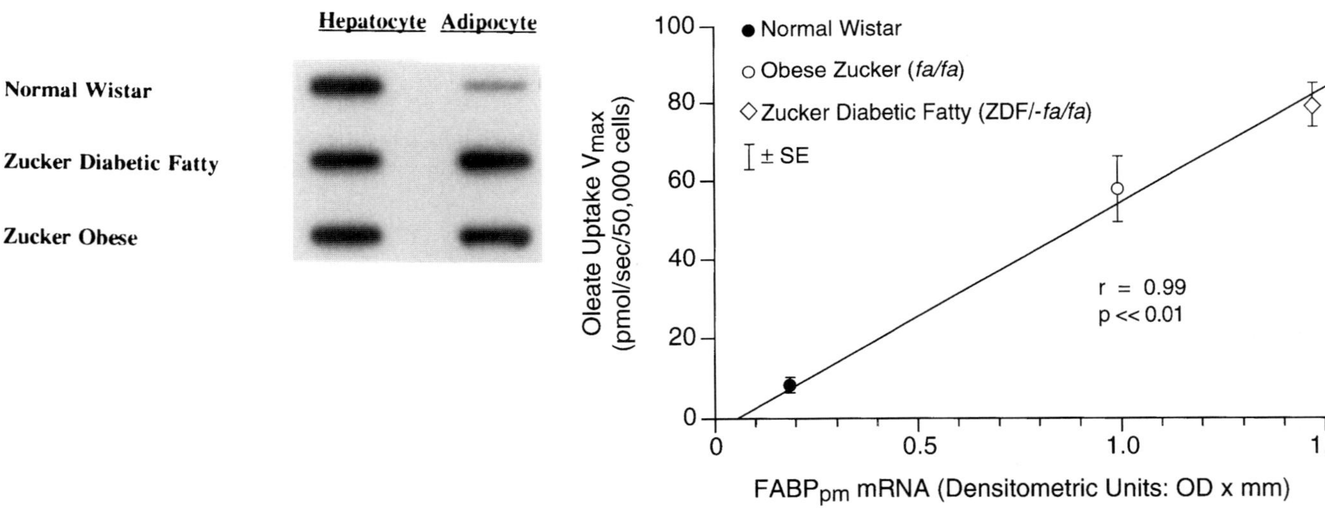

Fig. 14. Relation of mAspAT expression and FFA uptake in Normal Wistar (+/+), homozygous Zucker fatty (*fa*/*fa*), and Zucker diabetic fatty *(ZDF)* rats. Left hand panel: PolyA$^+$ RNA slot blots probed with a [^{32}P]-labeled probe for the 5′ end of the mAspAT message. After normalization for the signal from the β-actin probe, which served as an internal control, relative mAspAT mRNA levels in adipocytes were increased 5.7 fold in *fa*/*fa* and 8.9 fold in *ZDF* rats compared to +/+ controls. No appreciable differences in hepatocyte mAspAT mRNA levels were detected among these three groups of animals. Right hand panel: Comparison of relative mAspAT mRNA levels in adipocytes with oleate uptake Vmax values (see Fig. 12) in the same cell populations. The two variables are highly correlated. Both mAspAT mRNA and V_{max} values were equivalent in hepatocytes from these strains of animals. Reproduced from [75].

on a high fat diet then did the S 5B/P1 animals. There was no change in adipocyte FFA uptake kinetics in the S 5B/P1 strain. However, the FFA uptake V_{max} increased significantly in Sprague-Dawley, and even further in the Osborne-Mendel animals. Among these three strains there was a highly significant correlation ($r = 0.91$, $p < 0.05$) between the weight gained on the high fat diet and the increase in adipocyte FFA uptake V_{max}. No changes were observed in hepatocyte FFA transport in these studies (unpublished). Thus, tissue specific changes in adipocyte FFA uptake occur not only in purely genetically determined obesity but also in obesity having a predominantly dietary etiology.

The tissue specific changes in FFA uptake observed in these studies have the net effect of shunting plasma FFA away from tissues where they would be oxidized as fuel and toward adipose tissue where they are stored as fat [81]. Such changes, established early in the development of obesity, would predispose toward further fat accumulation. They would also explain the common observation that, once obesity is established, its reversal by dietary means, at least initially, proves difficult.

The observation that tissue specific physiologic alterations in FFA uptake occur in obesity raises the possibility that equally tissue specific pharmacologic manipulations could reverse these changes. Pursuit of this approach should, logically, be based on an understanding of the factors that regulate FFA transport in the relevant tissues. Such information is currently lacking. Nevertheless, documentation that FFA enter cells by regulatable, facilitated transport mechanisms has important pathophysiologic and therapeutic implications for human disease.

Acknowledgements

Supported in part by NIH Grants DK-26438 and DK-52401 from the National Institute of Diabetes, Digestive and Kidney Disease, and by generous gifts from the Liver Diseases Research Fund.

References

1. Distel RJ, Robinson GS, Spiegelman BM: Fatty acid regulation of gene expression. J Biol Chem 267: 5937–5941, 1992
2. Amri E-Z, Ailhaud G, Grimaldi PA: Fatty acids as signal transducing molecules: Involvement in the differentiation of preadipose to adipose cells. J Lip Res 35: 930–937, 1994
3. Noy N, Zakim D: Physical chemical basis for the uptake of organic compounds by cells. In: N. Tavoloni and P.D. Berk (eds). Hepatic Transport and Bile Secretion, Raven Press, New York, 1993, pp 313–335
4. Zakim D: Fatty acids enter cells by simple diffusion. Proc Soc Exptl Biol Med 212: 5–14, 1996
5. Kamp F, Hamilton JA: pH gradients across phospholipid membranes caused by fast flip-flop of un-ionized fatty acids. Proc Natl Acad Sci USA 89: 11367–11370, 1992
6. Kamp F, Hamilton JA: Movement of fatty acids, fatty acid analogues, and bile acids across phospholipid bilayers. Biochem 32: 11074–11086, 1993
7. Kamp F, Zakim D, Zhang F, Noy N, Hamilton JA: Fatty acid flip-flop in phospholipid bilayers is extremely fast. Biochem 34: 11928–11937, 1995
8. Hamilton JA, Civelek VN, Kamp F, Tornheim K, Corkey BE: Changes in internal pH caused by movement of fatty acids into and out of clonal pancreatic β-cells (HIT). J Biol Chem 269: 20852–20856, 1994
9. Civelek VN, Hamilton JA, Tornheim K, Kelly KL, Corkey BE: Intracellular pH in adipocytes: Effects of free fatty acid diffusion across the plasma membrane, lipolytic agonists, and insulin. Proc Natl Acad Sci USA 93: 10139–10144, 1996
10. Kleinfeld AM, Storch J: Transfer of long chain fluorescdnt fatty acids between small and large unilamellar vesicles. Biochem 32: 2053–2061, 1993
11. Sorrentino D, Berk PD: Free fatty acids and the sinusoidal plasma membrane: Concepts, trends and controversies. In: N. Tavoloni and P.D. Berk (eds). Hepatic Transport and Bile Secretion, Raven Press, New York, 1993, pp 197–210
12. Berk PD, Bradbury M, Zhou S-L, Stump D, Han NI: Characterization of membrane transport processes: Lessons from the study of BSP, bilirubin and fatty acid uptake. Semin Liv Dis 16: 107–120, 1996
13. Goresky CA, Stremmel W, Rose CP, *et al.*: The capillary transport system for free fatty acids in the heart. Circ Res 74: 1015–1026, 1994
14. Vyska K, Meyer W, Stremmel W, Notohamiprodjo G, Minami K, Machulla H-J, Gleichmann U, Meyer H, Körfer R: Fatty acid uptake in normal human myocardium. Circ Res 69: 857–880, 1991
15. Weisiger RA, Gollan J, Ockner R: Receptor for albumin on the liver cell surface may mediate uptake of fatty acids and other albumin bound substances. Science 211: 1048–1051, 1981
16. Sorrentino D, Robinson RB, Kiang C-L, Berk PD: At physiologic albumin/oleate concentrations oleate uptake by isolated hepatocytes, cardiac myocytes and adipocytes is a saturable function of the unbound oleate concentration. J Clin Invest 84: 1325–1333, 1989
17. Sorrentino D, Zifroni A, Van Ness K, Berk PD: Unbound ligand concentrations drive the hepatocellular uptake of taurocholate and sulfobromophthalein at physiologic albumin concentrations. Am J Physiol 266: G425–G432, 1994
18. Sorrentino D, Van Ness K, Stump DD, Berk PD: Oleate uptake kinetics in the perfused rat liver are consistent with pseudofacilitation by albumin. J Hepatol 21: 551–559, 1994
19. Spector AA, Fletcher JE, Ashbrook JD: Analysis of long chain free fatty acid binding to bovine serum albumin by determination of stepwise equilibrium constants. Biochemistry 10: 3229–3232, 1971
20. Richieri GV, Anel A, Kleinfeld AM: Interactions of long chain fatty acids and albumin: Determination of free fatty acid levels using the fluorescent probe ADIFAB. Biochem 32: 7574–7580, 1993
21. Rose H, Conventz M, Fischer Y, Ringling E, Hennecke T, Kammermeier H: Long-chain fatty acid-binding to albumin: Re-evaluation with directly measured concentrations. Biochim Biophys Acta 1215: 321–326, 1994
22. Bojesen IN, Bojesen E: Water-phase palmitate concentrations in equilibrium with albumin-bound palmitate in a biological system. J Lipid Res 33: 1327–1334, 1992
23. Bojesen IN, Bojesen E: Binding of arachidonate and oleate to bovine serum albumin. J Lipid Res 35: 770–7784, 1994
24. Abumrad NA, Perkins RC, Park JH, Park CR: Mechanism of long chain fatty acid permeation in the isolated adipocyte. J Biol Chem 256: 9183–9191, 1981

25. Abumrad NA, Park JH, Park CR: Permeation of long-chain fatty acid into adipocytes. J Biol Chem 259: 8945–8953, 1984
26. Nunes R, Kiang C-L, Sorrentino D, Berk PD: 'Albumin-receptor' kinetics do not require an intact lobular architecture and are not specific for albumin. J Hepatol 7: 293–304, 1988
27. Stump, DD, Nunes RM, Sorrentino D, Berk PD: Characteristics of oleate binding to liver plasma membrane and its uptake by isolated hepatocytes. J Hepatol 16: 304–315, 1992
28. Schwieterman W, Sorrentino D, Potter BJ, Rand J, Kiang C-L, Stump D, Berk PD: Uptake of oleate by isolated rat adipocytes is mediated by a 40-kDa plasma membrane fatty acid binding protein closely related to that in liver and gut. Proc Natl Acad Sci USA 85: 359–363, 1988
29. Berk PD, Zhou S-L, Kiang C-L, Stump D, Bradbury M, Isola LM: Uptake of long chain free fatty acids is selectively up-regulated in adipocytes of Zucker rats with genetic obesity and non-insulin-dependent diabetes mellitus. J Biol Chem 272: 8830–8835, 1997
30. Schmider W, Fahr A, Voges R, Gerok W, Kurz G: Irreversible inhibition of fatty acid salt uptake by photoaffinity labeling with 11,11-azistearate. J Lipid Res 37: 739–753, 1996
31. Stremmel W, Berk PD: Hepatocellular influx of [^{14}C]oleate reflects membrane transport rather than intracellular metabolism or binding. Proc Natl Acad Sci USA 83: 3086–3090, 1986
32. Stremmel W, Strohmeyer G, Berk PD: Hepatocellular uptake of oleate is energy dependent, sodium linked, and inhibited by an antibody to a hepatocyte plasma membrane fatty acid binding protein. Proc Natl Acad Sci USA 83: 3584–3588, 1986
33. Stremmel W: Translocation of fatty acids across the basolateral rat liver plasma membrane is driven by an active potential-sensitive sodium-dependent transport system. J Biol Chem 262: 6284–6289, 1987
34. Sorrentino D, Stump D, Potter BJ, Robinson RB, White R, Kiang CL, Berk PD: Oleate uptake by cardiac myocytes is carrier mediated and involves a 40-kD plasma membrane fatty acid binding protein similar to that in liver, adipose tissue and gut. J Clin Invest 82: 928–935, 1988
35. Stremmel W: Fatty acid uptake by isolated rat heart myocytes represents a carrier-mediated transport process. J Clin Invest 81: 844–852, 1988
36. Stremmel W: Uptake of fatty acids by jejunal mucosal cells is mediated by a fatty acid binding membrane protein. J Clin Invest 82: 2001–2010, 1988
37. Trimble ME: Mediated transport of long chain fatty acids by rat renal basolateral membranes. Am J Physiol F539–F546, 1989
38. Storch J, Lechene C, Kleinfeld A: Direct determination of free fatty acid transport across the adipocyte plasma membrane using quantitative fluorescence microscopy. J Biol Chem 266: 13473–13476, 1991
39. Turcotte LP, Kiens B, Richter EA: Saturation kinetics of palmitate uptake in skeletal muscle. FEBS Lett 279: 327–329, 1991
40. Sorrentino D, Stump DD, Van Ness K, Simard A, Schwab AJ, Zhou S-L, Goresky CA, Berk PD: Oleate uptake by isolated hepatocytes and the perfused rat liver is competitively inhibited by palmitate. Amer J Physiol 270: G385–G392, 1996
41. Zhou S-L, Stump DD, Sorrentino D, Potter BJ, Berk PD: Adipocyte differentiation of 3T3-L1 cells involves augumented expression of a 43 kDa plasma membrane fatty acid binding protein. J Biol Chem 267: 14456–14461, 1992
42. Zhou S-L, Stump DD, Kiang C-L, Isola LM, Berk PD: Mitochondrial aspartate aminotransferase expressed on the surface of 3T3-L1 adipocytes, mediates saturable fatty acid uptake. Proc Soc Exptl Biol Med 208: 263–270, 1995
43. Isola LM, Zhou S-L, Kiang C-L, Stump DD, Bradbury MW, Berk PD: 3T3 fibroblasts transfected with a cDNA for mitochondrial aspartate aminotransferase express plasma membrane fatty acid-binding protein and saturable fatty acid uptake. Proc Nati Acad Sci USA 92: 9866–9870, 1995
44. Daniels C, Noy N, Zakim D: Rates of hydration of fatty acids bound to unilamellar vescicles of phosphatidyl choline or to albumin. Biochem 24: 3286–3292, 1985
45. Storch J, Kleinfeld AM: Transfer of long chain fluorescent free fatty acids between unilamellar vesicles. Biochem 25: 1717–1726, 1986
46. Sorrentino D, Zhou S-L, Kokkoutou E, Berk PD: Sex differences in hepatic fatty acid uptake reflect a greater affinity of the membrane transport system in females. Am J Physiol 263: G380–G385, 1992
47. Skulachev VP: Fatty acid circuit as a physiological mechanism of un-coupling of oxidative phosphorylation. FEBS Letts 294: 158–162, 1991
48. Garlid KD, Orosz DE, Modriansky M, Vassanelli S, Jezek P: On the mechanism of fatty acid induced proton transport by mitochondrial uncoupling protein. J Biol Chern 271: 2615–2620, 1996
49. Stremmel W, Strohmeyer G, Borchard F, Kochwa S, Berk PD: Isolation and partial characterization of a fatty acid binding protein in rat liver plasma membranes. Proc Natl Acad Sci USA 82: 4–8, 1985
50. Fujii S, Kawaguchi H, Yasuda H: Isolation and partial characterization of an amphiphilic 56-kDa fatty acid binding protein from rat renal basolateral membrane. J Biochem Biophys Res 101: 679–684, 1987
51. Fujii S, Kawaguchi H, Yasuda H: Purification of high affinity fatty acid receptors in rat myocardial sarcolemmal membranes. Lipids 22: 544–546, 1987
52. Trigatti BL, Mangaroo D, Gerber GE: Photoaffinity labeling and fatty acid permeation in 3T3-L1 adipocytes. J Biol Chem 266: 22621–22625, 1991
53. Abumrad NA, Raafat El-Maghrabi M, Amri E-Z, Lopez E, and Grimaldi PA: Cloning of a rat adipocyte membrane protein implicated in binding or transport of long-chain fatty acids that is induced during preadipocyte differentiation. J Biol Chem 268: 17665–17668, 1993
54. Schaffer JE, Lodish HF: Expression cloning and characterization of a novel adipocyte long chain fatty acid transport protein. Cell 79: 427–436, 1994
55. Stump DD, Zhou S-L, Berk PD: Comparison of the plasma membrane FABP and the mitochondrial isoform of aspartate amino-transferase from rat liver. Am J Physiol 265: G894–G902, 1993
56. Potter BJ, Berk PD: The liver plasma membrane fatty acid binding protein. In: N. Tavoloni and P.D. Berk (eds). Hepatic Transport and Bile Secretion, Raven Press, New York, 1993, pp 253–267
57. Stremmel W, Theilmann L: Selective inhibition of long chain fatty acid uptake in short-term cultured rat hepatocytes by an antibody to the rat liver plasma membrane fatty acid binding protein. Biochim Biophys Acta 877: 191–197, 1986
58. Stremmel W, Diede HE, Rodilla-Sala E, Vyska K, Schrader M, Fitscher B, Passarella S: The membrane fatty acid binding protein is not identical to mitochondrial glutamic oxaloacetic transaminase (mGOT). Mol Cell Biol 98: 191–199, 1990
59. Berk PD, Wada H, Horio Y, Potter PJ, Sorrentino D, Zhou S-L, Isola LM, Stump D, Kiang C-L, Thung S: Plasma membrane fatty acid binding protein and mitochondrial glutamic-oxaloacetic transaminase of rat liver are related. Proc Natl Acad Sci USA 87: 3484–3488, 1990
60. Zhou S-L, Kiang C-L, Bradbury M, Isola LM, Stump DD, Berk PD: Up-regulation of mitochondrial aspartate aminotransferase mRNA parallels fatty acid uptake in 3T3-L1 cells and adipocytes from obese and diabetic rats (Abstract). Hepatol 22: 313A, 1995
61. Guarnieri F, Stump DD, Roboz J, Yu Q, Berk PD: Mitochondrial aspartate aminotransferase contains a 500 Å^3 hydrophobic groove representing a putative fatty acid binding site (Abstact). Hepatol 22: 424A, 1995
62. Zhou S-L, Kiang C-L, Bradbury M, Gordon R, Stump DD, Han NI, Berk PD: Effects of ETOH on mitochondrial aspartate aminotransferase expression and cellular leakage, and fatty acid uptake in HepG2 cells (Abstract). Hepatol 22: 241A, 1995

63. Zhou S-L, Gordon R, Kiang C-L, Stump DD, Bradbury M, Isola LM, Berk PD: Intracellular movement and cellular export of mitochondrial aspartate aminotransferase from hepatoma cell lines (Abstract). Hepatol 24: 128A, 1996
64. Soltys BJ, Gupta RS: Immunoelectron microscopic localization of the 60 kDa heat shock chaperonin protein (Hsp60) in mammalian cells. Exper Cell Res 222: 16–27, 1996
65. Dil M, Cascante M, Cortes A, Bozal J: Intramitochondrial location and some characteristics of chicken liver aspartate aminotransferase. Int J Biochem 19: 355–363, 1987
66. Kava R, Greenwood MRC, Johnson PR: Zucker (*fa/fa*) rat. ALAR News 32: 4–8, 1990
67. Chua SC Jr, Chung WK, Wu-Peng S, Zhang Y, Liu SM, Tartaglia L, Leibel RL: Phenotypes of mouse *diabetes* and rat *fatty* due to mutations in the OB (leptin) receptor. Science 271: 994–996, 1996
68. Phillips MS, Liu Q, Hammond HA, Dugan V, Hey PJ, Caskey CT, Hess JF: Leptin receptor missense mutation in the fatty Zucker rat. Nat Genet 13: 18–19, 1996
69. Iida M, Murakami T, Ishida K, Mizuno A, Kuwajima M, Shima K: Substitution at codon 269 (glutamine→proline)on the leptin receptor (OB-R) is the only mutation found in the Zucker fatty (*fa/fa*) rat. Biochem Biophys Res Comun 224: 597–604, 1996
70. Chua SC Jr, White DW, Sharon Wu-Peng X, Liu SM, Okada N, Kershaw EE, Chung WK, Power-Kehoe L, Chua M, Tartaglia LA, Leibel RL: Phenotype of *fatty* due to G1n269Pro mutation in the leptin receptor *(Lepr)*. Diabetes 45: 1141–1143, 1996
71. Pelleymounter MA, Cullen MJ, Baker MB, Hecht R, Winters T, Boone T, Collins F: Effects of the *obese* gene product on body weight regulation in *ob/ob* mice. Science 269: 540–543, 1995
72. Halaas JL, Gajiwala KS, Maffei M, Cohen SL, Chaoit BT, Rabinowitz D, Lallone RL, Burley SK, Friedman M: Weight-reducing effects of the plasma protein encoded by the *obese* gene. Science 269: 543–546, 1995
73. Campfield LA, Smith FJ, Guisez Y, Devos R, Burn P: Recombinant mouse OB protein: Evidence for a peripheral signal linking adiposity and central neural networks. Science 269: 546–549, 1995
74. Peterson RG, Shaw WN, Neel MA, Little LA, Eichberg J: Zucker diabetic fatty rat as a model for non-insulin-dependent diabetes mellitus. ALAR News 32: 16–19, 1990
75. Berk PD, Zhou S-L, Bradbury M, Stump D, Kiang C-L, Isola LM: Regulated membrane transport for free fatty acids in adipocytes: Role in obesity and non-insulin dependent diabetes mellitus. Trans Amer Clin Clim Assoc 108: 26–43, 1997
76. Zhang Y, Proenca R, Maffel M, Barone M, Leopold L, Friedman M: Positional cloning of the mouse *obese* gene and its human homologue. Nature 372: 425–431, 1994
77. Lee GH, Proenca R, Montez M, Carroll KM, Darvishzadeh JG, Lee JI, Friedman JM: Abnormal splicing of the leptin receptor in *diabetic* mice. Nature 379: 632–635, 1996
78. Chen H, Charlat O, Tartaglia LA, Woolf EA, Weng X, Ellis SJ, Lakey ND, Culpepper J, Moore M, Breitbart RE, Duyk GM, Tepper RI, Morgenstern JP: Evidence that the diabetes gene encodes the leptin receptor: Identification of a mutation in the leptin receptor gene in *db/db* mice. Cell 84: 491–495, 1996
79. Chua SC Jr, Chung WK, Wu-Peng S, Zhang Y, Liu SM, Tartaglia L, Leibel RL: Phenotypes of mouse *diabetes* and rat *fatty* due to mutations in the OB (leptin) receptor. Science 271: 994–996, 1996
80. Schemmel R, Mickelsen J, Gill JL: Dietary obesity in rats: Body weight and body fat accretion in seven strains of rats. J Nutr 100: 1041–1048, 1990
81. Bessesen DH, Rupp CL, Eckel RH: Dietary fat is shunted away from oxidation, toward storage in obese Zucker rats. Obes Res 3: 179–189, 1995

Molecular and Cellular Biochemistry **192:** 33–40, 1999.

Biochemical and biophysical analysis of the intracellular lipid binding proteins of adipocytes

Melanie A. Simpson, Vince J. LiCata, Natalie Ribarik Coe and David A. Bernlohr

Department of Biochemistry, University of Minnesota, St. Paul, MN, USA

Abstract

Adipocytes express two lipid-binding proteins; the major one termed the adipocyte lipid-binding protein or aP2 (ALBP/aP2) and a minor one referred to as the keratinocyte lipid-binding protein (KLBP). In order to evaluate the potential physiological roles for these proteins, their biochemical and biophysical properties have been analyzed and compared. ALBP/aP2 and KLBP exhibit similar binding affinities for most long-chain fatty acids; however, ALBP/aP2 exhibits a two to three-fold increased affinity for myristic, palmitic, oleic and linoleic acids, the predominant fatty acids of adipocytes. As measured by guanidinium hydrochloride denaturation, the stability of ALBP/aP2 is nearly 3 kcal/mol greater than that of KLBP. While the pI of ALBP/aP2 was determined to be 9.0, that of KLBP is 6.5 suggesting differing net charges at physiological pH. Analysis of surface electrostatic properties of ALBP/aP2 and KLBP revealed similar charge polarity, although differences in the detailed charge distribution exist between the proteins. The distribution of hydrophobic patches was also different between the proteins, ALBP/aP2 has only scattered hydrophobic surfaces while KLBP has a large hydrophobic patch near the ligand portal into the binding cavity. In sum, these results point out that despite the striking similarity between ALBP/aP2 and KLBP in tertiary structure, significant differences in ligand binding and surface properties exist between the two proteins. Hence, while it is tempting to speculate that ALBP/aP2 and KLBP are metabolically interchangeable, careful analysis suggests that the two proteins are quite distinct and likely to play unique metabolic roles. (Mol Cell Biochem **192**: 33–40, 1999)

Key words: fatty acids, adipocytes, binding proteins, electrostatics

Introduction

Lipid-binding proteins (LBPs) are small, abundantly distributed, cytosolic polypeptides which bind hydrophobic ligands [for reviews see [1–3]. Over 20 different LBP family members have been identified in vertebrates alone. Although named frequently for the tissue from which they were initially isolated, most LBPs are expressed in several cell or tissue types, and any given tissue may express multiple LBPs. Crystal structures have revealed that members of the family share a superimposable tertiary structure despite having primary sequence identity that varies from 20–70%. The tertiary structure is characterized by a flattened ten-stranded β-barrel that encompasses a water-filled internal ligand-binding cavity secluded from solvent by a somewhat flexible helix-turn-helix cap. Despite their relative abundance, solubility, ease of purification, and a wealth of structural data, the specific physiological functions of these proteins are unknown. The presence of a LBP in lipid metabolizing tissues is likely necessitated by cellular demand for fatty acids at a level beyond their inherent cytosolic solubility. This is especially true in adipocytes, where massive fluxing of lipids occurs continually.

The adipocyte member of the LBP family (ALBP or aP2) is found exclusively in adipocytes or adipogenic cell lines and was, until recently, thought to be the only LBP in fat cells. It has since been shown that murine adipocytes express a minor LBP, the keratinocyte lipid-binding protein (KLBP) as well, albeit at very low levels relative to ALBP/aP2 (approximately 1% of ALBP/aP2 protein levels) [4]. KLBP was originally

Address for offprints: D.A. Bernlohr, Department of Biochemistry, University of Minnesota, St. Paul, MN 55108, USA

identified as an over-expressed mRNA in murine squamous cell carcinomas and other transformed skin lines [5, 6]. Examination of adipose tissue from transgenic mice null for ALBP/aP2 revealed that when ALBP/aP2 is not expressed, KLBP becomes significantly upregulated. However, neither protein nor message levels of KLBP achieve the concentration of ALBP/aP2 in normal adipocytes [4].

Initial characterization of ALBP/aP2 null mice on a standard lab chow diet containing 4% fat demonstrated few metabolic abnormalities. Hence, it was speculated, KLBP must function so similarly within the adipocyte that it is able to compensate on a molecular level for the loss of ALBP/aP2. However, a strikingly different interpretation arose when ALBP/aP2 null mice were maintained on a high fat diet. Hotamisligil *et al.*, observed that although there were no obvious outward phenotypic differences, the ALBP/aP2 mice failed to develop obesity-related non insulin dependent diabetes mellitus [7]. That is, when fed a diet high in fat (40% of calories from fat), wild type mice became somewhat obese, hyperglycemic, hyperinsulinemic, and responded poorly to insulin or glucose tolerance tests. In contrast, ALBP/aP2 null mice became quite obese, but maintained low circulating levels of insulin and glucose, and responded well to insulin or glucose tolerance tests. Relative expression levels of several other adipocyte mRNAs, including enzymes involved in fat cell lipid metabolism, were unchanged, suggesting that essential fat cell metabolism was largely unaltered by the ALBP/aP2 null status. However, whereas obese wild type mice expressed high levels of tumor necrosis factor α (TNFα), neither lean wild type nor obese ALBP/aP2 disrupted mice harbored significant quantities of this diabetes-associated cytokine.

Ribarik Coe *et al.* [4], further observed that the efflux of fatty acids from adipocytes of null mice was impaired relative to wild type and that free fatty acids were elevated in such nulls. Together these findings suggest severe alterations in lipid trafficking resulting from loss of ALBP/aP2 expression. Since adipocytes express more than one lipid-binding protein, the failure to develop obesity-linked non-insulin dependent diabetes mellitus could derive from loss of ALBP/aP2 and/or increase in KLBP. Altered efflux efficacy and subsequent accumulation of fatty acids could result from less productive interaction between KLBP and cytoplasmic lipids. Alternatively, metabolic abnormalities may arise due to inefficient interactions between KLBP and intracellular proteins or membranes. Differences in the mechanism of ligand transfer from protein to phospholipid vesicles have been documented between LBPs, and in some cases, have been attributed to specific charged residues [8–11]. These observations have prompted us to explore the biochemical and biophysical characteristics of the two proteins.

We present here a detailed comparison of the intrinsic biochemical, ligand-binding, and electrostatic properties of ALBP/aP2 and KLBP. Although binding properties are very similar, there are notable differences. Additionally, the proteins were found to have significantly different isoelectric points, surface properties, and chemical stabilities. We discuss these differences within the context of altered lipolytic capacities in ALBP/aP2-disrupted mice and speculate about implications for non-insulin dependent diabetes mellitus.

Materials and methods

Purification of ALBP/aP2 and KLBP protein

Recombinant proteins were expressed in *E. coli* and purified essentially as described [12]. Briefly, both purifications employ pH 5 acetate precipitation and gel filtration chromatography. Following gel filtration, ALBP/aP2 or KLBP is loaded on a BioS anion exchange column (BioRad) in 50 mM NaOAc at pH 5.2 and eluted with a gradient of 0–1 M NaCl (ALBP/aP2) or 0–0.6 M NaCl (KLBP) in 50 mM acetate (pH 5.2).

Isoelectric focusing of ALBP/aP2 and KLBP

Standard two-dimensional electrophoresis techniques were employed to determine the isoelectric points of ALBP and KLBP. Isoelectric focusing calibration markers (pI range 3–10, Pharmacia) in the presence and/or absence of 20 μg of purified KLBP were run in an acrylamide (5.5%) vertical tube gel (400 V, 4°C, 18 h) from basic (sodium hydroxide) to acidic conditions (13 mM phosphoric acid). 40 μg of purified ALBP/aP2 was run in a denaturing acrylamide tube gel (8.4% acrylamide, 9.25 M urea, 4% Nonidet P-40) under identical conditions. The resulting tube gels were analyzed in the second dimension by SDS-PAGE (12.5% acrylamide).

1,8-ANS binding and competition assays

In vitro binding data for various putative ligands of ALBP/aP2 or KLBP were measured by displacement of LBP-bound 1-anilinonaphthalene-8-sulfonic acid (1,8-ANS) as previously described [13]. Briefly, increasing competitor ligand concentrations were added to ANS-bound LBP ([ANS] = 500 nM, [ALBP] = 540 nM, [KLBP] = 390 nM) in 50 mM $NaPO_4$ (pH 7.4). The decrease in fluorescence was plotted as a function of competitor concentration and used to calculate competitor constants. Excitation and emission wavelengths were 368 and 465 nm, respectively, for ANS/ALBP or 375 and 473 nm for ANS/KLBP.

Stability of ALBP/aP2 and KLBP proteins

Relative stability of ALBP/aP2 and KLBP was assessed by monitoring the shift in maximum tryptophan emission wavelength as a function of increasing guanidine HCl concentration as previously described [14]. Denaturation curves were analyzed to determine Cm, the concentration of guanidine HCl at 50% denaturation, and free energies of unfolding were calculated by the linear extrapolation method of Pace [15].

Electrostatic and hydrophobic surfaces of ALBP/aP2 and KLBP

Crystal coordinates for ALBP/aP2 were used to model its surface electrostatics (Brookhaven Protein Data Bank code 1LIB). The crystal structure for KLBP is not yet available so it was modeled using Swiss-Model, an automated modeling package implemented by internet (http://www.expasy.hcuge.ch/swissmod/SWISS-MODEL.html) [16, 17]. KLBP was modeled on the basis of its similarity to homologous structures existing in the Brookhaven Protein Data Bank: 10PA (apo cellular retinol binding protein II), 1ADL (ALBP/aP2-arachidonate), 1AB0 (C1G, V32D, F57H apoALBP, crystallized at pH 4.5), 1ACD (C1G, V32D, F57H apoALBP, crystallized at pH 6.4), 1PMP (myelin P2 protein), 1HMR (HFABP-elaidic acid), 1CBI (cellular retinoic acid binding protein I), and 1CBQ (cellular retinoic acid binding protein II). After primary modeling, the structure was energy minimized using CHARMm. Hydrogen atoms were added to structures using Insight II (Biosym, Inc.). Electrostatic calculations were carried out with the program GRASP [18]. Program defaults were used for all adjustable parameters except ionic strength, which was set at 0.145 M. Electrostatic potential contours shown were generated with GRASP. Hydrophobic residue distributions were generated with the program Rasmol [19].

Results

To explore the potential molecular mechanisms by which ALBP/aP2 and KLBP could impact fat cell metabolism, several biochemical and biophysical properties of ALBP/aP2 and KLBP were compared. Table 1 summarizes the molecular weights, tissue distributions [5, 20], and relative abundances in adipose [4]. ALBP/aP2 and KLBP are found in very different tissues, which might imply different functions for the proteins (ALBP/aP2 in adipose exclusively, KLBP low in intestine and kidney, some in heart, brain, liver, spleen, muscle, lung, adipose, mammary, and lens/tongue/epidermal epithelial cells). ALBP/aP2 is extremely abundant in adipocytes, constituting 1–5% of total soluble protein, whereas KLBP levels are about 1% those of ALBP/aP2. Relative abundance in adipose may be related to metabolic role; however, identification of functional differences between ALBP/aP2 and KLBP in adipose cells is more likely to occur through comparative ligand binding analysis. Richieri *et al.* have reported a systematic ligand-binding comparison between several lipid-binding proteins which demonstrated significant variability in both affinities and specificities [21]. We present here a comparative study of *in vitro* ligand-binding properties of ALBP/aP2 and KLBP.

Table 2 summarizes binding studies comparing affinities of ALBP/aP2 and KLBP for potential fatty acid ligands, fatty

Table 1. General properties of ALBP/aP2 and KLBP.

Property	ALBP/aP2	KLBP
number of amino acids	131	135
molecular mass (Da)	14,578	15,137
tissue distribution	adipose	skin, adipose, lens epithelium
relative abundance in fat	60 mg/g total fat protein	0.6 mg/g total fat protein
pI (predicted/experimental)	8.55/≈9	6.14/≈6.5

Molecular masses and predicted isoelectric points were obtained from the internet Swiss-Model server (http:llexpasy.hcuge.ch/swissmod/SWISS-MODEL.html). Empirical isoelectric points were estimated by comparison with standards on isoelectric focusing gels as illustrated in Fig. 2.

Table 2. In vitro binding of putative ligands by ALBP and KLBP

Ligand		ALBP	KLBP
decanoate	10:0	2800 ± 400	4300
myristate	14:0	506 ± 63	1873 ± 489
palmitate	16:0	390 ± 30	1087 ± 63
oleate	18:1	215 ± 20	320 ± 11
linoleate	18:2	368 ± 1	499 ± 40
linolenate	18:3	553 ± 8	495 ± 28
arachidonate	20:4	284 ± 21	412 ± 4
docosahexaenoate	22:6	198 ± 22	450
5-HPETE	20:4	317 ± 31	1100 ± 200
15-HPETE	20:4	412	600 ± 240
homogamma linolenate	20:3	832 ± 24	219 ± 6
eicosatrienoate	20:3	655 ± 17	186 ± 18
conjugated linoleate	18:2	149 ± 7	204 ± 14
15-deoxy-Δ12,14-PGJ_2		1910 ± 110	>10,000
LY-171883		2100	1860 ± 800
Wy-14643		≈10,000	≈10,000

In vitro binding data for various putative ligands of ALBP or KLBP were measured by displacement of 1-anilinonaphthalene-8-sulfonic acid (1,8-ANS) as described in Materials and methods. Briefly, increasing competitor ligand concentrations were added to protein and bound ANS in 50 mM $NaPO_4$ (pH 7.4). Excitation and emission wavelengths for ANS/ALBP were 368 and 465 nm, respectively. Corresponding wavelengths used for ANS/KLBP were 375 and 473. The decrease in fluorescence was plotted as a function of competitor concentration and used to calculate competitor constants.

acid analogs and several other lipids. Lipid competitor constants were measured based upon each ligand's ability to compete with the fluorescent probe 1-anilinonaphthalene-8-sulfonic acid for binding to ALBP/aP2 or KLBP. Affinities and specificities show few major differences, except for oleate, myristate, palmitate, and linoleate, all of which showed 2-3 fold lower affinity for KLBP than for ALBP/aP2. Since these are major constituents of the cellular fatty acid pool, a two-fold difference might affect the metabolic availability of substrate in ALBP/aP2 nulls relative to wild type cells.

The low levels of KLBP expression in adipose tissue relative to ALBP/aP2 prompted a comparison of the relative chemical stabilities of the two proteins. Shifts in tryptophan emission maxima resulting from denaturation were plotted as a function of denaturant concentration. Figure 1 illustrates a dramatic difference in the midpoint of the denaturation curve for KLBP vs. ALBP/aP2, which translates to a free energy of unfolding almost 3 kcal/mol lower for KLBP than for ALBP/aP2 ($\Delta G = -2.3$ kcal/mol for KLBP vs. $\Delta G = -5.3$ kcal/mol for ALBP). Hence, the relatively low intracellular levels of KLBP may be partially determined by intrinsic stability. This is supported by the observation that KLBP message levels in aP2 disrupted mice are elevated 40-fold relative to wild type, whereas steady state protein levels are only about 7-fold higher.

Storch and colleagues have demonstrated that collisional transfer of fatty acids from LBPs to phospholipid vesicles is dependent upon phospholipid charge composition [8, 9]. Specific positively charged residues in ALBP/aP2 and HFABP may mediate this dependence [10, 11]. While information on KLBP interaction with membrane vesicles is not available, potential explanations for membrane interaction specificity between LBPs include: (1) specific surface charge location; (2) overall protein charge density and (3) altered surface charge distribution. To analyze possible interactive differences between ALBP/aP2 and KLBP and membranes, we examined the electrostatic properties of ALBP/aP2 and KLBP. As presented in Fig. 2, isoelectric focusing resolved net charge differences between ALBP/aP2 and KLBP. ALBP/aP2 has a pI of about 9.0, whereas KLBP has a pI of about 6.5. Hence at intracellular pH, ALBP/aP2 will be more basic than KLBP. The primary sequences of ALBP/aP2 and KLBP are aligned in Fig. 3. All surface accessible lysines analogous to those mutagenized and shown to be involved in the transfer mechanism for heart FABP are conserved between ALBP/aP2 and KLBP suggesting that although these might regulate the mechanism of transfer, they probably do not confer any specificity of interaction.

Surface charge and hydrophobic amino acid distributions were examined for potential surface differences that might relate to function (Fig. 4A; details of the electrostatics of ALBP will be discussed elsewhere: [28]). Both ALBP/aP2 and KLBP are charge polarized, with positive dipoles for the two proteins oriented out the helix-turn-helix cap. The details of the electrostatic distribution differ for the two molecules, however. ALBP/aP2 has a conspicuous positive ridge across the top of the molecule, which is less pronounced in KLBP.

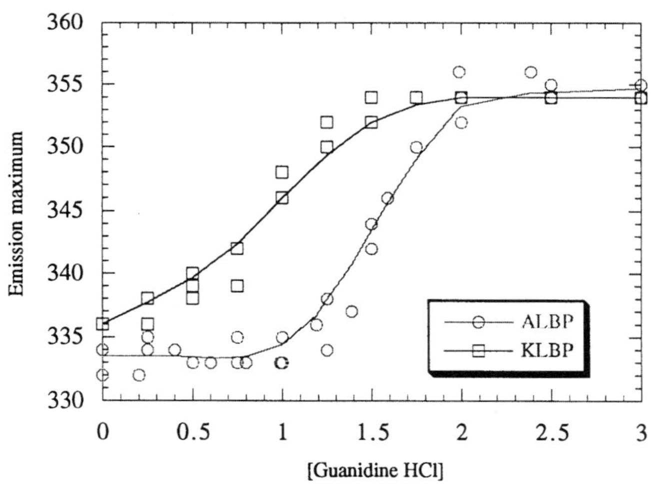

Fig. 1. Guanidine HCl denaturation of ALBP and KLBP. Relative stability of ALBP and KLBP was assessed by monitoring the shift in maximum tryptophan emission wavelength as a function of increasing guanidine HCl concentration as described in Materials and methods. Denaturation curves were fitted to determine Cm, the concentration of guanidine HCl at 50% denaturation, and free energies of unfolding were calculated as described.

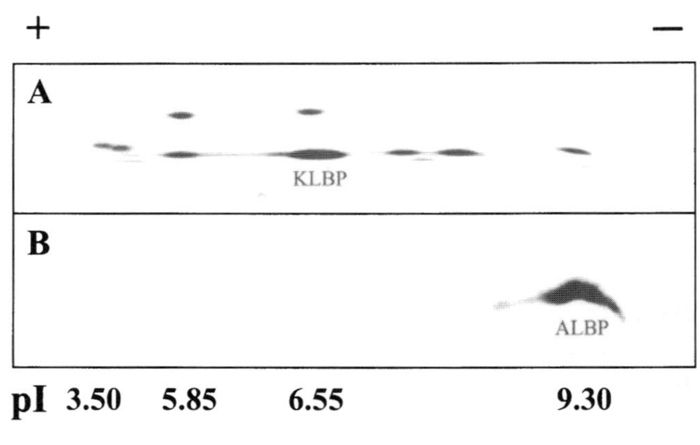

Fig. 2. Isoelectric focusing of ALBP and KLBP. (A) Standard isoelectric focusing calibration markers (pI range 3–10, Pharmacia) in the presence and/or absence of 20 μg of purified KLBP were run in an acrylamide vertical tube gel from basic to acidic conditions. The resulting tube gel was analyzed in the second dimension by SDS-PAGE. By comparison with marker pI values ranging from 3.5 to 9.3, the pI of KLBP was determined to be approximately 6.5. (B) 40 μg of purified ALBP was run in an acrylamide-urea tube gel from basic to acidic conditions (first dimension) prior to standard SDS-PAGE analysis (second dimension). The pI of ALBP under denaturing conditions is approximately 9.

```
               10         20         30         40         50
                   *            †          †
ALBP   ...CDAFVGT WKLVSSENFD DYMKEVGVGF ATRKVAGMAK PNMIISVNGD
KLBP   MASLKDLEGK WRLMESHGFE EYMKELGVGL ALRKMAAMAK PDCIITCDGN

               60         70         80         90        100
                  †      *               * *               *
ALBP   LVTIRSESTF KNTEISFKLG VEFDEITADD RKVKSIITLD GGALVQVQKW
KLBP   NITVKTESTV KTTVFSCNLG EKFDETTADG RKTETVCTFQ DGALVQHQQW

              110        120        130   135             Ref
         *    * *     *        *
ALBP   DGKSTTIKRK RDGDKLVVEC VMKGVTSTRV YERA.            [27]
KLBP   DGKESTITRK LKDGKMIVEC VMNNATCTRV YEKVQ             [5]
```

Fig. 3. Primary sequence alignment of murine ALBP and KLBP. Published cDNA sequences were translated and aligned using GCG program defaults. ALBP surface lysine residues are designated with an asterisk [26]. Those surface lysines analogous to site-specifically mutagenized residues in HFABP [11] as discussed in the text are denoted by a dagger.

Roughly corresponding areas of positive and negative potential exist in all views of the two molecules, but KLBP shows consistently smaller positive patches and larger negative patches. These differences will contribute to interactions between the two proteins and any potential partner, although the corresponding patterns suggest that the two proteins may be able to interact electrostatically with the same partners but with different affinities.

Comparison of the hydrophobic surfaces of the two proteins also reveals some potentially pertinent differences (Fig. 4B). Hydrophobic residues are nearly randomly distributed over the surface of ALBP/aP2, with no obvious large patches. KLBP, however, has a relatively large hydrophobic patch near the top of the molecule, on the portal or front face. Overall, the distribution of hydrophobic residues in KLBP is more patchy or clustered, while ALBP/aP2's surface hydrophobicity is more dispersed, lending strength to a model of ALBP/aP2 and KLBP interacting differentially.

Discussion

Despite intensive scrutiny from diverse experimental angles, it has remained difficult to assign a particular function to any of the LBPs (though several recent reviews thoroughly discuss possibilities, [1–3]). ALBP/aP2 null mice have afforded another opportunity to ask physiological questions. Because ALBP/aP2 is specifically expressed in adipose, disruption of its gene should have distinct and measurable effects. The loss of ALBP/aP2 is partially compensated on a molecular level by the upregulation of KLBP [4, 7], a lipid-binding protein normally found in epithelial-type cells. Although the phenotypes of wild type and ALBP/aP2 null mice are virtually identical on a diet of standard lab chow, important metabolic differences exist between the two strains which become more dramatic on a diet which stresses the adipocyte [7]. Unlike their wild type counterparts, null mice have higher free fatty acid levels [4], lower resting and stimulated lipolytic rates [4], and fail to become insulin resistant when fed a diet high in fat [7]. These differences prompted a systematic comparison of the biochemical properties of ALBP/aP2 and KLBP.

Table 2 summarizes dissociation constants for ALBP/aP2 and KLBP binding to a series of potential physiological ligands. While similar constants are observed in most cases, binding of oleate, palmitate, linoleate, and myristate differed by 2–3 fold. Interestingly, analysis of the free fatty acid composition of adipocytes from wild type or ALBP/aP2 disrupted mice revealed 2–3 fold increases in the levels of those fatty acids [4], perhaps resulting from an inability of KLBP to direct them appropriately for metabolism. Denaturation of the two proteins reveals inherent differences in stability. Northern analysis demonstrated 40-fold upregulation of KLBP message in nulls relative to wild type while protein levels only rose 7-fold [4], leading to significantly diminished total LBP levels in nulls. This discrepancy may be the result of the reduced intrinsic stability of KLBP. Furthermore, the isoelectric points for the two proteins are quite different, as are the patterns of surface charges and hydrophobic patches.

LBPs are frequently hypothesized to serve either as passive providers of fatty acid buffering capacity to the aqueous cytosol, in which lipids are poorly soluble, or as more active fatty acid chaperones, responsible for trafficking of lipids between various intracellular locales. Lipids may exist at low levels free in the cytosol, but non-esterified fatty acid concentrations in adipocytes are many fold higher [4] than in vitro solubility measurements would suggest possible [22]. Fatty acids may additionally equilibrate within the plasma membrane, various organelle membranes, or the surface of the triglyceride droplet, but elevated levels will exert a disruptive micellar effect on membranes and proteins. Hence, it has often been proposed that the purpose of the abundant LBPs in lipid-active tissues is to permit free fatty acids to exist at appropriate metabolic levels within the cell while preventing the deleterious effects of such high concentrations.

LBP overexpression has been shown to facilitate cellular uptake of exogenously added fatty acids and dispersion among organelles [23, 24]. One might predict, therefore, that total nonesterified fatty acid levels would decrease in proportion to a decrease in LBP concentration. However, ALBP/aP2 null mice seem to contradict this supposition. The concentration of nonesterified fatty acid is inversely proportional to the level of LBP [4]. If ALBP/aP2 were simply a buffer for fatty acids, merely sequestering and solubilizing them, transgenic mice should have lower non-esterified fatty acid levels than wild type, since the overall LBP concentration is lower. In this case, elevation of fatty acids would saturate the buffering capacity of the LBP pool, leading to product inhibition of hormone-sensitive lipase and decreased liberation of fatty acids from lipolysis. Lipolysis in ALBP/aP2

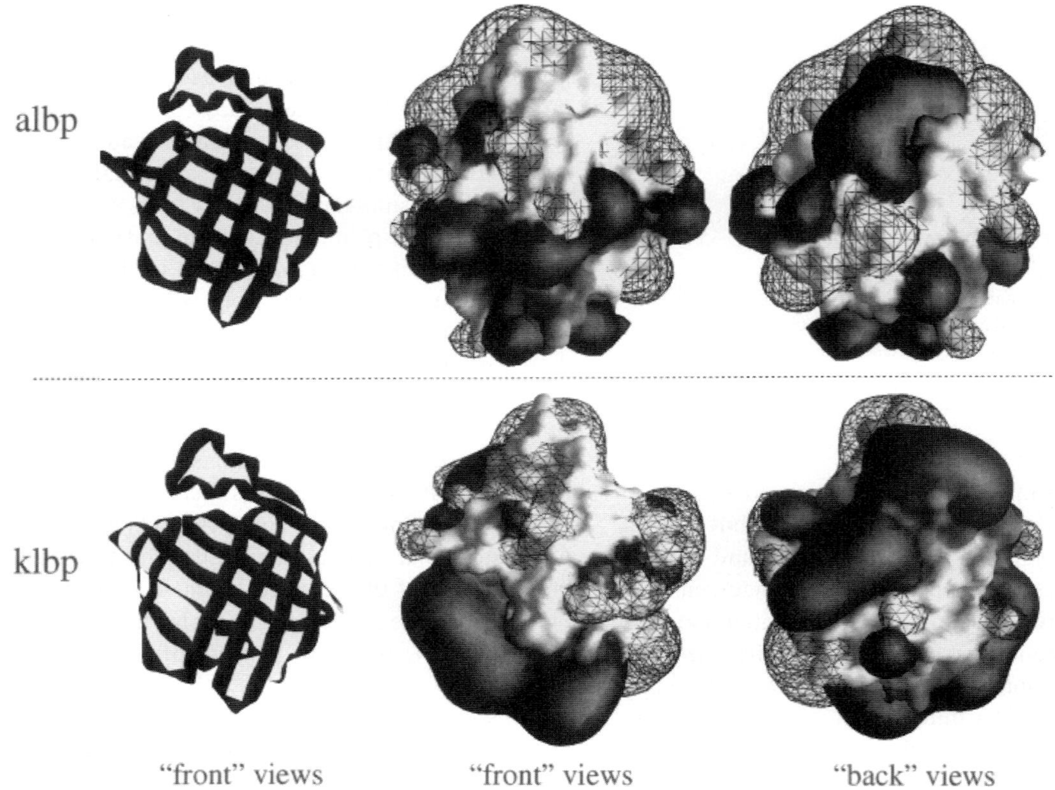
albp
klbp
"front" views
"front" views
"back" views

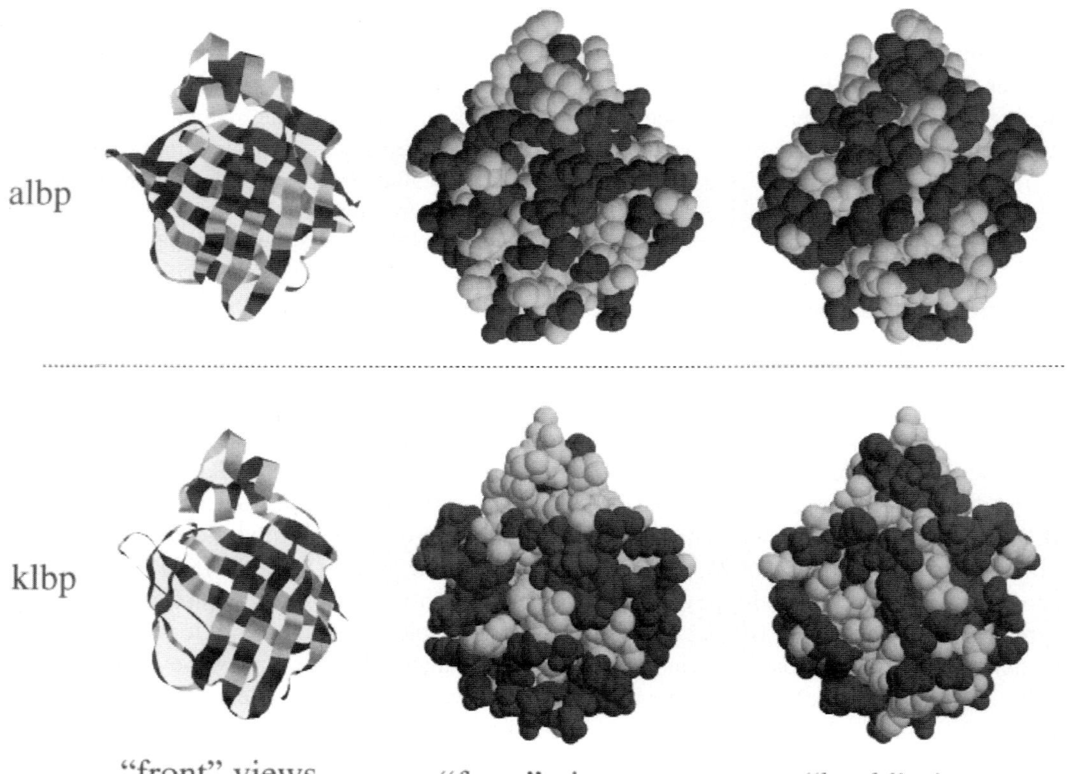
albp
klbp
"front" views
"front" views
"back" views

null mice is compromised in both resting and catecholamine-stimulated adipocytes [4], but because stimulation is equally effective in both types of adipocytes, product inhibition of hormone-sensitive lipase is unlikely to be the causative agent. This line of reasoning leads to a model in which ALBP/aP2 is an active shuttle of lipids, whose specific intracellular interactions are required for efflux of fatty acids. ALBP/aP2 is 55% identical and 70% similar to KLBP; the differences in overall charge and surface charge distribution could determine the unique interactions of the two proteins, particularly in lipolysis.

Storch and colleagues have examined the mechanisms of ligand transfer from LBPs to synthetic membranes. In a fluorescence assay designed to measure kinetics of fatty acid dissociation, ALBP/aP2 transferred bound fluorescent fatty acid analogs to phospholipid vesicles in a concentration dependent manner [8]. This implies that dissociation of lipids from ALBP/aP2 occurs preferentially through collision with vesicles rather than by random diffusion. In diffusional transfer, as in the case of liver FABP [25], fatty acid dissociation from the LBP would precede its insertion in the phospholipid vesicle, the kinetics of which would not change with vesicle concentration. The rate of collisional transfer from ALBP/aP2 was further shown to be dependent upon surface electrostatic interactions [9, 10]. ALBP/aP2 has several lysine residues near the site of ligand entry/exit (the portal). Transfer was found to occur more rapidly to vesicles containing a higher fraction of negatively charged, rather than predominantly neutral or positively charged, phospholipids. When all surface lysines were neutralized by acetylation, the mechanism of transfer became diffusional, regardless of phospholipid composition. We can speculate that KLBP, on the basis of its lower pI (6.5) and its differing electrostatic contour distribution, may transfer fatty acids at rates distinctly different from ALBP/aP2. Comparison of ALBP/aP2 and KLBP transfer rates may help address the importance of specific lysines vs. overall charge or charge distribution.

In conclusion, metabolic dysfunction exhibited by the ALBP/aP2 knock-out mouse model has underscored the importance of lipid-binding proteins in intracellular lipid trafficking. The biochemical and biophysical analyses described in this study characterize ALBP/aP2 and KLBP as separate and distinct members of the LBP multigene family. While it is tempting to conclude that the upregulation of KLBP in ALBP/aP2 null mice is an example of molecular compensation, careful analysis reveals that the properties of KLBP are not identical to those of ALBP/aP2. A more comprehensive understanding of those differences may provide clues to the role of fatty acids in the development of non-insulin dependent diabetes mellitus.

Fig. 4. Electrostatic and hydrophobic surfaces of ALBP and KLBP. Crystal structure coordinates were used for ALBP (PDB code 1LIB). The crystal structure for KLBP is not yet available so it was modeled using the Swiss-Model internet modeling facility. KLBP was modeled on the basis of its similarity to eight structurally homologous templates existing in the Brookhaven Protein Data Bank: five fatty acid-binding proteins, two retinoic acid-binding proteins, and one retinol-binding protein. (A). Top panel. Electrostatic calculations were carried out with the program GRASP. Electrostatic potential contours shown were generated with GRASP. Negative potential is displayed as dark gray, positive as wire mesh, and neutral in white. (B). Lower panel. Hydrophobic residue distributions were generated with the program Rasmol. Lighter regions show non-polar residues; darker residues are polar.

Acknowledgments

The authors would like to thank the members of the Bernlohr laboratory for helpful discussions during the preparation of this manuscript. This work supported by NSF MCB 9506088 to DAB.

References

1. Banaszak LJ, Winter N, Xu Z, Bernlohr DA, Cowan S, Jones TA: Lipid-binding proteins: a family of fatty acid and retinoid transport proteins. Adv Prot Chem 45: 89–151, 1994
2. Bernlohr DA, Simpson MA, Hertzel AV, Banaszak LJ: Intracellular lipid-binding proteins and their genes. Annu Rev Nutr 17: 277–303, 1997
3. Veerkamp JH, Maatman RGHJ: Cytoplasmic fatty acid-binding proteins: Their structure and genes. Prog Lipid Res 34: 17–52, 1995
4. Ribarik Coe N, Simpson MA, Johnston Smith A, Bernlohr DA: Targeted disruption of the adipocyte lipid binding protein impairs fat cell lipolysis and increases cellular fatty acid levels. J Biol Chem, submitted
5. Krieg P, Feil S, Furstenberger G, Bowden GT: Tumor-specific overexpression of a novel keratinocte lipid-binding protein: Identification and characterization of a cloned sequence activated during multistage carcinogenesis in mouse skin. J Biol Chem 238: 17362–17369, 1993
6. Kane CD, Ribarik Coe N, Vanlandingham B, Krieg P, Bernlohr DA: Expression, purification, and ligand-binding analysis of recombinant keratinocyte lipid-binding protein (MAL-1), an intracellular lipid-binding protein found overexpressed in neoplastic skin cells. Biochemistry 35: 2894–2900, 1996
7. Hotamisligil GS, Johnson RS, Distel RJ, Ellis R, Papaioannou VE, Spiegelman BM: Uncoupling of obesity from insulin resistance through a targeted mutation in aP2, the adipocyte fatty acid binding protein. Science 274: 1377–1379, 1996
8. Wootan MG, Bernlohr DA, Storch J: Mechanism of fluorescent fatty acid transfer from adipocyte fatty acid binding protein to membranes. Biochemistry 32: 8622–8627, 1993
9. Wootan MG, Storch J: Regulation of fluorescent fatty acid transfer from adipocyte and heart fatty acid binding proteins by acceptor membrane lipid composition and structure. J Biol Chem 269: 10517–10523, 1994
10. Herr FM, Matarese V, Bernlohr DA, Storch J: Surface lysine residues modulate the collisional transfer of fatty acid from adipocyte fatty acid binding protein to membranes. Biochemistry 34: 11840–11845, 1995

11. Herr FM, Aronson J, Storch J: Role of portal region lysine residues in electrostatic interactions between heart fatty acid binding protein and phospholipid membranes. Biochemistry 35: 1296–1303, 1996
12. Xu Z, Buelt MK, Banaszak U, Bernlohr DA: Expression, purification, and crystallization of the adipocyte lipid binding protein. J Biol Chem 266: 14367–14370, 1991
13. Kane CD, Bernlohr DA: A simple assay for intracellular lipid-binding proteins using displacement of 1-anilinonaphthalene-8-sulfonic acid. Anal Biochem 233: 197–204, 1996
14. Buelt MK, Xu Z, Banaszak W, Bernlohr DA: Structural and functional characterization of the phosphorylated adipocyte lipid-binding protein (pp 15). Biochemistry 31: 3493–3499, 1992
15. Pace CN: The stability of globular proteins. Crit Rev Biochem 3: 1–43, 1975
16. Peitsch MC: Protein modeling by E-mail. Bio/Technology 13: 658–660, 1995
17. Peitsch MC: ProMod and Swiss-Model: Internet-based tools for automated comparative protein modelling. Biochem Soc Trans 24: 274–279, 1996
18. Nicholls A, Sharp KA, Honig B: Protein folding and association: Insights from the interfacial and thermodynamic properties of hydrocarbons. Proteins 11: 281–296, 1991
19. Sayle RA, Milner-White EJ: RASMOL: Biomolecular graphics for all. TIBS 20: 374, 1995
20. Bernlohr DA, Doering TL, Kelly U Jr, Lane MD: Tissue specific expression of p422 protein, a putative lipid carrier, in mouse adipocytes. Biochem Biophys Res Comm 132: 850–855, 1985
21. Richieri GV, Ogata RT, Kleinfeld AM: Equilibrium constants for the binding of fatty acids with fatty acid-binding proteins from adipocyte, intestine, heart, and liver measured with the fluorescent probe ADIFAB. J Biol Chem 269: 23918–23930, 1994
22. Vorum H, Brodersen R, Kragh-Hansen U, Pedersen AO: Solubility of long-chain fatty acids in phosphate buffer at pH 7.4. Biochim Biophys Acta 1126: 135–142, 1992
23. Prows DR, Murphy EJ, Schroeder F: Intestinal and liver fatty acid binding proteins differentially affect fatty acid uptake and esterification in L-cells. Lipids 30: 907–910, 1995
24. Murphy EJ, Prows DR, Jefferson JR, Schroeder F: Liver fatty acid-binding protein expression in transfected fibroblasts stimulates fatty acid uptake and metabolism. Biochim Biophys Acta - Lipids and Lipid Metab 1301: 191–198, 1996
25. Hsu K-T, Storch J: Fatty acid transfer from liver and intestinal fatty acid-binding proteins to membranes occurs by different mechanisms. J Biol Chem 271: 13317–13323, 1996
26. Xu Z, Bernlohr DA, Banamak U: Crystal structure of recombinant murine adipocyte lipid-binding protein. Biochemistry 31: 3484–3492, 1992
27. Bernlohr DA, Angus CW, Lane MD, Bolanowski MA, Kelly U Jr: Expression of specific mRNAs during adipose differentiation: Identification of an mRNA encoding a homologue of myelin P2 protein. Proc Natl Acad Sci USA 81: 5468–5472, 1984
28. LiCata VJ, Bernlohr DA: Surface properties of adipocyte lipid binding protein. Response to lipid binding and comparison with homologous proteins. Proteins (in press)

Molecular and Cellular Biochemistry **192**: 41–52, 1999.

Long-chain fatty acid transport in bacteria and yeast. Paradigms for defining the mechanism underlying this protein-mediated process

Concetta C. DiRusso, Paul N. Black

Department of Biochemistry and Molecular Biology, The Albany Medical College, Albany, New York, USA

Abstract

Protein-mediated transport of exogenous long-chain fatty acids across the membrane has been defined in a number of different systems. Central to understanding the mechanism underlying this process is the development of the appropriate experimental systems which can be manipulated using the tools of molecular genetics. *Escherichia coli* and *Saccharomyces cerevisiae* are ideally suited as model systems to study this process in that both [1] exhibit saturable long-chain fatty acid transport at low ligand concentration; [2] have specific membrane-bound and membrane-associated proteins that are components of the transport apparatus; and [3] can be easily manipulated using the tools of molecular genetics. In *E. coli*, this process requires the outer membrane-bound fatty acid transport protein FadL and the inner membrane associated fatty acyl CoA synthetase (FACS). FadL appears to represent a substrate specific channel for long-chain fatty acids while FACS activates these compounds to CoA thioesters thereby rendering this process unidirectional. This process requires both ATP generated from either substrate-level or oxidative phosphorylation and the proton electrochemical gradient across the inner membrane. In *S. cerevisiae,* the process of long-chain fatty acid transport requires at least the membrane-bound protein Fat1p. Exogenously supplied fatty acids are activated by the fatty acyl CoA synthetases Faa1p and Faa4p but unlike the case in *E. coli*, there is not a tight linkage between transport and activation. Studies evaluating the growth parameters in the presence of long-chain fatty acids and long-chain fatty acid transport profiles of a *fat1Δ* strain support the hypothesis that Fat1p is required for optimal levels of long-chain fatty acid transport. (Mol Cell Biochem **192**: 41–52, 1999)

Key words: fatty acid transport, fatty acyl CoA synthetase, long chain fatty acids

Introduction

Long-chain fatty acids represent important sources of metabolic energy. In addition, these compounds or their derivatives are involved in a wide variety of cellular processes including phospholipid synthesis [1–4], protein export [5, 6], protein modification [7, 8], organelle biogenesis [9], enzyme activation or deactivation [10–13], cell signaling [14–17], membrane permeability [18] and transcriptional control [19–23]. The mechanisms which mediate the uptake of exogenous long-chain fatty acids across the plasma membrane are incompletely understood. Most cell types have the capacity to take up exogenous long-chain fatty acids by a diffusional mechanism yet many different cell types show high level, specific and regulated transport. It is now generally accepted that specific membrane-bound proteins participate in the transport of exogenous long-chain fatty acids across the membrane (e.g. see refs [24–33]). The findings that specific membrane-bound proteins participate in transport argues that fatty acid entry into the cell is highly specific and subject to regulation.

A number of different membrane-bound proteins have been identified that are thought to represent long-chain fatty acid transporters [24–34]. Over the past twelve years we have been defining the mechanism underlying this protein-mediated process by investigating structure-function relationships of the long-chain fatty acid transport protein FadL and associated fatty acyl CoA synthetase (FACS) found in

Address for offprints: P.N. Black, Department of Biochemistry and Molecular Biology, A10, The Albany Medical College, Albany, New York 12208, USA

Echerichia coli [32–40]. More recently we have begun investigations on the fatty acid transporter Fat1p and the fatty acyl CoA synthetases Faa1p and Faa4p found in *Saccharomyces cerevisieae* [41]. In *E. coli*, FadL is specifically involved in the binding and translocation of these compounds across the outer membrane of the cell envelope. Exogenous long-chain fatty acids destined for β-oxidation also requires FACS which is associated with the inner membrane [42]. In the context of transport, there is a tight linkage between FadL and FACS indicating that these two proteins must act in concert in the facilitated transport and activation of these hydrophobic compounds. The fatty acid transport apparatus in yeast is quite distinct from that defined in *E. coli* but may emulate that being defined in higher eukaryotic cells. This process requires the fatty acid transport protein Fat1p which is presumed to reside in the plasma membrane. The localization of the fatty acyl CoA synthetases Faa1p and Faa4p have not been defined, although these two enzymes function to activate imported fatty acids. In this context, Faa1p and Faa4p may act with Fat1p in the facilitated transport and activation of exogenous long-chain fatty acids. A central theme for both the bacterial and yeast fatty acid transport systems is the generation of long-chain fatty acyl CoA. This compound represents a pivotal regulatory metabolite having effects as diverse as membrane biogenesis and transcriptional regulation [9, 19–22].

The use of the bacterial and yeast model systems to explore the problem of fatty acid transport across a biological membrane has broad implications. These two systems are amenable to detailed molecular-genetic dissections that will allow us to define the mechanism underlying this process. The genes encoding specific components of the fatty acid transport apparatus have been identified in both model systems which has facilitated the generation of strains containing deletions in one or more genes whose protein products are involved in this process.

Components of the bacterial fatty acid transport apparatus

The cell envelope of gram negative bacteria represents a formidable barrier for long-chain fatty acids. It is composed of two structurally and functionally distinct membrane systems. The outer membrane is composed of an external layer of lipopolysaccharide and an internal layer of phospholipid. The external layer of lipopolysaccharide is refractory towards hydrophobic compounds thereby providing a protective shield for the cell while the internal layer of phospholipid is associated with a layer of peptidoglycan. Outer membrane proteins involved in the acquisition of nutrients fall into three general classes: (1) non-specific porins; (2) substrate-specific porins; and (3) high affinity, substrate-specific transport proteins. The inner membrane is a more typical phospholipid bilayer and contains proteins involved in nutrient transport, energy production, and phospholipid biosynthesis. The two membranes are separated by the aqueous periplasmic space which is rich in protein (some of which function in nutrient transport) and membrane-derived oligosaccharides. In order for *E. coli* to utilize exogenous long-chain fatty acids, these compounds must first traverse the three layers of the cell envelope.

Exogenous long-chain fatty acids traverse the cell envelope of *E. coli* by a highly specific, concentrative process that minimally consists of the outer membrane-bound fatty acid transport protein FadL and the inner membrane associated fatty acyl CoA synthetase (FACS) (Fig. 1). The periplasmic protease Tsp is required for optimal levels of transport although its precise role in this process remains undefined [43]. There is evidence which supports the existence of a specific fatty acid/H^+ cotransporter in the inner membrane; the structural gene encoding this protein has not been identified [44, 45].

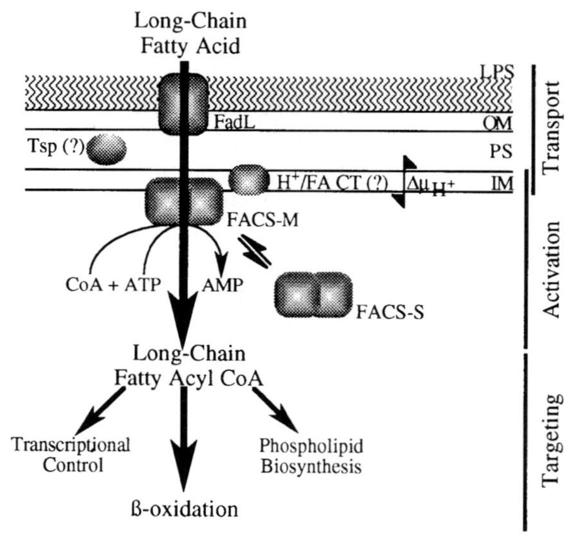

Fig. 1. Components of the long-chain fatty acid transport apparatus in *Escherichia coli.* Exogenous long-chain fatty acids bind to the outer membrane-bound fatty acid transport protein FadL which subsequently facilitates their transport across the outer membrane. It is unknown how long-chain fatty acids traverse the periplasmic space although the periplasmic protease Tsp appears to play a peripheral role (noted by the ?). As this process is osmotic shock sensitive, it may require a specific periplasmic fatty acid binding protein. The proton electrochemical gradient ($\Delta\mu_{H+}$) across the inner membrane is required for optimal transport. An oleic acid binding protein has been identified in the membrane that may function as a H^+/fatty acid cotransporter (noted by the ?). Fatty acids are activated to CoA thioesters by fatty acyl CoA synthetase which we presume partitions into the inner membrane in the presence of fatty acid ligands. There is no apparent efflux of long-chain acyl CoA which renders this a unidirectional process. The process is divided into Transport, Activation, and Targeting as noted by the bars on the right. In *E. coli*, there is linkage between transport and activation as indicated by the overlapping bars.

The process of fatty acid transport in *E. coli* was originally described in work characterizing the *fadD* gene, encoding FACS [42]. In these seminal papers, FACS was proposed to vectorially acylate fatty acids concomitant with transport thereby rendering this process unidirectional. On the basis of kinetic data, these early studies predicted that at least one additional protein is involved in the facilitated transport of long-chain fatty acids in *E. coli*. This prediction was confirmed in the late 1970s by Nunn and Simons [46] who identified and mapped the *fadL* gene, encoding the long-chain fatty acid transport protein FadL. The subsequent studies describing the kinetics of fatty acid transport in *E. coli* from Nunn's laboratory are consistent with the postulate that FadL is specifically involved in the transport of long-chain fatty acids [47, 48]. More recent work from our laboratory defining fatty acid-FadL binding properties demonstrate that this protein has specificity towards long-chain fatty acids and that the carboxylate of the fatty acid is essential for ligand binding [33, 40]. FadL spans the outer membrane where it functions to bind and transport long-chain fatty acids. This protein also serves as the primary receptor for bacteriophage T2 [49]. The localization of a specific long-chain fatty acid transport protein within the outer membrane is physiologically relevant. The outer layer of lipopolysaccharide renders this membrane refractory towards hydrophobic compounds (including long-chain fatty acids). Therefore, it is reasonable to predict these compounds would specifically interact with an outer membrane-bound protein that facilitates their transport across this layer. FadL fulfills these criteria: this protein binds long-chain fatty acids with a high affinity and specifically facilitates their transport across a membrane barrier which is otherwise impermeable.

How does FadL mediate long-chain fatty acid binding and transport? This protein spans the outer membrane and is composed of two discrete domains that are functionally distinct [38, 39, 49]. We postulate that the amino terminal proximal domain is exposed at the cell surface, is protease-sensitive and is involved in ligand binding. The carboxyl terminal domain of the protein is embedded within the membrane and is predicted to form a long-chain fatty acid-specific channel. Our working hypothesis is that long-chain fatty acid binding to FadL results in a conformational change thereby exposing the transport channel and facilitating transport. We have identified two mutations in the FadL structural gene which we suggest are defective in this conformational change. Both of these mutations apparently keep FadL in an open, or channel exposed, conformation [38, 39]. How fatty acids specifically traverse the outer membrane via the FadL channel is not known, but on the basis of data generated from a collection of *fadL* mutants is presumed to involve both hydrophobic and charged amino acid residues within the carboxyl-terminal region of the protein [39]. Once fatty acids traverse the outer membrane in a FadL-dependent manner, they must also traverse the aqueous periplasmic space. We have shown that the process of long-chain fatty acid transport is shock sensitive (supporting the existence of a periplasmic component) and that the periplasmic protease Tsp is peripherally involved in this process [43, 50]. Tsp has sequence similarities to the family of interphotoreceptor retinoid binding proteins and thus may function to bind hydrophobic ligands including long-chain fatty acids although this has not been experimentally defined. To date, no specific periplasmic fatty acid binding proteins have been identified which are players in the process of longchain fatty acid transport. The transport of these compounds across the inner (cytoplasmic) membrane is somewhat of an enigma. An oleate binding protein has been purified from the membrane and shown to increase the maximal rate of oleate transport in spheroplasts loaded with FACS [44]. In addition there is evidence which supports the proposal that fatty acids are transported across this membrane layer by a H^+/fatty acid cotransport process [45]. It is not known whether the oleate-binding protein identified represents the postulated H^+/fatty acid cotransporter.

As free fatty acid can not be detected in the cell following transport, FACS is presumed to represent a component of the transport apparatus and thus from an operational standpoint must necessarily function in concert with FadL and other components of the long-chain fatty acid transport apparatus. This enzyme is routinely isolated from the cytosolic fraction although there is evidence that it becomes membrane-bound in response to specific physiological conditions [51]. The precise mechanism which facilitates the movement of this enzyme into the membrane is not known. We have obtained data that suggests long-chain fatty acids in the growth media are required for this process (Weimar and Black, unpublished). The *E. coli* FACS belongs to the family of adenylate-forming enzymes specifically by containing a presumptive ATP/AMP binding site [37]. This enzyme also belongs to a smaller family of enzymes which includes other fatty acyl CoA synthetases, coumurate CoA ligases and firefly luciferases by containing a twenty-five amino acid segment and which bind hydrophobic ligands, including fatty acid, coumurate and lucifern respectively [37]. Data from our laboratory demonstrates that this region of the FACS is involved in conferring fatty acid chain length specificity [37].

The energetics of long-chain fatty acid transport in *E. coli*

The energy which drives transport systems in *E. coli* comes from the hydrolysis of ATP and from the respiration-derived proton electrochemical gradient (proton motive force;

PMF) [52]. Central to both sources of energy is the proton-translocating enzyme ATP synthase [53]. ATP synthase requires the proton electrochemical gradient for the synthesis of ATP from ADP and inorganic phosphate via an irreversible reaction. We have defined the contribution of ATP and PMF in the uptake of long-chain fatty acids in studies using wild-type and membrane-bound ATP synthase-deficient (Δ*atp*) strains and a battery of different metabolic inhibitors and exogenous energy sources. Oleate transport increases 3- to 4-fold in the presence of glucose and D-lactate in the wild-type strain but does not respond to D-lactate in the Δ*atp* strain (Fig. 2A). Oleate transport in the presence of arsenate is significantly reduced ($p \leq 0.01$) in the Δ*atp* and wild-type strains under starved and D-lactate and glucose energized conditions (Fig. 2B). The conclusion from these data is that oleate transport requires ATP generated from substrate-level or oxidative phosphorylation. We presumed *a priori* that oleate uptake would require ATP due to the requirements of the fatty acyl CoA synthetase as a component of the fatty acid transport apparatus.

Oleate uptake is also sensitive to the uncoupler carbonyl cyanide *m*-chlorophenylhydrazone (CCCP) in both D-lactate and glucose energized cells suggesting this process also responds to the proton motive force (Fig. 3). The treatment of wild-type cells with CCCP yields essentially the same level of oleate transport as measured in strains defective for ATP synthase. Treatment of the Δ*atp* strain with CCCP reduces the level of oleate transport even further and in an additive manner. While the absolute levels of oleate uptake are higher in both wild-type and Δ*atp* strains, these results are mirrored in the D-lactate energized cells. Of particular importance is the finding that glucose fails to restore maximal levels of oleate uptake in the wild-type and Δ*atp* strains treated with CCCP supporting the conclusion that the proton electrochemical gradient across the inner membrane plays a crucial role in energizing the long-chain fatty acid transport apparatus. Cyanide also decreases oleate uptake under D-lactate and glucose energized condition's supporting data obtained using CCCP further supporting a role of an energized membrane in this process (Fig. 3).

The data described above is somewhat of a paradox as both ATP and the PMF are involved in the uptake of long-chain fatty acids. This is in contrast to other nutrient transport systems in gram-negative bacteria which are powered by either ATP or PMF but not both. Yet if one considers the unique nature of the fatty acid transport apparatus in *E. coli*, the contribution of both sources of metabolic energy makes sense. These studies apparently distinguish the fatty acid transport activity from the fatty acid activation activity. Disruption of either the electrochemical gradient or the intracellular ATP pools compromises the cell's ability to transport these hydrophobic compounds. Long-chain fatty acid ligands bind to FadL with high affinity; this is consistent with the involvement of the PMF which may act to facilitate ligand release. Additionally, the PMF may be required for the activity of the presumed fatty acid/H^+ cotransporter. The activity of FACS requires ATP. In this

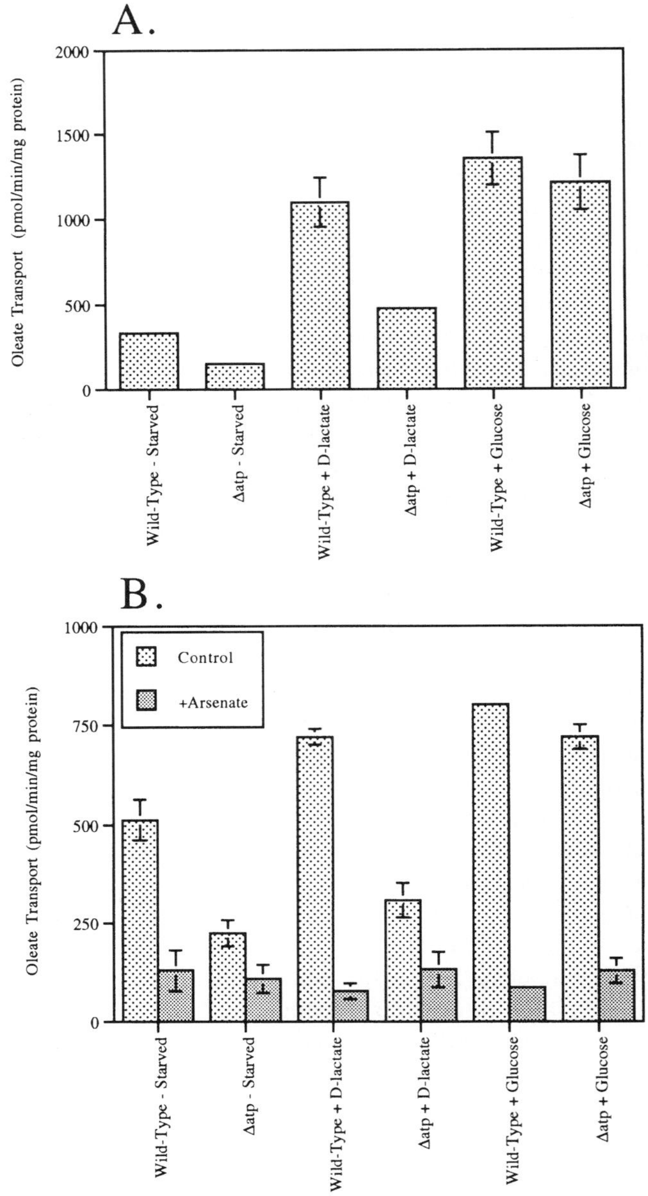

Fig. 2. Long-chain fatty acid transport in *E. coli* responds to the energized state of the cell. (A) Oleate transport in wild-type and ATP synthase-deficient cells (Δ*atp*) in starved cells and cells energized with D-lactate or glucose; (B) Oleate transport in wild-type and Δ*atp* cells in the presence of 100 μM arsenate. Cells were starved to deplete endogenous energy stores for 30 min in minimal salts containing 100 μg/ml chloramphenicol. Starved cells were assayed using 100 μM oleate and where noted included 1 mM D-lactate or 1 mM glucose and an exogenous energy source. Details of the transport assay are described by Kumar and Black [38].

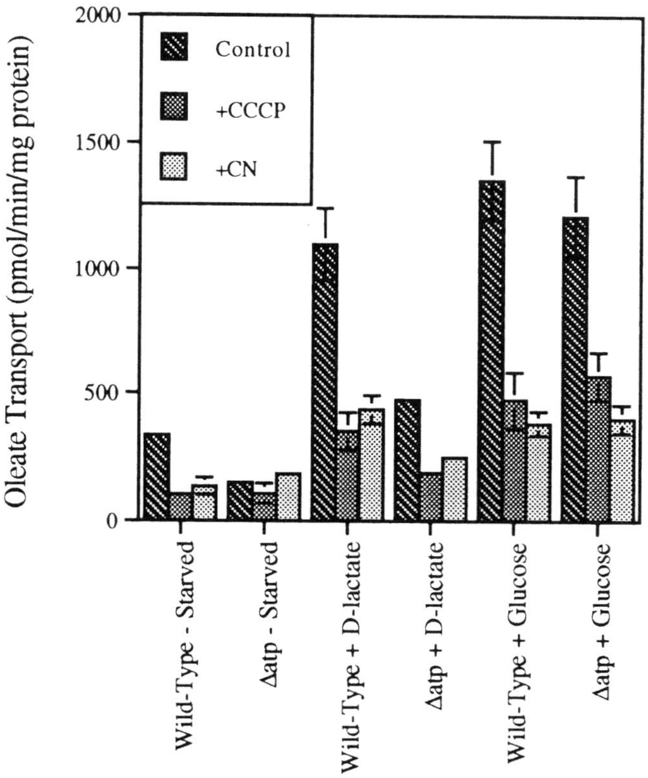

Fig. 3. Disruption of the proton electrochemical gradient compromises long-chain fatty acid transport in *E. coli*. Oleate transport was measured in cells (wild-type and Δ*atp*; starved and energized with D-lactate or glucose) as noted in Fig. 2 in the presence of 10 μM carbonyl cyanide m-chlorophenylhydrazone or 1 mM KCN.

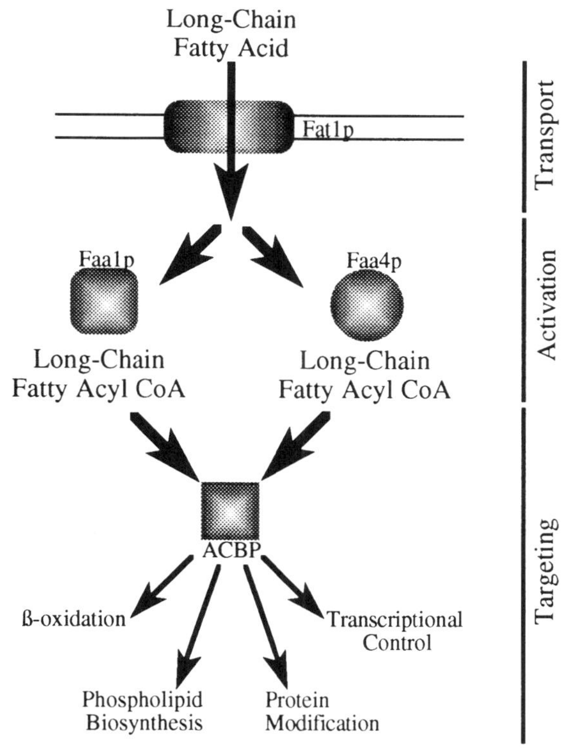

Fig. 4. Components of the long-chain fatty acid transport apparatus in *Saccharomyces cerevisiae*. Exogenous long-chain fatty acids traverse the membrane via the membrane-bound fatty acid transport protein Fat1p. Fatty acids are activated to CoA thioesters by the fatty acyl CoA synthetases Faa1p and Faa4p. No information is presently available on the compartmentalization of these two enzymes. The activated fatty acids are bound by acyl CoA binding protein (ACBP) which is presumed to act as a delivery vehicle to sites of utilization as well as buffering the cell against the detergents effects of long-chain fatty acyl CoA. The process is divided into Transport, Activation, and Targeting as noted by the bars on the right.

regard the transport of long-chain fatty acids specifically requires the PMF while the activation of long-chain fatty acids requires ATP (Fig. 1).

The components of the fatty acid transport system in yeast

The process of long-chain fatty acid transport is saturable in yeast including *Saccharomyces uvarum*, *Saccharomycopsis lipolytica*, and *Saccharomyces cerevisiae* suggesting that this process is protein-mediated [54, 55]. The components required for fatty acid transport and activation in *S. cerevisiae* are illustrated in Fig. 4. We have identified the protein Fat1p (encoded within the *FAT1* gene) which appears to represent a plasma membrane-bound long-chain fatty acid transporter. Fat1p has significant homology to the murine fatty acid transport protein FATP identified from 3T3-L1 adipocytes [41]. Once transported across the membrane, long-chain fatty acids are activated to CoA thioesters by the fatty acyl CoA synthetases Faa1p and Faa4p [56, 57]. The activated fatty acyl CoA thioesters become bound by the intracellular fatty acyl CoA binding protein (ACBP) [58]. ACBP must necessarily act to buffer the detergent effects of these compounds in addition to providing a vehicle for the intracellular trafficking of these compounds including targeting for phospholipid biosynthesis, protein acylation, and energy production. It is unknown if ACBP has a direct role in the transport of exogenous long-chain fatty acids. There does not appear to be as tight of linkage between the transport and activation of long-chain fatty acids in yeast which makes this process different from that defined in bacteria.

Cells which contain a deletion in *FAT1* are essentially unable to grow on YPD media containing the fatty acid synthase inhibitor cerulenin even when supplemented with long chain fatty acids [41]. Furthermore *fat1*Δ mutants have an impaired ability to accumulate the fluorescent fatty acid analogue boron dipyrromediane difluoride dodecanoic acid

(BODIPY-3823) and to transport the long-chain fatty acids myristate, palmitate and oleate [41].

Like the fatty acyl CoA synthetases, Fat1p belongs to the family of adenylate-forming enzymes by containing the same presumptive ATP/AMP binding site [41]. However Fat1p lacks the 25 amino acid residue FACS signature motif common to the family of acyl CoA synthetases. The role of the presumptive ATP/AMP binding site has not been defined but it is reasonable to predict that this region of the protein is of functional or structural significance. On the basis of computer modeling Fat1p is predicted to span the membrane four times with the ATP/AMP binding site residing on the cytoplasmic side (Fig. 4).

Gordon and coworkers have demonstrated that exogenous long-chain fatty acids are activated by the fatty acyl CoA synthetases Faa1p and Faa4p with Faa1p being the principal enzyme responsible for this function [56, 57]. Thus in the context of long-chain fatty acid transport, these two enzymes appear to be functionally interchangeable. Strains carrying deletions in both *FAA1* and *FAA4* are not viable on YPD plates containing cerulenin and long-chain fatty acid (myristate, palmitate or oleate) demonstrating a defect in the uptake pathway. Thus from a phenotypic standpoint, strains carrying deletions in *FAT1* or *FAA1* and *FAA4* appear to be quite similar [41, 56, 57]. The prediction from these types of studies is that mutations in *FAT1, FAA1* and *FAA4* compromise the cells ability to take up exogenous long-chain fatty acids. Although long-chain fatty acid transport and activation do not appear to be linked, these data provide compelling evidence that the fatty acyl CoA synthetases Faa1p and Faa4p must act in concert with Fat1p in the context of long-chain fatty acid transport. From a functional standpoint, this appears to be the case at least in adipocytes [27]. The fatty acyl CoA synthetase from 3T3-L1 adipocytes was identified along with the fatty acid transport protein FATP using expression cloning attesting to the functional importance of this enzyme in the context of long-chain fatty acid transport [27].

Physiological role of Fat1p and fatty acid transport in yeast

When *S. cerevisiae* is grown anaerobically or when fatty acid synthesis is compromised either using antibiotics (e.g. cerulenin) or by mutation in the fatty acid synthase structural gene, there is a requirement for exogenously supplied long-chain fatty acids. Under anaerobic conditions, there is a specific requirement for unsaturated long-chain fatty acids and for sterol since the oxygen-dependent enzymes of unsaturated fatty acid and sterol synthesis are inactive [59]. A specific transporter for sterol, Sut1p, has recently been identified and its structural gene (*SUT1*) cloned [60]. We hypothesize that Fat1p is a central component in the acquisition of fatty acids under these conditions. As noted above, strains carrying a deletion in *FAT1* cannot grow on YPD plates containing cerulenin and long-chain fatty acid. Furthermore, the growth of the *fat1*Δ strain in YNB-oleate media is severely compromised (Fig. 6). To test the requirement for Fat1p during anaerobic growth, we are currently evaluating the growth of both the wild-type *FAT1* and *fat1*Δ strains on YNB containing increasing concentrations of oleate. Our preliminary data indicate that cells harboring a mutation in *FAT1* are unable to grow under anaerobic conditions with oleate supplementation. These data attest to the physiological importance of Fat1p in the transport of these compounds under these conditions.

Prospective

The transport and activation of exogenous long-chain fatty acids allows the cell to use these compounds in lipid biosynthesis, protein acylation, organellar biogenesis and energy production. The mechanisms underlying this process appear to be complex and involve both specific membrane-bound

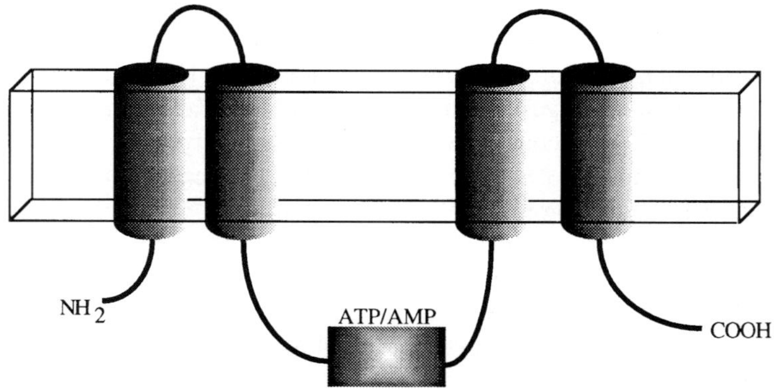

Fig. 5. Predicted membrane topology of Fat1p. ATP/AMP refers to the presumptive ATP/AMP signature common to adenylate-forming enzymes.

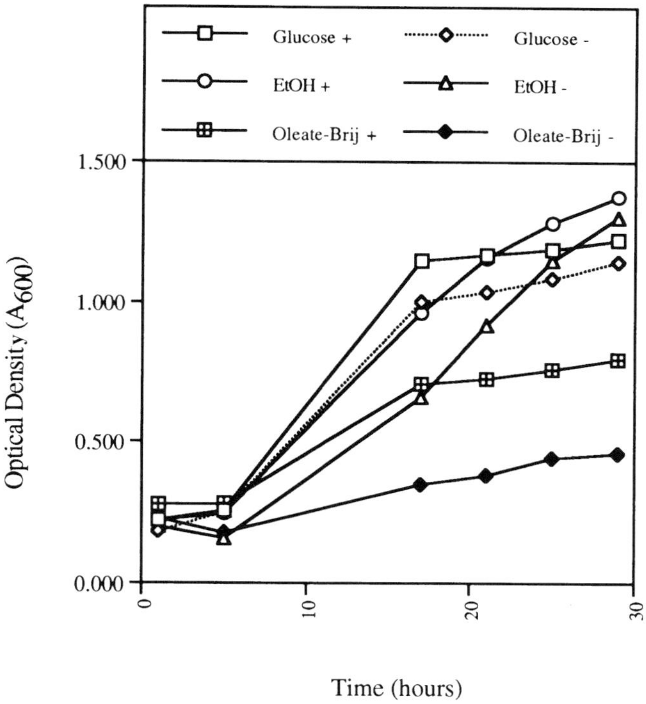

Fig. 6. Growth kinetics of *FAT1* (+) and *fat1*Δ (–) strains in YNB supplemented with the carbon sources noted.

proteins and fatty acid activating enzymes. Long-chain fatty acid transport is highly regulated and linked to the energized state of the cell. We are confronted with the challenge to define how specific membrane-bound proteins and fatty acid activating enzymes function in concert to facilitate the specific and regulated transport of long-chain fatty acids using well-defined genetic model systems.

Long-chain fatty acid transport and activation in bacteria – implications in early states of infection

The genes required for acquisition of long-chain fatty acids in *E. coli* are regulated at the level of transcription. Under high nutrient conditions the genes required for fatty acid transport, activation and β-oxidation are repressed by the global transcription factor FadR. When the cell encounters long-chain fatty acids, FadL and FACS function to transport and activate these compounds into intracellular pools of long-chain acyl CoA. These compounds represent the effectors that when in sufficiently high intracellular concentrations bind to FadR resulting in the derepression of the genes involved in fatty acid transport, activation and β-oxidation. This sets forth a cascade of events that allow the cell to utilize long-chain fatty acids as a carbon and energy source. DNaseI footprinting has identified two FadR binding sites for both the *fadL* (encoding the transporter) and *fadD* (encoding FACS) genes [21, 35]. The precise roles of each of these operator sites in transcriptional repression is under investigation. Furthermore we are currently measuring the intracellular pools of long-chain fatty acyl CoA to assess the conditions that result in FadR binding and transcriptional derepression. Both the *fadL* and *fadD* genes are normally expressed at basal levels under high nutrient growth conditions. In the presence of long-chain fatty acids as the primary carbon source these two genes become induced 2- to 3-fold [19, 61]. In addition to being regulated through long-chain fatty acyl CoA-FadR-mediated process, the *fadL* gene is also regulated in response to changes in osmolarity through OmpR while the *fadD* gene is also regulated by catabolite repression through cAMP-CRP (e.g. see refs [19, 35, 36, 62]. These multiple levels of transcriptional regulation of *fadL* and *fadD* must be necessary for the cell given the physiological consequences of high levels of long-chain fatty acyl CoA. In addition to regulation at the level of transcription, FACS is apparently regulated both at the levels of translational initiation [36] and membrane association [51]. The long-chain fatty acid transport system is also responsive to the energized state of the cell adding yet another dimension of regulation.

Given the multiple levels of regulation which we have defined for the long-chain fatty acid transport system in *E. coli*, one might question why such a system has evolved. The answer must lie in the opportunistic nature of these organisms. The outer membrane of *E. coli* and other gram negative bacteria has an outer leaflet of lipopolysaccharide which renders it refractory towards hydrophobic compounds (including long-chain fatty acids). Long-chain fatty acids represent important sources of metabolic energy and carbon for macromolecular synthesis and therefore must be specifically and efficiently transported across the cell envelope. As FadL and FACS are present at basal levels under nutrient-rich conditions, they can function to specifically transport and activate long-chain fatty acids when they become encountered in the environment. Under conditions where there are high levels of long-chain fatty acids are encountered in the environment, there is a specific induction of the fatty acid transport apparatus which occurs by way of long-chain fatty acyl CoA-FadR-mediated derepression of the *fadL* and *fadD* genes.

The transport of long-chain fatty acids in gram negative bacteria also appears to have pathophysiological implications. Using *in vivo* expression technology (IVET), Mahan *et al.* have shown that the *fadB* gene is specifically induced during early infection of mice by *Salmonella typhimurium* [63]. The *fadB gene* is part of the *fadBA* operon which encodes the β-oxidation multienzyme complex. As noted above, the expression of the genes involved in β-oxidation of long-chain fatty acids (including *fadBA*) are controlled at the level of transcription by FadR in response to intracellular long-chain fatty acyl CoA. High level expression of FadL and FACS is dependent upon FadR-mediated transcriptional control. The finding that *fadB* becomes induced implies that the intracellular levels of long-chain fatty acyl CoA rise which must necessarily be the result of the concerted activity of FadL and FACS in the transport and activation of exogenous long-chain fatty acids. High concentrations of long-chain fatty acids, including arachidonate, are found in the extracellular inflammatory milieu as the result of phagocytosis and the action of specific classes of phospholipases [64]. The result of transport, activation and β-oxidation of long-chain fatty acids such as arachidonic acid by gram negative bacteria would result in suppressing a local inflammatory response. This may provide the cells an advantage during early stages of colonization as well as a means of detoxification of high concentrations of long-chain fatty acids that may be encountered under such conditions. Therefore the expression of *fadB* along with the other fatty acid transport, activation and degradative genes may contribute to the metabolism of bactericidal or proinflammatory host fatty acids.

Our present challenge it to define the mechanism by which FadL and FACS interact to facilitate transport. Our approach is to define the structure and topology of FadL and FACS; define the fatty acid binding domains within FadL and FACS; and define whether there are additional protein players in the transport apparatus. Our presumption is that FadL and FACS are the major components of the fatty acid transport apparatus and that the periplasmic protease Tsp is peripheral. We have limited data to support the existence of the reported fatty acid/H^+ cotransporter. If the hypothesis is correct that FACS partitions into the inner membrane in response to fatty acid ligands, part of the challenge will be to define how FadL delivers fatty acids to the inner membrane and how this signal facilitates membrane association. Given the pathophysiological implications of fatty acid metabolism in enterotoxigenic gram negative bacteria noted above, an additional challenge that we face is to define whether this process provides cells a protective advantage during early stages of inflammation.

Fatty acid transport and trafficking in yeast

In yeast, long-chain fatty acids imported by Fat1p are converted to CoA thioesters by the fatty acyl CoA synthetases Faa1p and Faa4p [41, 56, 57]. The long-chain fatty acyl-CoAs are incorporated into phospholipids, used as a substrate in protein acylation and used as carbon and energy source. Intracellular long-chain acyl CoAs are presumed bind ACBP which acts to buffer the cell from the detergent properties of these compounds as well as providing a means for intracellular targeting [58]. While this scheme of fatty acid trafficking appears to be reasonably simple, there are a number of unanswered questions. For example, we do not yet understand the precise roles of the activating enzymes Faa1p and Faa4p. Are they essentially redundant or are they differentially compartmentalized or expressed? The role of ACBP is presumed to be a long-chain fatty acyl CoA pool former but can this protein also target these compounds to specific sites of utilization? Lastly, it is possible that other proteins interact with ACBP or bind long-chain fatty acyl CoA directly to specify intracellular targeting?

β-oxidation of long-chain fatty acids occurs exclusively in the peroxisomes, of yeast [65]. Growth of yeast in the presence of long-chain fatty acids induces the proliferation of peroxisomes, and results in the activation of genes whose protein products are involved in β-oxidation [66]. For example, the *POX1* gene (encoding fatty acid oxidase) and the *FOX2* and *FOX3* genes (encoding a multifunctional enzyme complex made up of a trifunctional polypeptide subunit and a subunit with thiolase activity) are specifically induced in the presence of long-chain fatty acids [67]. Therefore these compounds or a derivative thereof must signal the transcriptional apparatus to specifically induce genes required for peroxisomal proliferation and β-oxidation. On the other hand, growth in the presence of long-chain fatty acid oleate results in the repression of *OLE1,* encoding fatty acid desaturase. In this regard, the intracellular trafficking of these compounds must be highly regulated in order to maintain cellular homeostasis.

On the basis of our work identifying Fat1p as a fatty acid transporter and that of Gordon and colleagues demonstrating Faa1p and Faa4p are involved in the activation of exogenous long-chain fatty acids, we suggest there must be a flow of long chain fatty acid metabolites proceeding from Fat1p at the plasma membrane to intracellular sites of metabolic utilization. ACBP must play a pivotal role in regulating the flux of long-chain fatty acyl CoA within the cell. In fact recent work from Knudsen's laboratory demonstrated that the disruption of the ACBP structural gene (*ACB1*) alters acyl CoA metabolism [68, 69]. We are now faced with the challenge of defining the flow of these fatty acid metabolites following transport and activation. Central to this is understanding [1] how Fat1p and Faa1p and Faa4p function to transport and activate long-chain fatty acids and [2] whether ACBP is specifically involved in intracellular targeting. At the level of the peroxisome, certain details of long-chain fatty acid targeting are now starting to be resolved. For example, two independent pathways describing long-chain fatty acid transport across the peroxisomal membrane have been defined [70]. The first is specific for long-chain fatty acyl CoAs and requires Pat1p (Pxa2p) and Pat2p (Pxa1p). The second is specific for medium-chain fatty acids and requires the fatty acyl CoA synthetase Faa2p and is apparently independent of Pat1p and Pat2p [70]. Pat1p and Pat2p were initially identified as peroxisomal integral membrane proteins each of which have six or seven predicted transmembrane domains and an ATP binding cassette characteristic of ATP-binding protein (ABC) transporters. These two proteins share sequence similarities to the protein product of the human adrenoleukodystrophy gene [71, 72]. Deletion of either Pat1p or Pat2p impairs growth of yeast on the long chain fatty acids palmitate and oleate but not myristate. The conclusion that these two protein are involved in fatty acid transport into the peroxisome is based upon studies showing that a deletion of either gene reduces β-oxidation of whole cells but not cell extracts [70]. A third putative peroxisomal fatty acid transporter, Fat2p (Psc60p) has been cloned by reverse genetics by selection for peroxisomal proteins induced by oleic acid [73]. Although no functional studies on Fat2p have been reported, we have found that this protein shares 23% amino acid identity and 45% amino acid similarity to Fat1p (DiRusso and Black, unpublished). Like Fat1p, Fat2p contains a presumptive ATP/AMP signature motif characteristic of the enzymes which form adenylated intermediates as part of their catalytic cycle (including the fatty acyl CoA synthetases) [73]. Yet, unlike these enzymes, Fat2p has no measurable fatty acyl-CoA synthetase activity. Deletion of Fat2p has no apparent effect on peroxisomal induction and proliferation or β-oxidation of long chain fatty acids which indicates either that the protein is not required for these functions or that there additional protein(s) that serve the same function.

Regulation of gene expression by fatty acids

Exogenous fatty acids have diverse effects on lipid metabolism and organelle structure, function and biogenesis. The activation of genes involved in fatty acid degradation and the repression of genes involved in fatty acid synthesis in response to exogenous fatty acids occurs widely in prokaryotes and eukaryotes (e.g. [3, 74–76]. One of the most pronounced effects of high levels of fatty acids is the proliferation of peroxisomes [65]. In mammalian systems, the transcription factors which effect activation include the peroxisomal proliferator family including particularly PPARγ and FAAR [77, 78]. In yeast, the factors Oaf1p (Pip1p) and Oaf2p (Pip2p) are required to activate many genes encoding enzymes required for synthesis of peroxisomal proteins [79]. Oaf1p and Oaf2p form heterodimers which interact specifically with promoter DNA containing an fatty acid response element. Fatty acids or FACoA are not required for DNA binding of Oaf1p/Oaf2p.

FRM2 also encodes a fatty acid-responsive regulatory protein, Frm2p, that is required for repression of *OLE1* [81]. It is not known whether Frm2p binds DNA and/or regulates *OLE1* transcription directly. Analysis of the cloned DNA sequence predicts a 21,000 molecular weight polypeptide with no homologies to other proteins in the database. It has been reported that fatty acid acid-mediated repression of *OLE1* requires the activity of fatty acyl-CoA synthetase *(FAM* or *FAM* [68]) suggesting that FACoA mediates regulation perhaps by modifying an additional protein that in turn regulates DNA binding. Binding of FACoA to the *Escherichia coli* transcription factor FadR, which our laboratory has extensively characterized, prevents DNA binding and transcription activation or repression [19–21].

While two specific fatty acid-responsive transcription factors have been identified in yeast, the mechanism of regulation of their activity by fatty acids remains undefined. Since fatty acids or fatty acyl CoAs do not affect transcription factor binding, these effectors must mediate control at a step prior to DNA binding. For example, fatty acid or a fatty acid derivative may stimulate a protein kinase which in turn regulates the factors activity by phosphorylation. The fatty acid effector could stimulate or repress the expression of the transcription factor itself. The effector might regulate the activity or expression of a protein which interacts with Frm2p and/or Oaf1p/Oaf2p by a mechanism similar to Ga180 regulation of Ga14.

Pathophysiological implications of deficiencies in long-chain fatty acid transport

Long-chain fatty acids represent the preferred source for energy production by the heart. Under physiological conditions it is estimated that circulating long-chain fatty acids provide 60–70% of the cardiac energy requirements [80–82]. The uptake of long-chain fatty acids into myocardial cells must therefore be efficient. Two proteins have been identified in cardiac cells that appear to function in the facilitated import of long chain fatty acids. The first, *F*atty *A*cid *T*ransport *P*rotein (FATP), may operate in conjunction with fatty acyl-CoA synthetase to promote unidirectional transport [27, 28]. As noted above Fat1p of yeast represents the FATP homologue defined in murine adipocytes [41]. The second, *F*atty *A*cid *T*ranslocase (FAT) may function with intracellular fatty acid binding proteins [24–26]. Both FATP and FAT mRNAs are found in cardiac cells suggesting these transport proteins are important in the transport of exogenous long-chain fatty acids and therefore in cardiovascular physiology. In this regard it appears that certain types of cardiomyopathies are the consequence of depressed levels of long-chain fatty acid uptake [28, 83, 84]. In addition to potential cardiac pathophysiologies, deficiencies in long-chain fatty acid uptake may have a pathophysiological role in type H diabetes and atherosclerosis [85, 86].

Acknowledgments

Work from our laboratories has been supported by the National Institutes of Health, The National Science Foundation, and The American Heart Association.

References

1. Carmen GM, Zeimetz GM: Regulation of phospholipid synthesis in the yeast *Saccharomyces cerevisiae*. J Biol Chem 271: 13293–13296, 1996
2. Greenberg ML, Lopes JM: Genetic regulation of phospholipid biosynthesis in *Saccharomyces cerevisiae*. Microbiol Rev 60: 1–20, 1996
3. Hillgartner FB, Salati LM, Goodridge AG: Physiological and molecular mechanisms involved in nutritional regulation of fatty acid synthesis. Physiol Rev 25: 47–76, 1995
4. Paltauf BL, Kohlwein SD, Henry SA: Regulation and compartmentalization of lipid synthesis in yeast. In: E.W. Jones, J.R. Pringle, J.R. Broach (eds). The Molecular and Cellular Biology of the Yeast *Saccharomyces cerevisiae*: Gene Expression. Cold Spring Harbor Press, Cold Spring Harbor, NY, 1992, pp 415–500
5. Glick BS, Rothman JE: Possible role for fatty acyl-coenzyme A in intracellular protein transport. Nature 326: 309–312, 1987
6. Pfanner N, Orci L, Glick BS, Amherdt M, Ardern SR, Malhotra V, Rothman JE: Fatty acyl-coenzyme A is required for budding of transport vesicles from Golgi cistemae. Cell 59: 95–102, 1989
7. Gordon JI, Duronio RS, Rudnick DA, Adams SP, Gokel GW: Protein N-myristoylation. J Biol Chem 266: 8647–8650, 1991
8. McLaughlin S, Aderem A: The myristoyl-electrostatic switch: A modulator of reversible protein-membrane interactions. TIBS 20: 270–276, 1995
9. Schneiter R, Kohlwein SD: Organelle structure, function and inheritance in yeast: A role for fatty acid synthesis? Cell 88: 431–434, 1997
10. Fujimoto Y, Tsunomori M, Sumiya T, Nishida H, Sakuma S, Fujita T: Effects of fatty acyl coenzyme A esters on lipoxygenase and cyclooxygenase metabolism of arachidonic acid in rabbit platelets. Prostaglandins Leuko Essen Fatty Acids 52: 255–258, 1995
11. Fulceri R, Gamberucci A, Scott HM, Giunti R, Burchell A, Benedetti A: Fatty acyl-CoA esters inhibit glucose-6-phosphatase in rat liver microsomes. Biochem J 307: 391–397, 1995
12. Lai JCK, Liang BB, Jarri EL, Cooper AJL, Lu DR: Differential effects of fatty acyl coenzyme A derivatives on citrate synthase and glutamate dehydrogenase. Res Com Chem Pathol Pharm 82: 331–338, 1993
13. Yamashita A, Watanabe M, Tonegawa T, Sugiura T, Waku K: Acyl-CoA binding and acylation of UDP-glucuronosyltransferase isoforms of rat liver: Their effect on enzyme activity. Biochem J 312: 301–308, 1995
14. Aihaud GR, Abumrad NA, Amri E-Z, Grimaldi PA: A new look at fatty acids as signal-transducing molecules. In: G.C. Simopoulos, E. Tremoli (eds). Fatty Acids and Lipids: Biological Aspects. Karger Press, Basel, Switzerland, 1994
15. Glatz JE, Borchers T, Spener E, van der Vusse, GJ: Fatty acids in cell signaling: Modulation by lipid binding proteins. Prostaglandins Leuko Essen Fatty Acids 52: 121–127, 1995
16. Korchak HM, Kane LH, Rossi MW, Corkey BE: Long-chain acyl coenzyme A and signaling in neutrophils. An inhibitor of acyl coenzyme A synthetase, triacsin C, inhibits superoxide anion generation and degranulation by human neutrophils. J Biol Chem 269: 30281–30287, 1994
17. Shrago E, Woldegiorgis G, Ruoho AE, DiRusso CC: Fatty acyl CoA esters as regulators of cell metabolism. Prostaglandins Leuko Essen Fatty Acids 52: 163–166, 1995
18. Rich GT, Comerford JG, Graham S, Dawson AP: Effects of CoA and acyl-CoA on Ca^{2+} permeability of endoplasmic-reticulum membranes from rat liver. Biochem J 306: 703–708, 1995
19. DiRusso CC: The role of long-chain acyl-CoA esters in the regulation of gene expression and metabolism. In: J.Y. Vanderhoek (ed). *Escherichia coli*. Frontiers in Bioactive Lipids. Plenum Press, NY, 1995, pp 15–22
20. DiRusso CC, Heimert TL, Metzger AK: Characterization of FadR, a global transcriptional regulator of fatty acid metabolism in *Escherichia coli*. J Biol Chem 267: 8685–8691, 1992
21. DiRusso CC, Metzger AK, Heimert TL: Regulation of transcription of genes required for fatty acid transport and unsaturated fatty acid biosynthesis in *Escherichia coli* by FadR. Mol Microbiol 7: 311–322, 1993
22. Raman N, DiRusso CC: Analysis of acyl coenzyme A binding to the transcription factor FadR and identification of amino acid residues required for ligand binding. J Biol Chem 270: 1092–1097, 1995
23. DiRusso CC, Nystom T: The fats of *Escherichia coli* during infancy and old age. Mol Microbiol 27: 1–8, 1998
24. Abumrad NA, Forest CC, Regen DM, Sanders S: Increase in membrane uptake of long-chain fatty acids early during preadipocyte differentiation. Proc Nat Acad Sci USA 88: 6008–6012, 1991
25. Abumrad NA, El-Maghrabi MR, Amri E-Z, Lopez E, Crimaldi PA: Cloning of a rat adipocyte membrane protein implicated in binding or transport of long-chain fatty acids that is induced during preadipocyte differentiation. J Biol Chem 268: 17665–17668, 1993
26. Harmon CM, Abumrad NA: Binding of sulfosuccinimidyl fatty acid to adipocyte membrane proteins: Isolation and amino-terminal sequence of an 88-kD protein implicated in transport of long-chain fatty acid. J Membr Biol 133: 43–49, 1993
27. Schaffer JE, Lodish HF: Expression cloning and characterization of a novel adipocyte long-chain fatty acid transport protein. Cell 79: 427–436, 1994
28. Schaffer JE, Lodish HF: Molecular mechanism of long-chain fatty acid uptake. Trends Cardiovasc Med 5: 218–224, 1995
29. Berk PD, Wada H, Horio Y, Potter BL, Sorrentino D, Zhou SL, Isola LM, Stump D, Kiang CL, Thung S: Plasma membrane fatty acid binding protein and mitochondrial glutamic-oxaloacetic transaminase of rat liver are related. Proc Nat Acad Sci USA 87: 3484–3488, 1990
30. Zhou S-L, Stump D, Isola LM, Berk PD: Constitutive expression of a saturable transport system for non-esterified fatty acids in *Xenopus laevis* oocytes. Biochem J 297: 315–319, 1994
31. Zhou S-L, Stump D, Kiang C-L, Isola LM, Berk PD: Mitochondrial aspartate aminotransferase expressed on the surface of 3T3-L1 adipocytes mediates saturable fatty acid uptake. Proc Soc Exp Biol Med 208: 263–270, 1995
32. Black PN: Primary sequence of the *Escherichia coli fadL* gene encoding an outer membrane protein involved in fatty acid transport. J Bacteriol 173: 535–442, 1991
33. Black PN: Characterization of FadL-specific fatty acid binding in *Escherichia coli*. Biochim Biophys Acta 1046: 97–105, 1990
34. Black PN, DiRusso CC: Molecular and biochemical analyses of fatty acid transport, metabolism, and gene regulation in *Escherichia coli*. Biochim Biophys Acta 1210: 123–145, 1994
35. Black PN, DiRusso CC, Metzger AK, Heimert TL: Cloning, sequencing, and characterization of the *fadD* gene of *Escherichia coli* encoding fatty acyl coenzyme A synthetase. J Biol Chem 267: 25513–25520, 1992
36. Black PN: Acyl-CoA synthetase and the transport of long-chain fatty acids. In: J.Y. Vanderhoek (ed). Frontiers in Bioactive Lipids. Plenum Press, NY, 1996, pp 7–14
37. Black PN, Zhang Q, Weimar JD, DiRusso CC: Mutational analysis of a fatty acyl-coenzyme A synthetase signature motif identifies seven amino acid residues that modulate fatty acid substrate specificity. J Biol Chem 272: 4896–4903, 1997

38. Kumar GB, Black PN: Linker mutagenesis of a bacterial fatty acid transport protein. Identification of domains with functional importance. J Biol Chem 266: 1348–1353, 1991
39. Kumar GB, Black PN: Bacterial long-chain fatty acid transport: Identification of amino acid residues within the outer membrane protein FadL required for activity. J Biol Chem 268: 15469–15476, 1993
40. Black PN, Zhang Q: Evidence that His[110] of protein FadL in the outer membrane of *Echerichia coli* is involved in the binding and uptake of long-chain fatty acids. Possible role of this residue in carboxylate binding. Biochem J 310: 389–394, 1995
41. Færgeman NL, DiRusso CC, Elberger A, Knudsen L, Black PN: Disruption of the *Saccharomyces cerevisiae* homologue to the murine fatty acid transport protein impairs uptake and growth on long-chain fatty acids. J Biol Chem 272: 8531–8538, 1997
42. Overath P, Pauli G, Schairer HU: Fatty acid degradation in *Escherichia coli*. An inducible acyl-CoA synthetase, the mapping of old mutants, and isolation of regulatory mutants. Eur J Biochem 7: 559–574, 1969
43. Azizan A, Black PN: Use of transposon Th*phoA* to identify genes for cell envelope proteins of *Escherichia coli* required for long-chain fatty acid transport: The periplasmic protein Tsp potentiates long-chain fatty acid transport. J Bacteriol 176: 6653–6662, 1994
44. Kameda K: Partial purification and characterization of fatty acid binding protein(s) in *Escherichia coli* membranes and reconstitution of fatty acid transport system. Biochem Int 13: 343–350, 1986
45. Kameda K, Suzuki LK, Imai Y: Transport of fatty acid is obligatory coupled with H^+ entry in pheroplasts of *Escherichia coli* K-12. Biochem Int 14: 227–234, 1987
46. Nunn WD, Simons RW: Transport of long-chain fatty acids by *Escherichia coli*: Mapping and characterization of mutants in the *fadL* gene. Proc Nat Acad Sci USA 75: 3377–3381, 1978
47. Nunn WD, Simons RW, Egan PA, Maloy SR: Kinetics of the utilization of medium and long chain fatty acids by a mutant of *Escherichia coli* defective in *fadL gene*. J Biol Chem 254: 9130–9134, 1979
48. Maloy SR, Ginsburgh CL, Simons RW, Nunn WD: Transport of long and medium chain fatty acids by *Escherichia coli* K12. J Biol Chem 256: 3735–3742, 1981
49. Black PN: The *fadL* gene product of *Escherichia coli* is an outer membrane protein required for uptake of long-chain fatty acids and involved in sensitivity to bacteriophage T2. J Bacteriol 170: 2850–2854, 1988
50. Azizan A, Sherin D, DiRusso CC, Black PN: Energetics underlying the process of long-chain fatty acid transport: The roles of ATP synthase and an energized membrane (submitted)
51. Mangroo D, Gerber GE: Fatty acid uptake in *Escherichia coli*: Regulation by recruitment of fatty acyl CoA synthetase to the plasma membrane. Biochem Cell Biol 71: 431–443, 1993
52. Berger EA: Different mechanisms of energy coupling for the active transport of proline and glutamine in *Escherichia coli*. Proc Nat Acad Sci USA 70: 1514–1518, 1973
53. Berger EA, Heppel LA: Different mechanisms of energy coupling for the shock-sensitive and shock-resistant amino acid permeases of *Escherichia coli*. J Biol Chem 249: 7747–7755, 1974
54. Kohlwein SD, Paltauf F: Uptake of fatty acids by the yeasts *Saccharomyces uvarum* and *Saccharomycopsis lipolytica*. Biochim Biophys Acta 792: 310–317, 1983
55. Knoll LL, Johnson DR, Gordon JI: Complementation of *Saccharomyces cerevisiae* strains containing fatty acid activation gene (FAA) deletions with a mammalian acyl CoA synthetase. J Biol Chem 270: 10861–10867, 1995
56. Johnson DR, Knoll LL, Levin DE, Gordon JI: *Saccharomyces cerevisiae* contains four fatty acid activation (FAA) genes: As assessment of their role in regulating protein N-myristoylation and cellular lipid metabolism. J Cell Biol 127: 751–762, 1994
57. Knoll LJ, Johnson DR, Gordon JI: Biochemical studies of three *Saccharomyces cerevisiae* acyl-CoA synthetases, Faa1p, Faa2p, and Faa3p. J Biol Chem 269: 16348–16356, 1994
58. Rasmusen JT, Færgeman NL, Kristiansen K, Knudsen J: Acyl-CoA binding protein (ACBP) can mediate intermembrane transport and donate acyl-CoA for beta-oxidation and glycerolipid synthesis. Biochem J 299: 165–170, 1994
59. Andreasen AA, Stier TJB: Anaerobic nutrition of *Saccharomyces cerevisiae*. II. Unsaturated fatty acid requirement for growth in a defined medium. J Cell Comp Physiol 41: 23–26, 1953
60. Bourot S, Karst F: Isolation and characterization of the *Saccharomyces cerevisiae SUT1* gene involved in sterol uptake. Gene 165: 97–102, 1995
61. Sallus L, Haselbeck RL, Nunn WD: Regulation of fatty acid transport in *Escherichia coli*: Analysis by operon fusion. J Bacteriol 155: 1450–1454, 1983
62. Higashitani HA, Nishimura Y, Hara H, Aiba H, Mizuno T, Horiuchi K: Osmoregulation of the fatty acid receptor gene *fadL* in *Escherichia coli*. Mol Gen Genet 240: 339–347, 1993
63. Mahan ML, Tobias JW, Slauch JM, Hanna PC, Collier RL, Mekalanos JJ: Antibiotic-based selection for bacterial genes that are specifically induced during infection of a host. Proc Nat Acad Sci USA 92: 669–673, 1995
64. Pace L, Hayman ML, Galan JE: Signal transduction and invasion of epithelial cells by *S. typhimurium*. Cell 72: 505–514, 1993
65. Paltauf BL, Kohlwein SD, Henry SA: Regulation and compartmentalization of lipid synthesis in yeast. In: E.W. Jones, J.R. Pringle, J.R. Broach (eds). The Molecular and Cellular Biology of the Yeast *Saccharomyces cerevisiae*: Gene expression. Cold Spring Harbor Press, Cold Spring Harbor, NY, 1992, pp 415–500
66. Elgersma Y, Tabak HF: Proteins involved in peroxisome biogenesis and functioning. Biochim Biophys Acta 1286: 269–283, 1996
67. Hiltunen JK, Wenzel B, Beyer A, Erdman R, Fossa A, Kunau W: Peroxisomal multifunctional beta-oxidation protein of *Saccharomyces cerevisiae*. Molecular analysis of the fox2 gene and gene product. J Biol Chem 267: 6646–6653, 1992
68. Karlskov Schjerling C, Hummel R, Hansen JK, Børsting C, Mikkelsen JM, Kristiansen K, Knudsen J: Disruption of the gene encoding the acyl CoA binding protein (*ACB1*) perturbs acyl-CoA metabolism in *Saccharomyces cerevisiae*. J Biol Chem 37: 22514–22521, 1996
69. Rasmusen JT, Rosendal J, Knudsen J: Interaction of acyl-CoA binding protein (ACBP) on processes for which acyl-CoA is a substrate, product, or inhibitor. Biochem J 292: 907–913, 1993
70. Hettema EH, van Roermund CWT, Distel B, van den Berg M, Vilela C, Rodrigues-Pousada C, Wanders RJA, Tabak HF: The ABC transporter proteins Pat1 and Pat2 are required for import of long-chain fatty acids into peroxisomes of *Saccharomyces cerevisiae*. EMBO J 15: 3813–3822, 1996
71. Lombard-Platet G, Savary S, Sarde CO, Mandel JL, Chimini G: A close relative of the adrenoleukodystrophy (ADL) gene codes for a peroxisomal protein with a specific expression pattern. Proc Nat Acad Sci USA 93: 1265–1269, 1996
72. Shani N, Valle D: A *Saccharomyces cerevisiae* homolog of the human adrenoleukodystrophy transporter is a heterodimer of two half ATP-binding cassette transporters. Proc Nat Acad Sci USA 93: 11901–11906, 1996
73. Blobel F, Erdman R: Identification of a yeast peroxisomal member of AMP binding proteins. Eur J Biochem 240: 468–476, 1996
74. Frayn KN, Shadid S, Hamlani R, Humphreys SM, Clark ML, Fielding BA, Boland O, Coppack SW: Regulation of fatty acid movement in human adipose tissue in the postabsorptive-to-postprandial transition. Am J Physiol 266: E308–E317, 1994
75. Gonzalez CI, Martin CE: Fatty Acid-responsive control of mRNA stability. J Biol Chem 271: 25801–25809, 1996

76. Jump DB, Clarke SD, Thelen A, Liimatta M: Coordinate regulation of glycolytic and lipogenic gene expression by polyunsaturated fatty acids. J Lipid Res 35: 1076–1084, 1994
77. Cornelius P, MacDougal OA, Lane MD: Regulation of adipocyte development. Annu Rev Nutr 14: 99–129, 1994
78. Grimaldi PA, Gaillard D, Inadera H, Teboul L, Ailhaud G, Amri E-Z: Molecular mechanisms involved in the adipogenic action of fatty acids. In: J.Y. Vanderhoek (ed). Frontiers in Bioactive Lipids. 1996, pp 1–6
79. Einerhand AW, Kos W, Distel D, Tabak HF: Characterization of a transcriptional control element involved in the proliferation of peroxisomes in yeast in response to oleate. Eur J Biochem 214: 323–331, 1993
80. Burczynski FL, Cai ZS, Moran JB, Geisbuhler T, Rovetto M: Palmitate uptake by cardiac myocytes and endothelial cells. Am J Physiol 268: H1659–H1666, 1995
81. Lopaschuk GD, Collins-Nakai RL, Itoi T: Developmental changes in energy substrate use by the heart. Cardio Res 26: 1172–1180, 1992
82. Schulz H: Regulation of fatty acid oxidation in the heart. J Nutr 124: 165–171, 1994
83. Van Nieuwenhoven FA, Verstijnen CP, Abumrad NA, Willemsen PF, Van Eys GL, Van der Vusse GL, Glatz JF: Putative fatty acid translocase and cytoplasmic fatty acid binding protein am co-expressed in rat heart and skeletal muscles. Biochem Biophys Res Comm 207: 747–752, 1995
84. Vork MM, Glatz JFC, van der Vusse: On the mechanism of long chain fatty acid transport in cardiomyocytes as facilitated by cytoplasmic fatty acid binding protein. J Theor Biol 160: 207–222, 1993
85. Franken PR, De Geeter E, Dendale P, Block P, Bossuyt A: Abnormal free fatty acid uptake in subacute myocardial infarction after coronary thrombolysis: Correlation with wall motion and inotropic reserve. J Nucl Med 35: 1758–1765, 1994
86. Siguel EN, Lerman RH: Altered fatty acid metabolism in patients with angiographically documented coronary artery disease. Metab Clin Exp 43: 982–993, 1994

Molecular and Cellular Biochemistry **192:** 53–61, 1999.

Cytochrome P450, peroxisome proliferation, and cytoplasmic fatty acid-binding protein content in liver, heart and kidney of the diabetic rat

Wim Engels,[1] Marc van Bilsen,[2] Bruce H.R. Wolffenbuttel,[1] Ger J. van der Vusse[2] and Jan F.C. Glatz[2]

[1]*Department of Internal Medicine and* [2]*Department of Physiology, Cardiovascular Research Institute Maastricht (CARIM), Maastricht University, Maastricht, The Netherlands*

Abstract

Diabetes mellitus generally results in an increased systemic fatty acid mobilization which can be associated with an increase in mitochondrial and peroxisomal β-oxidation of fatty acids in selected tissues. The latter is usually accompanied by a concomitant increase in the tissue content of cytoplasmic fatty acid-binding protein (FABP) which functions in the intracellular translocation of fatty acids. It was previously found that in liver clofibrate-induced proliferation of peroxisomes and increase in FABP expression each are dependent on the induction by cytochrome P4504A1 -mediated (CYP4A1) formation of dicarboxylic acids. We studied whether peroxisome proliferation and an increase of FABP contents in liver, heart and kidney of streptozotocin-induced diabetic rats are also accompanied by an increase of CYP4A1 activity, as this would indicate a possible regulatory role for dicarboxylic acids in peroxisome proliferation and FABP induction in diabetic organs other than liver. In livers of the diabetic rat, a concomitant increase was observed of the activities of CYP4A1 and the peroxisomal key enzyme fatty acyl-CoA oxidase (FACO) and of the FABP content. In the diabetic heart FACO activity and FABP content also increased, but there was no induction of CYP4A1 activity. Conversely, in diabetic kidney there was no increase in FACO activity nor FABP content in spite of a marked induction of CYP4A1 activity. It is concluded that streptozotocin-induced diabetes leads to increased peroxisome proliferation and increased levels of FABP in both liver and heart, which only in liver is accompanied by an induction of the cytochrome P450 system. Consequently, it is not likely that dicarboxylic acids are involved in the induction of peroxisome proliferation in the heart. (Mol Cell Biochem **192**: 53–61, 1999)

Key words: diabetes mellitus, cytochrome P450 4A, fatty acid-binding protein, peroxisome proliferation, streptozotocin, rat

Abbreviations: EC – ethoxycoumarin; FACO – fatty acyl CoA oxidase; FABP – fatty acid-binding protein; MES – β-morpholino-ethanesulfonic acid; MOPS – 3-[N-Morpholino] propane sulfonic acid; PPAR – peroxisome proliferator-activated-receptor; PPRE – peroxisome proliferator reponse element

Introduction

Diabetes mellitus is well known to lead to profound changes in whole body fatty acid metabolism. In both insulin-dependent and non-insulin dependent diabetes mellitus, glucose uptake and/or metabolism by extra-hepatic tissues is decreased, causing an increased mobilization of fatty acids from adipose tissue and a shift towards increased utilization of fatty acids by liver, heart and skeletal muscles [1]. Uncontrolled diabetes mellitus leads to a fatty acid overload, which is accompanied by an increased capacity for oxidation of fatty acids by the mitochondrial,

Address for offprints: W. Engels, Department of Internal Medicine, Maastricht University, P.O. Box 616, 6200 MD Maastricht, The Netherlands

peroxisomal and microsomal pathways of liver and heart [2].

Under normal circumstances, the peroxisomal β-oxidation is only a minor pathway for fatty acid oxidation relative to the mitochondrial counterpart, although it appears to be of particular importance for the β-oxidation of very-long-chain fatty acids which are poor substrates in the mitochondrial oxidative pathway [3–5]. Diets containing (very) long-chain fatty acids not only induce in liver the enzymes of both peroxisomal and microsomal oxidation [6], but also the cytoplasmic liver-type fatty acid-binding protein (L-FABP) [7]. This protein exhibits marked affinity for long-chain fatty acids and is considered to play a significant role in the cellular uptake and intracellular targeting of fatty acids [8]. Therefore, in the transport and metabolism of long-chain fatty acids during fatty acid overload, the expression of FABP and the activity of peroxisomal enzymes, respectively, each play a crucial role. However the precise mechanisms underlying the regulation of FABP expression and peroxisome proliferation are still unknown.

In liver the proliferation of peroxisomes and the induction of FABP have been demonstrated to be invariably dependent on the cytochrome P450 system localized at the endoplasmic reticulum [5, 9]. This was seen both under conditions of fatty acid overload and after administration of so-called peroxisome proliferators, such as clofibrate [9]. The mechanism has been proposed to be an adaptive response to altered intracellular fatty acid fluxes, mediated by dicarboxylic fatty acids formed via the cytochrome P4504A1 (CYP4A1) ω-oxidation pathway [9]. In this pathway long-chain fatty acids are first hydroxylated on the omega position, then oxidized via an intermediary aldehyde to a dicarboxylic acid [10] and finally converted into their CoA derivative [6, 10]. Recently, a soluble receptor termed peroxisome-proliferator-activated receptor (PPAR) belonging to the nuclear hormone receptor superfamily has been isolated [11]. These receptors are ligand-activated transcription factors and can bind to specific DNA sequences, i.e. the so-called peroxisome proliferator response elements (PPRE). The detection by Northern blotting of PPAR-specific mRNA in liver, heart, kidney, gut and brown adipose tissue of the rat, suggests its involvement in regulating aspects of peroxisomal function and fatty acid homeostasis [12]. However, none of the peroxisomal proliferators tested thus far has been shown to bind to PPAR while the endogenous ligands of this receptor remain unknown [13].

Under starved and diabetic conditions, co-induction in liver of peroxisomal fatty acid β-oxidation and microsomal fatty acid ω-oxidation has also been established [14, 15]. Conversely, conditions of hyperinsulinemia inhibit hepatic peroxisomal β-oxidation in rats [16]. Therefore, it is reasonable to hypothesize that in diabetes mellitus peroxisomal fatty acid β-oxidation and microsomal ω-oxidation in rat liver are cooperatively regulated by PPAR in order to achieve an optimal utilization of fatty acids. With respect to kidney and heart, no data are available. The notion that dicarboxylic acids might play a role in these inductive processes in diabetes is substantiated by the observation that in diabetic rats the urinary excretion of dicarboxylic acids is enhanced [10]. However, Gustafsson *et al.* [17] demonstrated in liver that the cytochrome P450 system as well as dicarboxylic acids do not play a role in peroxisome proliferation. The assembled findings indicate that the precise mechanism of the induction of peroxisome proliferation and of FABP in liver, heart and kidney is not yet fully understood.

The aim of the present study was to determine whether in experimental diabetes, as induced by streptozotocin injection, CYP4A1 activity is implicated in peroxisome proliferation, in the induction of FABP, and/or in the induction of CYP4A1 itself, as was postulated by Kaikaus *et al.* [9, 18, 19] for clofibrate administration and high-fat feeding, and, if so, whether possible organ-specific mechanisms are involved. For this, we measured in rat liver, heart, and kidney not only CYP4A1 activity, but also the activities of CYP1A1 as a generally inducible cytochrome P450-related enzyme and of CY1P2El, a cytochrome P450 subspecies which is induced particularly under diabetic conditions [20]. Peroxisome proliferation was assessed by measuring the activity of fatty acyl-CoA oxidase (FACO). FABP was determined as L(iver) and H(eart)-type FABP by ELISA's type-specific antibodies.

Our data presented in this study indicate marked differences in the response of these organs towards conditions of streptozotocin-induced hyperglycemia with respect to peroxisome proliferation, FABP levels and CYP4A1 induction, thereby questioning the general role of cytochrome P450 related metabolites, such as dicarboxylic acids, in the process of a coordinated induction.

Materials and methods

Chemicals

Streptozotocin, clofibrate, 7-ethoxycoumarin, bovine serum albumin (fraction V), lauric acid, 11-hydroxy and 12-hydroxy lauric acid, 110-hydroxy caproic acid, arachidonic acid, MES (β-morpholino-ethanesulfonic acid), MOPS (3-[N-Morpholino]propane sulfonic acid) and Tris-HCl were obtained from Sigma Chemicals (St. Louis, MO, USA). Acetonitrile, diethyl ether, and tetrahydrofuran were purchased from Rathburn (Walkerburn, UK). Panacyl bromide [p-(9-anthroyl)phenacyl bromide] was from Molecular Probes Inc. (Eugene, OR, USA). NADH, NADPH, $NADP^+$, glucose-6-phosphate and glucose-6-phosphate dehydrogenase were obtained from Boehringer Mannheim (Mannheim, Germany). All other chemicals were at least reagent grade (Merck, Darmstadt, Germany).

Animals

Experiments were performed with adult male Wistar rats (Winkelmann, Borchen, FRG) (body wt 250–300 g). Animals were fed *ad libitum* (Diet SRIVI-A, Hope Farms, Woerden, The Netherlands), had free access to water, and were kept under an artificial dark-light cycle of 12 h. The experiments were approved by the Institutional Animal Care and User Committee of Maastricht University.

Treatment

Insulin-dependent diabetes was evoked by a single intravenous injection of streptozotocin (55 mg/kg body wt) as described in detail by Glatz *et al.* [21]. The rats were sacrificed after 3 weeks.

In a subset of experiments (n = 8) clofibrate was used as peroxisome proliferator. Hereto, clofibrate was administered daily for 7 days via an oral cannula. A solution of 80 mg of the drug per ml glycerol (60%) was given so that each rat received 250 mg/kg per day of clofibrate.

Blood glucose

Blood glucose concentrations were measured with a glucose sensor electrode (Medisense Inc., Waltham, MA, USA) in whole blood obtained by orbital punction.

Tissue homogenization and isolation of microsomes

Liver, heart and kidney were dissected and treated as described by McCallum *et al.* [22]. Briefly, rats were heparinized by an intraperitoneal injection of 1000 U Thromboliquine (Organon Technika, Oss, The Netherlands). After 10 min, the animals were decapitated and liver, heart, and kidney were quickly removed and weighed. Immediately thereafter, the organs were freeze-clamped and stored at –80°C until further analysis. Tissue was pulverized in an aluminum mortar cooled with liquid nitrogen and subsequently homogenized in 0.15 M KCl and 50 mM potassium phosphate, pH 7.4, at 0–4°C using a Potter-Elvehjem glass-Teflon homogenizer. Thereafter, microsomes were prepared by differential centrifugation as described by Rutten *et al.* [23]. The pellet was resuspended carefully with a 5 ml glass-Teflon homogenizer in ice-cold 0.15 M KCl in a volume equal to the original organ weight.

For the determination of FABP freeze-clamped tissue was homogenized as a 10% (w/v) solution in 50 mM Tris-HCl, pH 7.4, with an Ultraturrax (50 Watt) during three periods of 5 sec with 10 sec intervals at 0°C.

Fatty acyl-CoA oxidase (FACO)

Fatty acyl-CoA oxidase activity was measured using palmitoyl-CoA as the substrate, as described by Vannecq [24]. Briefly, the homogenates were freshly prepared by homogenizing the freeze-clamped organ tissue in Mannaerts buffer (10 mM tetrasodium pyrophosphate (pH 9.0), 10 μM FAD^+ and 1 mM EDTA) [25]. Then, 10 μl of the homogenate (containing 50–100 μg total protein) was added to 2 ml of a reaction mixture containing 100 mM HEPES (pH 7.6), 1 mM homovanillic acid, 1 μg horse radish peroxidase, 5 μM FAD^+, 1 mM NaN_3, 240 μg bovine serum albumin (fatty acid free), and 0.06% Triton X-100. The reaction temperature was 37°C, and the cuvet was equipped with a stirring bar at the bottom. The reaction was started by addition of 100 μl of a palmitoyl-CoA solution (1 mM in 20 mM MES buffer). The increase in fluorescence was measured during 15 min on a Shimadzu RF-5001PC fluorescence spectrophotometer (settings: excitation wavelength 327 nm, and emission wavelength 428 nm).

FABP assay

Tissue contents of L(iver-type)-FABP and H(eart-type)-FABP each were measured by specific affinity-enzyme-linked immuno-sorbent assays of the antigen capture type (sandwich-ELISAs), using purified IgG antibodies directed against purified rat L- or H- FABP and the streptavidin-biotin detection system as described by Vork *et al.* [26] and Van Nieuwenhoven *et al.* [27]. The cross-reactivities of H-FABP in the assay for L-FABP, and conversely, were found to be less than 0.002%.

Spectroscopic determination of cytochrome P450

Determinations of cytochrome P450 content of the microsomal preparations was based on the reduced carbon monoxide difference spectrum, using sodium dithionite as the reducing agent, according to Onnura and Sato [28], as described by Stegeman *et al.* [29]. Microsomal suspensions containing about 1 mg of protein per ml were used. The cytochrome P-450 content was calculated from the absorbance difference between 450 and 490 nm, using a molar extinction coefficient of 91 mM^{-1} cm^{-1} [28].

CYP4A1 activity

This activity was measured as the omega-hydroxylation of lauric acid. Hereto we developed a new specific HPLC method, using panacyl bromide as the fluorescent derivative.

Reaction mixtures containing microsomes (1–2 mg of protein), 100 nM lauric acid (added as a stock solution of 10 mM in ethanol), 0.1 ml of a NADPH regenerating system (10 mg NADH, 7.9 mg $NADP^+$, 6.8 mg glucose-6-phosphate, 14 μl glucose-6-phosphate dehydrogenase, and 6.8 mg $MgCl_2$ in 1 ml of 100 mM potassium phosphate buffer, pH 7.4) in a final reaction volume of 1 ml of 100 mM potassium phosphate buffer, pH 7.4, were incubated at 37°C for 10 min in an orbital shaking water bath (200 rpm). The reaction was stopped by the addition of 50 μl of acetic acid. The formation of the hydroxy derivatives of lauric acid was linear with time until at least 60 min. After addition of the internal standard 10-hydroxy caproic acid, the samples were extracted twice with 3 ml of diethyl ether. The organic phases were collected and dried under a stream of nitrogen at 40°C. The residue was dissolved in diethyl ether-acetic acid (100:0.5, v/v) and consecutively fractionated by silica gel chromatography, purified by reverse-phase chromatography, and derivatized with panacyl bromide as described earlier [30]. The panacyl esters were solubilized in 100 μl of acetonitrile and were analyzed by C_{18} reversed-phase HPLC with fluorimetric detection. The HPLC system consisted of a Lichrocart Supersphere 100-RP-18 column (250 × 4 mm ID, 4 μm particles with a precolumn guard cartridge) (Merck Darmstadt, Germany), a Spectroflow 400 pump (Kratos Analytical, Ramsey, NJ, USA) with a Rheodyne 7125 loop injector (Cotati, CA, USA), and a Jasco fluorescence detector (Jasco Corporation, Tokyo, Japan). The settings were: Excitation wavelength 360 nm and emission wavelength 480 nm. The mobile phase was acetonitrile-water (80:20, v/v). The injection volume was 20 μl and the flow rate 0.5 ml/min.

CYP1A1 activity

This activity was measured as the deethylation of ethoxycoumarin (EC) using a 1.0 ml of a reaction mixture containing 2 μM EC, 0.1 mM Tris (pH 8.0), 0.1 ml M NaCl, and 20 μl microsomes (3–8 mg of microsomal protein per ml). The reaction was started at 37°C by the addition of NADPH (final concentration 0.5 mM), and the appearance of coumarin was monitored with a fluorescence spectrophotometer (Shimadzu RF-5001PC). The fluorimeter settings were: Excitation wavelength 530 nm and emission wavelength 585 nm.

CYP2E activity

This activity was measured as the conversion of p-nitrophenol into 4-nitrocatechol according to Reinke [31] using a 2 ml incubation mixture containing 1.58 ml incubation buffer (50 mM Tris, pH 7.4, 5 mM $MgSO_4$), 0.20 ml of a NADPH regenerating system (see above) and 0.20 ml of microsome suspension. The reaction was started by addition of 20 μl of a p-nitrophenol solution (20 mM) and the mixture was incubated at 37°C for 30 min. The reaction was stopped by addition of 0.5 ml 0.6 M $HClO_4$. After centrifugation at 2000 × g for 10 min, 150 μl 10 M KOH was added to 1.5 ml supernatant. The tubes were placed in ice for 30 min and subsequently centrifuged at 2000 × g for 10 min. The extinction of the supernatants was measured spectrophotometrically at 546 nm. Hydroxylase activity was calculated based on the formation of 4-nitrocatechol, for which a molecular extinction coefficient of 10.28 $mM^{-1}.cm^{-1}$ was used.

Protein determination

Protein concentrations were determined by the Micro BCA (Bicinchoninic acid) assay (Pierce Chem Co., Rockford, IL, USA), using bovine serum albumin (fraction V, Sigma Chemicals) as the standard.

Statistics

Data are presented as mean values ± S.D. Comparisons between groups were made using one-way analysis of variance followed by Tukey's method for multiple comparisons [32]. P values ≤ 0.05 were considered to be statistically significant.

Results

General

The insulin-dependent diabetic state of the streptozotocin-treated adult rats manifested as a high plasma glucose concentration (20–25 mM) and a considerable loss (–27%) of body weight (Table 1). An increase of organ weight was observed only for kidney (+20%), probably due to hypertrophy [33]. In contrast, the absolute heart weight of the diabetic rats decreased significantly (–21%) when compared to the controls. No significant change in liver weight was observed.

Peroxisome proliferation

Generally, diabetes mellitus is characterized by a condition of increased mobilization of fatty acids which might lead to an increase of peroxisomal activity. Therefore, we measured peroxisomal activity based on the peroxisomal enzyme fatty acyl-CoA-oxidase (FACO) (Fig. 1). A significant increase of

Table 1. Body weight, blood plasma glucose concentration and organ weights in control and streptozotocin-induced diabetic rats

Condition	Body weight (g)	Glucose (mM)	Liver (g)	Heart (g)	Kidney (g)
Control	329 ± 25	7.0 ± 0.7	13.3 ± 1.5 (4%)	1.03 ± 0.11 (0.31%)	1.22 ± 0.10 (0.37%)
Diabetes	240 ± 39*	22.5 ± 2.7*	11.9 ± 1.7 (5%)	0.81 ± 1.7 (0.33%)	1.47 ± 0.17* (0.61%)

Values are means ± S.D. (n = 6). Values in parentheses refer to mean organ weight expressed as percentage of total body weight. Statistical significance was determined by Tukey's test. *p < 0.05.

FACO was observed in both liver (+62%) and heart (+265%). However, in kidney the FACO activity of the diabetic group did not differ from the control.

FABP

The diabetic animals showed an increase by 40–50% in the content of both L-FABP in liver and H-FABP in the heart (Table 2). In kidney from the control rats the amounts of L- and H-FABP were low when compared to those in liver and heart (Table 2). The 5–10 fold higher kidney content of H-FABP compared to L-FABP in control animals agrees with an earlier report [34]. A small but significant decrease of the L-FABP content was observed in kidney the diabetic animals, whereas the contents of H-FABP did not differ significantly between diabetic and control rats.

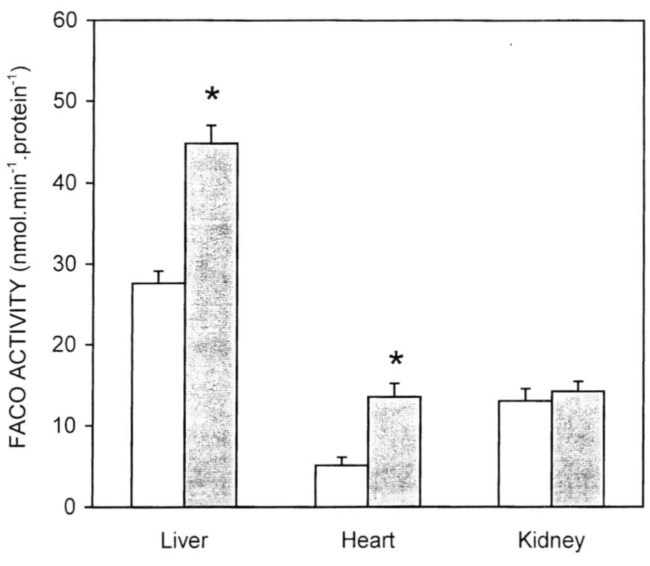

Fig. 1. Activities of the peroxisomal enzyme fatty acyl-CoA oxidase (FACO) in liver, heart and kidney of control and streptozotocin-induced diabetic rats. Activities were measured in tissue whole homogenates and are represented as means ± S.D. (n = 6). Statistical significance was determined by Tukey's test. *p < 0.05, compared with the control group. Open bars, control animals; dark bars, streptozotocin-induced diabetic animals.

Cytochrome P450

To explore whether peroxisome proliferation is a cytochrome P450-dependent process, in which the formation of dicarboxylic acids is crucial, we measured in addition to the total content of cytochrome P450 also the activity of CYP4A1 (based on ω- and ω-1 hydroxylation of lauric acid). Moreover, to assess the specificity of the cytochrome P450 induction, the activity of CYP1A1 and CYP2E was measured (Table 3). In both liver and kidney a 2 to 3-fold increase in the total content of cytochrome P450 was observed. In contrast, in heart cytochrome P450 or CYP4A1 activity (even after prolonged incubation times up to 1 h) could not be demonstrated. CYP4A1 activities were increased in liver and kidney of diabetic animals, but not in the heart. Ethoxycoumarin deethylase (CYP1A1) and p-nitrophenol (CYP2E) hydroxylase could be demonstrated only in liver. In kidney and heart these activities were not detectable, neither in control, nor in streptozotocin-treated animals.

Since streptozotocin might be a weak inducer of the cytochrome P450 system in heart, in a separate set of animals (n = 8) we used clofibrate, a compound commonly used to

Table 2. Content of liver-type(L-) and heart-type (H-) FABP in liver, heart and kidney of streptozotocin-induced diabetic and control rats

Tissue	Condition	L-FABP (μg.mg protein^{-1})	H-FABP (μg.mg protein^{-1})
Liver	Control	8.8 ± 2.5	< 0.0002
	Diabetes	12.4 ± 2.2*	< 0.0002
Heart	Control	0.005 ± 0.012	5.1 ± 0.6
	Diabetes	0.002 ± 0.012	7.4 ± 0.8*
Kidney	Control	0.031 ± 0.010	0.12 ± 0.02
	Diabetes	0.012 ± 0.007*	0.14 ± 0.05

FABP contents were measured in tissue whole homogenates and are expressed as means ± S.D. (n = 6). Statistical significance was determined by Tukey's test. *p < 0.05.

Table 3. Cytochrome P450 content and activities of ω- and ω-1-hydroxylation, 7-ethoxycoumarine deethylase, and p-nitrophenol in microsomal preparation in liver, heart and kidney of streptozotocininduced diabetic rats.

Tissue	Condition	Cytochrome P450[a]	Omega hydroxylation[b] (CYP4A1)	Omega hydroxylation[b] (CYP4A1)	Ethoxycournarin deethylase[b] (CYP1Al)	p-Nitrophenol hydroxylase[c] (CYP2E)
Liver	control	0.60 ± 0.17	1.17 ± 0.37	0.89 ± 0.12	0.91 ± 0.20	0.65 ± 0.15
	diabetes	1.69 ± 0.37*	3.18 ± 0.51	1.41 ± 0.10*	1.79 ± 0.42*	1.82 ± 0.12*
Heart	control	< 0.01	< 0.01	< 0.01	< 0.01	< 0.01
	diabetes	< 0.01	< 0.01	< 0.01	< 0.01	< 0.01
Kidney	control	0.20 ± 0.10	1.32 ± 0.27	0.53 ± 0.10	< 0.01	< 0.01
	diabetes	0.37 ± 0.12*	2.06 ± 0.54*	0.65 ± 0.25	< 0.01	< 0.01

Content/activities were measured in tissue whole homogenates and are expressed as [a]ng.mg total protein^{-1}; [b]nmol.min^{-1}.mg total protein^{-1}; [c]pmol.min^{-1}.mg total protein^{-1}, respectively. Values are means ± S.D. (n = 6). Statistical significance was determined by Tukey's test. *p < 0.05.

induce peroxisome proliferation [18]. In liver a significant increase in cytochrome P450 content (3-fold) and in CYP4A1 activity (4-fold) occurred. In heart no increased activity could be demonstrated. Nevertheless, peroxisomal β-oxidation, as assessed by FACO activity, increased 14-fold in liver and 3-fold in heart (data not shown). Additional experiments with β-naphtoflavone and phenobarbital as potent inducers of cytochrome P4501A1 in heart [22, 35] neither resulted in detectable amounts of cytochrome P450, nor of CYP4A1 activity in the heart (data not shown). In contrast, in the present study these compounds gave in liver a considerable and comparable induction of omega hydroxylation (+420%), ethoxycournarin deethylation (+160%), aniline hydroxylation (+300%) and 6β-testosterone hydroxylation (+240%).

Discussion

In the present study we explored the effect of streptozotocin-induced diabetes mellitus on peroxisome proliferation, cytochrome P450 activity and the content of cytoplasmic FABP in various tissues of the rat. In liver there was a co-induction of all three parameters, which finding extends the previous observations of Kaikaus *et al.* [18, 19] in clofibrate-treated rats, and suggests that also in the liver of diabetic animals the increase in peroxisome proliferation and L-FABP content is induced by dicarboxylic acids formed in the cytochrome P450 pathway. However, in the diabetic heart we found both an increased peroxisome proliferation and H-FABP content, but no detectable CYP4A1 activity, while in kidney an induction of CYP4A1 was seen in combination with a decreased L-FABP content, an unchanged H-FABP content and peroxisomal activity. These findings strongly suggest that peroxisome proliferation and H-FABP induction in the heart take place via a mechanism not involving CYP4A1 metabolites (such as dicarboxylic acids) and, conversely, that the enhanced CYP4A1 activity, as seen in kidney, does not necessarily lead to an induction of peroxisome proliferation and induction of either cytoplasmic L- or H-type FABP.

A potential role for dicarboxylic fatty acids, produced from long-chain monocarboxylic fatty acids via CY1P4A1, as mediators of the peroxisome proliferator response and possible ligands of PPAR, has been demonstrated in rat liver [18]. In contrast, Gustafson and coworkers demonstrated that the induction of peroxisome proliferation in rat liver by long-chain fatty acids was not dependent of cytochrome P450 and that dicarboxylic acids did not activate PPAR [17]. At present, no explanation can be given for these contradictory results. Additionally, it was shown by Göttlicher *et al.* [36, 37] that PPAR can be activated by long-chain fatty acids *per se*. These workers also demonstrated that β-oxidation of the inducing fatty acid is not required and that blocking of the β-oxidative pathway increases the potency of fatty acids or of an, as yet, unidentified metabolite for activating PPAR, most likely due to enhanced levels of unmetabolized long-chain fatty acids [37]. The latter would be in agreement with the suggestion made by Isseman *et al.* [7] that a rapid metabolism of β-oxidizable fatty acids within the cell requires high concentrations of these fatty acids to activate PPAR. This notion is also supported by the observation of Kaikaus *et al.* [19] that oleic acid alone was unable to stimulate peroxisomal β-oxidation or the expression of FAI3P, but in the presence of an inhibitor of carnitine palmitoyl transferase 1, induced peroxisomal β-oxidation and the expression of FABP pre-translationally.

In general, streptozotocin-induced diabetes not only leads in liver to the induction of CYP4A1, but also of subspecies, such as CYP1A1, CYP2E [38, 39]. The most well-defined change in livers of diabetic rats with respect to cytochrome P450 is the induction of CYP2E1 and this elevation can be reversed by treatment with insulin [39]. Data from the study of Shimojo [39] indicate that during diabetes most probably the enhanced production of ketone bodies is primarily responsible for the induction of CYP1A1 and CYP2E, whereas fatty acid overload results in an increased expression of CYP4A1.

Although it is well documented that administration of peroxisome proliferators, such as clofibrate, results in a markedly increased content of L-FABP in rat liver [18], data on tissue FABP levels under diabetic conditions are limited. In this study a significant increase in L-FABP content could be demonstrated, which most likely is related to the increased fatty acid utilization by the tissue [8]. In contrast, however, Veerkamp *et al.* [40] reported decreased L-FABP contents in the liver of streptozotocin-induced diabetic rats when compared to the control animals. Possibly, the duration of streptozotocin treatment (12 vs. 21 days) and differences in the assay procedure of FABP may have contributed to the discrepancy.

Peroxisome proliferation measured as an increase of FACO activity was relatively more pronounced in the heart than in liver (Fig. 1). This supports the marked increase in density of myocardial peroxisomes in experimental diabetes observed by others [41, 42]. Although the myocardial peroxisomal β-oxidative activity might be relatively low, when compared to that of the liver [43], it has been demonstrated to be significant for the β-oxidation of very-long chain fatty acids [44, 45]. Interestingly, no induction of the activities of cytochrome P450 subspecies (CYP1A1, CYP2E, and CYP4A1) could be demonstrated in this organ. Although, a number of other studies has provided evidence that very low levels of cytochrome P450, in particular CYP1A1, are present in the hearts of rabbit, [35], guinea pig [22] and rat [46], at present no other data are available on the occurrence of CYP4A1 activity in the heart. Since the microsomal P450 monooxygenase system is composed of a flavoprotein (NADPH-P450 reductase) and multiple isoenzymes of the hemoprotein P450 [47, 48], the observed low activities might be due not only to a low content of cytochrome P450 but also to low activities of P450-reductase in heart. This notion is substantiated by the observation that reconstitution of cardiac microsomal fractions with purified NADPH-cytochrome P450 reductase dramatically increases cytochrome P4501A activity *in vitro* [22, 49]. Therefore, it is possible that the low activity of cytochrome P450 monooxygenase in the heart, even after CYP1A1 induction, is based on a deficiency of NADPH-P450 reductase activity. The present study suggests that in cardiac cells the cytochrome P450 system at best may play a minor role in the formation of hydroxy fatty acids. It should be noted that we were not able to detect in rat heart, in contrast to liver and kidney, alcohol dehydrogenase activity (unpublished results). Thus, in cardionnyocytes the conversion of hydroxy fatty acids into the aldehyde and subsequently into the dicarboxylic form is unlikely.

The induction of CYP4A1 activity in kidney supports the observation that in this organ arachidonic acid can be converted not only into 20-HETE but also into 20-COOH-arachidonic acid [50]. It should be noticed that the cytochrome P450 system is capable of producing a variety of arachidonic acid metabolites which play an important role in electrolyte homeostasis and in the regulation of renal blood flow [50]. However, in contrast to heart and liver, no increase of FACO activity could be observed in kidney of streptozotocin-treated animals. In this respect, our data support the findings of Asayama *et al.* [41] that streptozotocin does not induce peroxisome proliferation in rat kidney. Also in mice feeding of high-fat diets failed to increase the number of peroxisomes and peroxisomal β-oxidation in kidney [51, 52].

The presence of both H- and L-FABP in kidney has been reported by various authors [8, 34], but the amounts are relatively low when compared to those observed in heart and liver, respectively. The small but significantly decreased L-FABP content of the diabetic kidney is difficult to explain, partly since the distinct functions of the two FABP types in kidney are not yet understood [8].

Our observations suggest that in kidney the role of the cytochrome P450 system is of greater importance for the conversion of fatty acids into bioactive substances regulating integrated renal function, such as participation in electrolyte homeostasis and regulation of tissue blood flow, than for the process of peroxisome proliferation which could help reducing fatty acid overload as associated with diabetes mellitus [50]. The role of FABP in the induction of these cytochrome P450-related activities has to be established.

The findings of the present study show that in the liver of diabetic rats a coordinated induction of CYP4A1, cytoplasmic FABP content and peroxisome proliferation occurs, similar to that observed in rats after administration with peroxisome proliferators (such as clofibrate) or feeding of high-fat diets. In heart, increased peroxisome proliferation and FABP content was not accompanied by an induction of any . Therefore, a role for cytochrome P4504A1 -mediated metabolites of fatty acids, such as dicarboxylic acids, in the induction of peroxisome proliferation and expression of FABP in the heart is unlikely. Finally, in kidney induction of cytochrome P450 activity was observed in absence of an increase in peroxisomal activity and in the content of FABP.

Acknowledgement

This study was supported by an Established Investigatorship of the Netherlands Heart Foundation to J.F.C.G. and a fellowship of the Royal Academy of Arts and Sciences to M. v. B.

References

1. DeFronzo RA: Pathogenesis of type 2 (non-insulin dependent) diabetes mellitus: A balanced overview. Diabetologica 35: 389–397, 1992

2. Mannaerts GP, van Veldhoven PP: Role of peroxisomes in mammalian metabolism. Cell Biochem Funct 10: 141–151, 1992
3. Lock EA, Mitchell AM, Elcombe CR: Biochemical mechanisms of induction of hepatic peroxisome proliferation. Annu Rev Pharmacol Toxicol 29: 145–163, 1989
4. Vamecq J, Draye JP: Peroxisomal and mitochondrial beta-oxidation of monocarboxyl-CoA, omega-hydroxymonocarboxyl-CoA and dicarboxyl-CoA esters in tissues from untreated and clofibrate-treated rats. J Biochem 106: 216–222, 1989
5. Osmundsen H, Bremer J, Pedersen J1. Metabolic aspects of peroxisomal β-oxidation. Biochim Biophys Acta 1085: 141–158, 1991
6. Reddy JK, Mannaerts GP: Peroxisomal lipid metabolism. Annu Rev Nutr 14: 343–370, 1994
7. Issemann I, Prince R, Tugwood J, Green S: A role for fatty acids and liver fatty acid binding protein in peroxisome proliferation? Biochem Soc Trans 20: 824–827, 1992
8. Glatz JFC, van der Vusse GJ: Cellular fatty acid-binding proteins: their function and physiological significance. Progr Lipid Res 35: 243–282, 1996
9. Kaikaus RM, Sui Z, Lysenko N, *et al.*: Regulation of pathways of extramitochondrial fatty acid oxidation and liver fatty acid-binding protein by long-chain monocarboxylic fatty acids in hepatocytes. Effect of inhibition of carnitine palmitoyltransferase 1. J Biol Chem 268: 26866–26871, 1993
10. Mortensen PB: Formation and degradation of dicarboxylic acids in relation to alterations in fatty acid oxidation in rats. Biochim Biophys Acta 1124: 71–79, 1992
11. Isseman I, Green S: Activation of a member of the steroid hormone receptor superfamily by peroxisome proliferators. Nature 347: 645–650, 1990
12. Beck F, Plummer S, Senior PV, Byrne S, Green S, Brammar W: The ontogeny of peroxisome-proliferator-activated receptor gene expression in the mouse and rat. Proc R Soc Lond B 247: 83–87, 1992
13. Mangelsdorf DJ, Thummel C, Beato M, *et al.*: The nuclear receptor superfamily: The second decade. Cell 83: 835–839, 1995
14. Osmundsen H, Brodal B, Hovik R: A luminometric assay for peroxisomal β-oxidation. Effects of fasting and streptozotocin-diabetes on β-oxidation. Biochern J 260: 215–220, 1989
15. Kozuka H, Watanabe T, Horie S, Yamada J, Suga T, Ikeda T: Characteristics of peroxisome proliferation: Co-induction of peroxisomal fatty acid oxidation-related enzymes with microsomal laurate hydroxyiase. Chem Pharm Bull Tokyo 39: 1267–1271, 1991
16. Xu L, Ash M, Abdelaleern S, Lowe JE, Badr M: Hyperinsulinemia inhibits hepatic peroxisomal beta-oxidation in rats. Horm Metab Res 27: 76–78, 1995
17. Gustafsson J-Å, Gearing K, Widmark E, *et al.*: Effects of fatty acids on gene expression mediated by a member of the nuclear receptor supergene family. Prog Clin Biol Res 387: 21–28, 1994
18. Kaikaus RM, Chan WK, Lysenko N, Ray R, Ortiz de Montellano PR, Bass NM: Induction of peroxisomal fatty acid beta-oxidation and liver fatty acid-binding protein by peroxisome proliferators - mediation via the cytochrome P-45OIVA1 omega-Hydroxylase pathway. J Biol Chem 268: 9593–9603, 1993
19. Kaikaus RM, Chan WK, Ortiz de Montellano PR, Bass NM: Mechanisms of regulation of liver fatty acid-binding protein. Mol Cell iochem 123: 93–100, 1993
20. Shimojo N, Ishizaki T, Imaoka S, Funae Y, Fujii S, Okuda K: Changes in amounts of cytochrome-P450 isozymes and levels of catalytic activities in hepatic and renal microsomes of rats with streptozocin-induced diabetes. Biochem Pharmacol 46: 621–627, 1993
21. Glatz JFC, van Breda E, Keizer HA, *et al.*: Rat heart fatty acid-binding protein content is increased in experimental diabetes. Biochem Biophys Res Commun 199: 639–646, 1994
22. McCallum GIP, Horton JE, Faikner KC, Bend JR: Microsomal Cytochrome-P450 1A1 dependent monooxygenase activity in guinea pig heart – induction, inhibition, and increased activity by addition of exogenous NADPH Cytochrome-P450 reductase. Can J Physiol Pharmacol 71: 151–156, 1993
23. Rutten AAJJL, Falke HE, Catsburg JF, *et al.*: Interlaboratory comparison of total cytochrome P-450 and protein determination in rat liver microsomes. Reinvestigation of assay conditions. Arch Toxicol 61: 27–33, 1987
24. Vannecq J: Fluorometric assay of peroxisomal oxidases. Anal Biochem 186: 340–349, 1990
25. Mannaerts GP, van Veldhoven P, van Broekhoven A, Vandebroek G, Debeer LJ: Evidence that peroxisomal acyl-CoA synthetase is located at the cytoplasmic side of the peroxisomal membrane. Biochem J 204: 17–23, 1982
26. Vork MM, Glatz JFC, Surtel DA, Knubben HJ, van der Vusse GJ: A sandwich enzyme linked immuno-sorbent assay for the determination of rat heart fatty acid-binding protein using the streptavidin-biotin system. Application to tissue and effluent samples from normoxic rat heart perfusion. Biochim Biophys Acta 1075: 199–205, 1991
27. Van Nieuwenhoven FA, Verstijnen CPHJ, van Eys GJJM, *et al.*: Fatty acid transfer across the myocardial capillary wall: No evidence of a substantial role for cytoplasmic fatty acid binding protein. J Mol Cell Cardiol 26: 1635–1647, 1994
28. Onnura T, Sato R. The carbon monoxide-binding pigment of liver microsomes. II. Solubilization, purification, and properties. J Biol Chem 239: 2379–2385, 1964
29. Stegeman JJ, Woodin BR, Klotz AV, Wolke RE, Orme-Johnson NR: Cytochrome P-450 and monooxygenase activity in cardiac microsomes from fish Stenotomus chrysops. Mol Pharmacol 21: 517–526, 1982
30. Engels W, Kamps MA, Lemmens PJ, van der Vusse GJ, Reneman RS: Determination of prostaglandins and thromboxane in whole blood by high-performance liquid chromatography with fluorimetric detection. J Chromatogr 427: 209–218, 1988
31. Reinke LA, Moyer MJ: p-Nitrophenol hydroxylation: A microsomal oxidation which is highly induceble by ethanol. Drug Metab Dispos 13: 548–552, 1985
32. Wallenstein S, Zucker C, Fleiss JL: Some statistical methods useful in circulation research. Circ Res 47: 1–9, 1980
33. Seyer-Hansen K. Renal hyperthrophy in experimental diabetes. Kidney Int 23: 643–646, 1983
34. Maatman RG, van de Westerlo EM, van Kuppevelt TH, Veerkamp JH: Molecular identification of the liver- and the heart-type fatty acid-binding proteins in human and rat kidney. Use of the reverse transcriptase polymerase chain reaction. Biochem J 288: 285–290, 1992
35. Abraham NG, Pinto A, Levere RD, Mullane K: Identification of heme oxygenase and cytochrome P-450 in the rabbit heart. J Mol Cell Cardiol 19: 73–81, 1987
36. Gottlicher M, Widmark E, Li Q, Gustafsson JA: Fatty acids activate a chimera of the clofibric acid-activated receptor and the glucocorticoid receptor. Proc Nati Acad Sci USA 89: 4653–4657, 1992
37. Göttlicher M, Demoz A, Svensson D, Tollet P, Berge RK, Gustafsson JA: Structural and metabolic requirements for activators of the peroxisome Proliferator-Activated receptor. Biochem Pharmacol 46: 2177–2184, 1993
38. Barnett CR, Flatt PR, Ioannides C: Modulation of the rat hepatic cytochrome p450 composition by long-term streptozotocin-induced insulin-dependent diabetes. J Biochem Toxicol 9: 63–69, 1994
39. Shimojo N: Cytochrome P450 changes in rats with streptozocin-induced diabetes. Int J Biochern 26: 1261–1268, 1994
40. Veerkamp JH, van Moerkerk HTB, van den Born J: No correlation between changes in fatty-acid binding content and fatty acid oxidation capacity of rat tissues in experimental diabetes. Int J Biochem Cell Biol 28: 473–478, 1996

41. Asayama K, Yokota S, Kato K. Peroxisomal oxidases in various tissues of diabetic rats. Diabetes Res Clin Pract 11: 89–94, 1991
42. Yokota S, Asayama K: Proliferation of myocardial peroxisomes in experimental rat diabetes: A biochemical and immunocytochemical study. Virchows Arch B 63: 43–69, 1992
43. Chu C, Mao L-F, Schultz H: Estimation of peroxisomal β-oxidation in rat heart by a direct assay of acyl-CoA oxidase. Biochem J 302: 23–29, 1994
44. Norseth J, Thomassen MS. Stimulation of microperoxisomal IS-oxidation in rat heart by high-fat diets. Biochim Biophys Acta 751: 312–320, 1983
45. van der Vusse GJ, Glatz JFC, Stam HCG, Reneman RS: Fatty acid homeostasis in the normoxic and ischemic heart. Physiol Rev 72: 881–940, 1992
46. Guengerich FP, Mason PS: Immunological comparison of hepatic and extrahepatic cytochromes P-450. Mol Pharmacol 15: 154–164, 1979
47. Nelson DR, Kamataki T, Waxman DJ, *et al.*: The P450 superfamily: Update on new sequences, gene mapping, accession numbers, early trivial names of enzymes, and nomenclature. DNA Cell Biol 12: 1–51, 1993
48. Porter TD, Coon MJ: Cytochrome P-450. Multiplicity of isoforms, substrates, and catalytic and regulatory mechanisms. J Biol Chem 266: 13469–13472, 1991
49. Wu S, Moomaw CR, Tomer KB, Faick JR, Zeldin DC: Molecular cloning and expression of CYP2J2, a human cytochrome P450 arachidonic acid epoxygenase highly expressed in heart. J Biol Chem 271: 3460–3468, 1996
50. McGiff JC, Quilley CP, Carroll MA: The contribution of cytochrome P450-dependent arachidonate metabolites to integrated renal function. Steroids 58: 573–579, 1993
51. Decraemer D, Vamecq J, Roels F, Vallee IL, Pauwels M, Vandenbranden C: Peroxisomes in liver, heart, and kidney of mice fed a commercial fish oil preparation: Original data and review on peroxisomal changes induced by high-fat diets. J Lipid Res 35: 1241–1250, 1994
52. Asayama K, Okada Y, Kato K: Peroxisomal beta-oxidation in liver and muscles of gold-thioglucose-induced obese mice: Correlation with body weight. Int J Obes 15: 45–49, 1991

Molecular and Cellular Biochemistry **192:** 63–68, 1999.

Long chain fatty acids as modulators of gene transcription in preadipose cells

Paul A. Grimaldi, Lydia Teboul, Danielle Gaillard, Anne V. Armengod and Ez Z. Amri

INSERM U 470, Centre de Biochimie, UFR Sciences, Université de Nice-Sophia Antipolis, Parc Valrose, Nice, France

Abstract

During the last years, it has been clearly established that long-chain fatty acids act as modulators of gene expression in various tissues, such as adipose tissue, intestine and liver. This transcriptional action of fatty acids explains in part adaptation mechanisms of tissues to nutritional changes and especially to high-fat diets by increasing expression of proteins involved in lipid catabolism in liver and fatty acid uptake and utilization in other tissues. It is now clearly demonstrated that some of these transcriptional effects of fatty acids are mediated by activation of specific nuclear hormone receptors, called peroxisome proliferator-activated receptors (PPARs). These findings will be discussed with a special reference to control of gene expression in preadipocytes and adipose tissue development. (Mol Cell Biochem **192**: 63–68, 1999)

Key words: long chain fatty acids, modulation of gene expression, fatty acid uptake

Abbreviations: ACS – acyl-CoA-synthetase; ALBP – adipocyte lipid-binding protein; FABP – fatty acid-binding protein; FAT – fatty acid transporter; L-FABP – liver-fatty acid-binding protein; GPDH – glycerophosphate dehydrogenase; GAPDH – glyceraldehyde-3-P-dehydrogenase; LCFA – long-chain fatty acid; PGJ2 – prostaglandin J2; PPAR – peroxisome proliferator-activated receptor; PPRE – peroxisome proliferator-responsive element; PUFA – polyunsaturated fatty acid

Introduction

Fatty acids are important energy storage substrates as precursors of triglycerides and phospholipids. In vertebrates, fatty acids are stored as triacylglycerol in adipose tissue, which is the major depot for energy storage, and released in blood circulation during periods of nutritional deprivation. In addition to their role as substrates for triglyceride and membrane synthesis, fatty acids play important regulatory functions in various tissues. Some effects of fatty acids are indirect and are exerted by arachidonic acid metabolites, such as prostaglandins and leukotrienes. Direct effects of fatty acids have also been described. For instance, fatty acids modulate ion channel permeability and activate guanylate cyclase or protein kinase C [1–3].

Fatty acids have been also suggested to play a regulatory role in the adaptation of various tissues to nutritional changes. For instance it has been reported that high-fat diets promote peroxisome proliferation and fatty acid oxidation in rodent liver [4] and obesity with hypertrophy and hyperplasia of adipose tissue in adult animals [5]. These effects of dietary fatty acids occur within days or weeks and it has been demonstrated that these cellular effects of fatty acids are related to an action of these molecules on the control of expression of lipid metabolism-associated genes. As described below, recent findings have established that these transcriptional actions of fatty acids are mediated by members of the nuclear hormone receptor superfamily which are expressed in various tissues and play a central role in lipid homeostasis.

Address for offprints: INSERM U 470, Génétique Moléculaire de la Différenciation et du Développement, Centre de Biochimie, UFR Sciences, Université de Nice-Sophia Antipolis, Parc Valrose, 06108 Nice, France

Fatty acids as modulators of gene expression

During the last years, it has been established that long-chain fatty acids (LCFAs) are directly involved in gene regulation in various tissues providing an explanation for the pleiotropic action of high-fat diets *in vivo*. Fatty acids exerted both positive and negative effects on gene expression. Transcriptional up-regulation of numerous genes by saturated and unsaturated LCFAs has been observed in various tissues including liver, adipose tissue and intestine whereas negative transcriptional effects of fatty acids were only described in liver. These repressive actions of fatty acids have been demonstrated *in vitro* as well as *in vivo* on transcription of genes encoding proteins involved in hepatic lipogenesis such as fatty acid synthase, stearoyl-CoA desaturase I or Spot-14 [6–8]. It is worth noting that such down-regulation of gene transcription is restricted to polyunsaturated long-chain fatty acids (PUFAs), whereas saturated and monounsaturated fatty acids as well as prostanoids are not active [9]. The molecular mechanisms involved in this negative modulation of gene expression by PUFAs remain unclear and are probably different from those involved in the positive transcriptional effects of LCFAs.

To date, it has been directly demonstrated by run-on experiments, that LCFAs up-regulate the transcription of several genes in liver, small intestine and preadipose cells. Interestingly, these genes are directly involved in lipid metabolism.

Direct transcriptional effects of LCFAs in adipose cells were originally described in the Ob1771 cell line. At the preadipose state, these cells do not express adipose-related genes and do not synthesize fatty acids or triglycerides. Exposure of these cells to LCFAs exerted potent effects on the transcriptional rate of genes such as FAT (Fig. 1), ALBP, ACS or lipoprotein lipase [10–12]. By contrast, this treatment did not activate the expression of other adipose-related genes, such as glycerophosphate dehydrogenase (Fig. 1B), adipsin or hormono-sensitive lipase, which remained transcriptionally inactive. Fatty acids containing 16 carbons and over are potent inducers whatever the unsaturation degree of the fatty acyl chain. By contrast, middle-, short- and very long-chain fatty acids were found to be poor inducers of gene expression in preadipose cells. To determine whether or not metabolism of LCFA was required, we investigated the capability of 2-bromopalmitate to induce gene expression in preadipose Ob1771 cells. Metabolic studies have shown that in preadipose Ob1771 cells 2-bromopalmitate is rapidly exchanged across the preadipocyte membrane but remained in the free form inside the cell. Further experiments demonstrated that the predominant acyl-CoA synthetase expressed in preadipose cells has a high affinity for arachidonate, a low affinity for saturated LCFA and is not at all active on 2-bromopalmitate [13]. As shown in Fig.1A, exposure of preadipose cells to linolenate or to 2-bromopalmitate resulted after one day of treatment in a tremendous accumulation of FAT mRNA. Run-on assays presented in Fig. 1B revealed that this increase of FAT mRNA amount in LCFA-treated preadipose cells was primarily due to transcriptional activation of the gene whereas transcriptional rate of GPDH remained unchanged. These findings demonstrated that the unprocessed form of fatty acid, rather than a fatty acid metabolite, was the actual inducer of gene expression in preadipose cells.

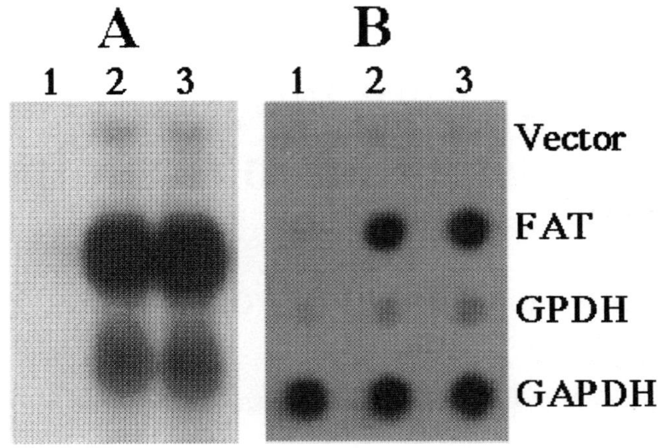

Fig. 1. Effect of long-chain fatty acids on FAT gene expression in Ob1771 preadipocytes. One-day post confluent cells were maintained for 24 h in medium containing 8% bovine serum in the absence (lanes 1) or presence of 200 μM linolenate (lanes 2) or 50 μM 2-bromopalmitate (lanes 3). At this time RNA and nuclei were prepared from the various cell conditions and northern-blot and run-on assay were performed as previously described [12]. (A) Northern blot (20 μg of RNA per lane) showing FAT mRNA; (B) Run-on transcription assay.

The induction of FAT mRNA by LCFAs appeared to be a relatively slow process, occuring within hours and reaching a maximum after 2 days of treatment. Interestingly, the induction of FAT mRNA was fully and rapidly reversible upon withdrawal of the LCFA from culture medium (Fig. 2A). It is noteworthy that a *de novo* protein synthesis is required for transcriptional induction of FAT and ALBP genes as inhibitor of protein synthesis such as cycloheximide completely impaired LCFA-induction of FAT gene (Fig. 2B). Similar pattern of LCFA-induction was found for the ALBP gene suggesting a common mechanism of regulation for both genes [11, 13].

More recently, it has been demonstrated or suggested that FAT and FABP genes are regulated in a way which is very similar in other tissues. For instance, FAT and L-FABP genes which are co-expressed in enterocytes are trancriptionally up-regulated by LCFAs. As described in preadipocytes, the

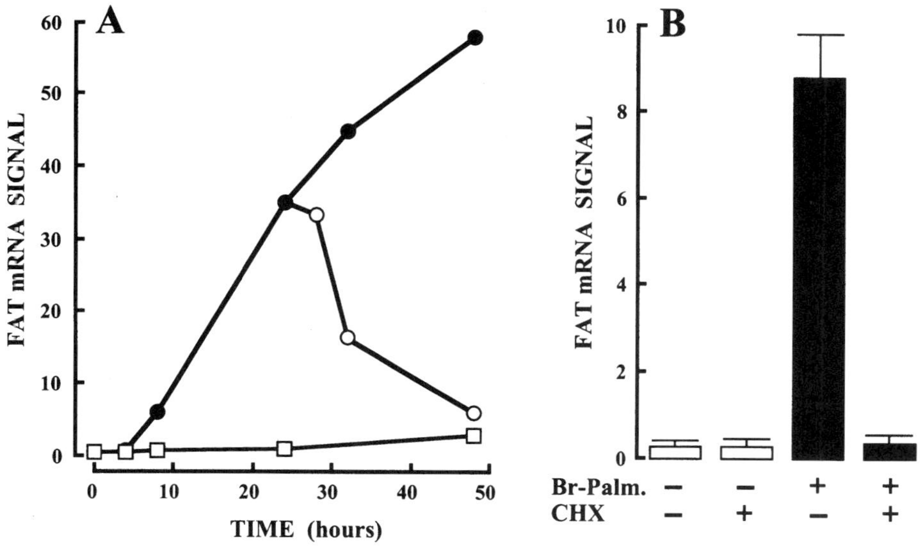

Fig. 2. (A) Time course for induction of FAT mRNA by 2-bromopalmitate. One-day post confluent Ob1771 cells were kept for the times indicated in medium containing 8% bovine serum without (□) or with (●) 50 μM 2-bromoplamitate. The fatty acid was removed (○) by two washes with fresh medium. FAT mRNA was quantitated by Northern-blot analysis as previously described [12]. (B) Effect of cycloheximide on LCFA-induction of FAT mRNA. Same cells as in A were kept for 8 h with or without 50 μM 2-bromopalmitate in the presence or absence of 10 μM cycloheximide as indicated. FAT mRNA was quantitated by Northern-blot analysis as previously described [12].

process of gene induction by LCFAs in enterocytes occurs within hours, requires on-going protein synthesis and shows the same inducer specificity including the fact that the non metabolizable fatty acid 2-bromopalmitate is a very strong inducer [14, 15]. It has been also demonstrated that in cultured hepatocytes LCFAs up-regulated L-FABP gene expression and that high-fat feeding led to overexpression of FAT and FABP genes in heart and mammary gland [16].

Taken together, these findings strongly suggest that LCFAs act as hormonal signal for the transcriptional expression of proteins postulated to play key roles in the control of intracellular free fatty acid concentration.

Molecular mechanisms of transcriptional effects of LCFA

It is now fully established that fatty acids and arachidonate metabolites exert some of their transcriptional effects through activation of members of the nuclear hormone receptor superfamily, called Peroxisome Proliferator-Activated Receptors (PPARs). The first PPAR was described by Issemann and Green [17], who have shown that it was activated in liver by peroxisome proliferators, hence the term PPAR. Two other rodent PPAR subtypes have been so far cloned from adipocyte and preadipocyte cDNA libraries and called PPARγ [18] and PPARδ or FAAR [19]. In rodents, PPAR isoforms have differential expression suggesting that they have specific regulatory roles in different organs. PPARγ is predominantly expressed in white and brown adipocytes and PPARα is mainly expressed in liver, kidney and brown adipose tissue. FAAR is more ubiquitously expressed but shows high levels of expression in adipose tissue, heart, skeletal muscle and intestine [20]. PPARγ and FAAR are induced during adipose differentiation. In Ob1771 cells, FAAR expression is a very early event of differentiation occuring in preadipose cells whereas PPARγ emerged lately during terminal differentiation [19].

It has been demonstrated by various authors that PPARs exert their transcriptional effects, after heterodimerization with RXRs, by binding to a specific element called Peroxisome Proliferator Responsive Element (PPRE). These regulatory sequences, first identified in the promoter of acyl-CoA oxidase gene [21], can be defined as DR-1 elements comprised of two TGACCT half-sites spaced by a single nucleotide. Additional PPREs have been found in regulatory sequences of various LCFA-responsive genes, such as ACS, L-FABP and ALBP (Fig. 3)[21–24]. We have recently characterized in the mouse FAT gene the regulatory sequence driving adipose-specific expression of FAT mRNA. Transient transfections in fibroblasts demonstrated that this minimal promoter is transactivated by PPARγ and FAAR. Computer analysis of this promoter revealed the presence of a PPRE-like sequence (Fig. 3) which specifically binds PPARγ/RXR and FAAR/RXR heterodimers. Scrambling mutations in this PPRE-like sequence, leading to the lack of binding to PPARs, completely

Acyl CoA oxidase	**TGACCT T TGTCCT**
Acyl CoA Synthetase	**TGACTG A TGCCCT**
L-FABP	**TGACCT A TGGCCT**
ALBP	**TGAACT C TGATCC**
FAT	**TGGCCT C TGACTT**

Fig. 3. Alignment of peroxisome proliferator-responsive elements of some LCFA-responsive genes.

abolished the PPAR-mediated transactivation of the FAT promoter in fibroblasts demonstrating the functionality of this responsive element. Interestingly, these mutations in the FAT promoter-PPRE led to a net decrease of FAT promoter activity during adipose differentiation. These findings illustrate the importance of the role of PPARs in specific gene expression in adipose cells (manuscript submitted for publication). As FAT gene is also expressed and sensitive to LCFAs in other tissues, such as small intestine and heart, it will be of interest to investigate the respective roles of the newly described FAT-PPRE and tissue-specific responsive elements in FAT gene regulation in this variety of expressing tissues.

Using transactivation assays, it has been shown that the three PPAR isoforms are activated by a broad variety of molecules including natural and non metabolized LCFAs, arachidonate metabolites and pharmaceutical compounds such as thiazolidinediones and fibrates. These studies performed in various recipient cells and with different reporter genes and/or expression vector coktails led to complex and often contradictory conclusions. However, the general picture emerging from these studies is that PPARα is activated preferentially by unsaturated LCFAs and fibrates, that PPARγ is activated by unsaturated LCFAs and thiazolidinediones whereas FAAR is equally sensitive to saturated and unsaturated LCFAs. More recently, binding assays demonstrated that thiazolidinediones and 15-deoxy-$\Delta^{12,14}$-PGJ_2, a PGJ_2 metabolite, are specific ligands for the γ isoform [25, 26], whereas PPARα specifically binds leukotriene B4 and fibrates [27]. Indomethacin was also found to bind to both PPAR isoforms [28]. So far, no specific ligands have been described for the third PPAR isoform.

The physiological roles of PPAR isoforms are not yet clearly established. However, the important role of the α isoform in liver catabolism of LCFAs is strongly supported by the finding that targeted disruption of this PPAR isoform in mice resulted in abolishment of liver peroxisomal proliferation and fatty acid oxidation [29]. The demonstration that retroviral expression of PPARγ in fibroblast cell lines induces adipogenesis is a strong evidence for a central role of this nuclear factor in the control of adipose-related gene expression during adipose differentiation [30]. We have recently demonstrated that stable transfection of 3T3C2 fibroblast with a FAAR expression vector confers LCFA-responsiveness to FAT and ALBP genes [19]. As shown in Fig. 4, induction of FAT gene by various LCFAs is very similar in these 3T3C2-FAAR expressing cells and in Ob1771 preadipocytes. In both cell lines, saturated and unsaturated LCFAs, including 2-bromopalmitate (lanes b-e), were strong activators whereas bromo-octanoate (lanes f) and clofibrate (lanes g) did not exert inducing effects. In order to determine whether or not prostanoid synthesis was required for induction of FAT mRNA in response to LCFAs, cells have been exposed to a combination of 2-bromopalmitate and cyclooxygenase inhibitors, i.e. indomethacin and aspirin. At the concentrations used, which led to a complete abolition of prostaglandin synthesis (not shown), both inhibitors were without effect on FAT mRNA expression in cells maintained in control medium (lanes h and i) and were not able to block the inducing effects of 2-bromopalmiate (lanes j and k) strongly suggesting that in preadipose cells as well as in FAAR-transfected fibroblasts, LCFAs are inducing FAT gene expression by a process which is not involving prostaglandin mediators. Altogether, these findings demonstrate that FAAR, which is expressed at high level in preadipocytes and in other cell types responsive to fatty acids or high-fat diets, is likely a good candidate as mediator of the effects of these molecules on gene expression.

Control of adipocyte proliferation and differentiation by LCFAs

In vivo studies have clearly demonstrated that adipose tissue development is regulated by nutritional status. For instance, it has been observed that high-fat and high-carbohydrate diets promote hypertrophy and hyperplasia of adipose tissue in adult rats [31]. More recently, it has been demonstrated that feeding rats a diet containing saturated fat lead to a 3-fold increase in the adipocyte number of retroperitoneal fat pad compared to animals receiving a diet containing poly-unsaturated fat [32]. Altogether, these findings suggested that saturated LCFAs exert a control on the proliferation and the differentiation of preadipose cells in adult rodents. In preadipose Ob1771 cells, we have demonstrated that LCFAs exerted a potent adipogenic action by increasing both the number of adipocytes and the level of expression of adipose-related genes. This adipogenic effect of LCFAs was not related to an increase of substrate availability for triacylglycerol synthesis as 2-bromopalmitate was found to be the most active fatty acid. Additional experiments have shown that LCFAs exerted their adipogenic action only during a critical period of time characterized as the first days of the differentiation

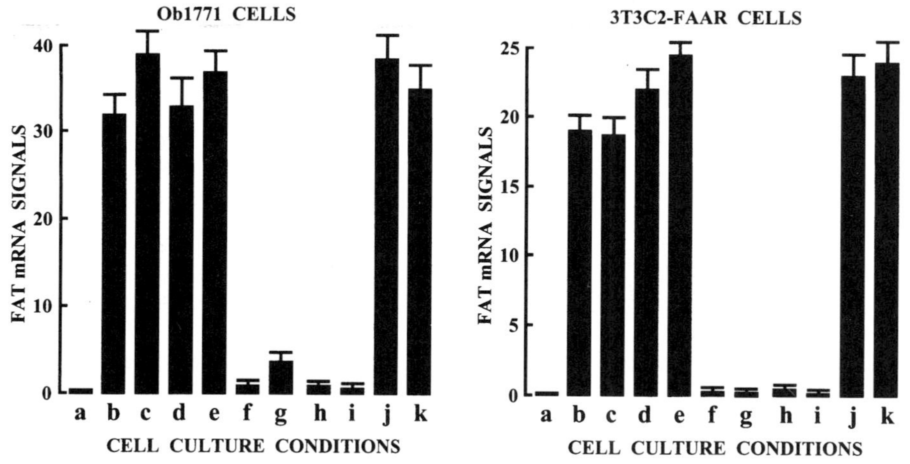

Fig. 4. LCFA-induction of FAT mRNA in Ob1771 and 3T3C2-FAAR cells. One-day post confluent cells were maintained for 24 h in medium containing 8% bovine serum in the absence (a), or presence of 200 μM palmitate (b), linoleate (c), or linolenate (d), 50 μM 2-bromopalmitate (e), or 2-bromooctanoate (f), 0.5 mM clofibrate (g). Lanes h to k correspond to cells exposed to 10 μM indomethacin (h, j) or 100 μM aspirin (i, k) in the absence (h, i) or presence of 50 μM 2-bromopalmitate (j, k). FAT mRNA was quantitated as in Fig. 2.

process corresponding to the preadipose state [33]. How LCFAs promote preadipocyte proliferation and differentiation still remains an open question. However, it is tempting to postulate that these cellular events are a consequence of the PPAR-mediated induction of gene expression in preadipose cells. With respect to the functions of proteins up-regulated in preadipocytes by LCFAs, i.e. FAT, ALBP and ACS, an increase of LCFA uptake and metabolism can be assumed with a subsequent accumulation of lipid-signalling molecules, such as prostanoids, which in turn could regulate cellular functions. Alternatively, some of the LCFA-induced proteins could be directly involved in cellular proliferation and differentiation. For instance, such roles have been suggested for FABPs in liver, heart and mammary gland [16]. In addition, it has been demonstrated that FAT is physically associated to protein kinases of the Src family which play important roles in the control of cellular proliferation. Further studies are clearly required for the understanding of these complex regulations. In that respect, gene disruption experiments, now in progress in several laboratories, would be helpful to characterize the respective roles of these proteins, i.e. nuclear receptors and LCFA-induced proteins, in the control of cell proliferation and differentiation of LCFAs.

In conclusion, fatty acids act as signal transducing molecules in addition to playing the role of an energy source and of being essential membranes components and eicosanoid precursors. In preadipose cells, LCFAs behave as hormones and stimulate gene expression by means of nuclear receptors of the PPAR subfamily. The characterization of these lipid-activated nuclear receptors which play a central role in the adaptation of various tissues to high-fat diets could be helpful to define new approaches to the treatment of nutritional obesities and associated metabolic disorders.

References

1. Braughler JM, Mittal CK, Murad F: Purification of soluble guanylate cyclase. Proc Natl Acad Sci USA 76: 219–222, 1979
2. McPhail LC, Clayton CC, Snyderman R: A potential second messenger role for unsaturated fatty acids: Activation of Ca-dependent protein kinase. Science 224: 622–626, 1984
3. Ordway RN, Singer JJ, Walsh JV: Direct regulation of ion channels by fatty acids. Trends Neurosci 14: 96–100, 1991
4. Flatmark T, Nilsson A, Krames J, Eikhom TS, Fukami MH, Kryvi H, Christiansen EN: On the mechanism of induction of the enzyme systems for peroxisomal β-oxidation of fatty acids in rat liver by diets rich in partially hydrogenated fish oil. Biochem Biophys Acta 962: 122–130, 1988
5. Faust IM, Johnson PR, Stern JS, Hirsch J: Diet-induced adipocyte number increase in adult rats: A new model of obesity. Am J Physiol 235: E279–E286, 1978
6. Clarke SD, Abraham S: Gene expression : nutrient control of pre-and post-transcriptional events. FASEB J 6: 3146–3152, 1992
7. Girard J, Perdereau D, Foufelle F, Prip-Buus C, Ferre P: Regulation of lipogenic enzyme gene expression by nutrients and hormones. FASEB J 8: 36–42, 1994
8. Landschulz KT, Jump DB, McDougald OA, Lane MD: Transcriptional control of stearoyl-CoA-desaturase-I gene by polyunsaturated fatty acids. Biochem Biophys Res Commun 200: 763–768, 1993
9. Clarke SD, Jump DB: Regulation of hepatic gene expression by dietary fats: A unique role for polyunsaturated fatty acids. In: C.D. Benardier and J.L. Hargrove (eds). Nutrition and Gene Expression. CRC Press, Boca Raton, 1993, p 227
10. Amri EZ, Bertrand B, Ailhaud G, Grimaldi P: Regulation of adipose cell differentiation: I. Fatty acids are inducers of the aP2 gene expression. J Lipid Res 32: 1449–1546, 1991

11. Amri EZ, Ailhaud G, Grimaldi P: Regulation of adipose cell differentiation: II: Kinetics of induction of the aP2 gene by fatty acids and modulation by dexamethasone. J Lipid Res 32: 1457–1463, 1991
12. Amri EZ, Teboul L, Vannier C, Grimaldi P, Ailhaud G: Fatty acids regulate the expression of lipoprotein lipase and activity in preadipose and adipose cells. Biochem J 314: 541–546, 1996
13. Grimaldi PA, Knobel SM, Whitesell RR, AbumradNA: Induction of the aP2 gene by nonmetabolized long-chain fatty acids. Proc Natl Acad Sci USA 89: 10930–10934, 1992
14. Poirier H, Degrace P, Niot I, Bernard A, Besnard P: Localization and regulation of the putative membrane fatty acid transporter (FAT) in the small intestine. Comparison with fatty acid-binding proteins (FABP). Eur J Biochem 238: 368–373, 1996
15. Niot I, Poirier H, Besnard P: Regulation of gene expression by fatty acids; special reference to fatty acid-binding proteins. Biochim (in press)
16. Glatz JFC, van der Vusse GJ: Cellular fatty acid-binding proteins: Their function and physiological significance. Prog Lipid Res 35: 243–282, 1996
17. Issemann I, Green S: Activation of a member of the steroid hormone receptor superfamily by peroxisome proliferators. Nature 347: 645–650, 1990
18. Tontonoz P, Hu E, Graves RA, BudavariAI, Spiegelman BM: PPARg2: Tissue-specific regulator of an adipocyte enhancer. Genes Dev 8: 1224–1234, 1994
19. Amri EZ, Bonino F, Ailhaud G, Abumrad NA, Grimaldi PA: Cloning of a protein that mediates transcriptional effects of fatty acids in preadipocytes. J Biol Chem 270: 2367–2371, 1995
20. Braissant O, Foufelle F, Scotto C, Dauca M, Wahli W: Differential expression of peroxisome proliferator-activated receptors: Tissue distribution of PPARα, β, and γ in the adult rat. Endocrinology 137: 354–366, 1996
21. Osumi T, Wen JK, Hashimoto T: Two cis-acting regulatory sequences in the peroxisome proliferator-responsive enhancer region of rat acyl-CoA oxidase gene. Biochem Biophys Res Commun 175: 866–871, 1991
22. Schoonjans K, Watanabe M, Suzuki H, Mahfoudi A, Krey G, Wahli W, Grimaldi P, Staels B, Yamamoto T, Auwerx J: Induction of the acyl-coenzyme A synthetase gene by fibrates and fatty acids is mediated by a peroxisome proliferator response element in the C promoter. J Biol Chem 270: 19269–19276, 1995
23. Issemann I, Prince RA, Tugwood JD, Green S: A role for fatty acids and liver fatty acid-binding protein in peroxisome proliferation. Biochem soc Trans 20: 824–827, 1992
24. Graves RA, Tontonoz P, Spiegelman BM: Analysis of a tissue-specific enhancer: ARF6 regulates adipogenic gene expression. Mol Cell Biol 12: 1202–1208, 1992
25. Forman BM, Tontonoz P, Chen J, Brun RP, Spiegelman BM, Evans RM: 15-deoxy-Δ12,14-prostaglandin J2 is a ligand for the adipocyte determination factor PPARγ. Cell 83: 803–812, 1995
26. Kliewer SA, Lenhard JM, Willson TM, Patel I, Morris DC, Lehmann JM: A prostaglandin J2 metabolite binds to PPARγ and promotes adipocyte differentiation. Cell 83: 813–819, 1995
27. Devchand PR, Keller H, Peters JM, Vazquez M, Gonzalez FJ, Wahli W: The PPARα-leukotriene B4 pathway to inflammation control. Nature 384: 39–43, 1996
28. Lehmann JM, Lenhard JM, Oliver BB, Ringold GM, Kliewer SA: Peroxisome proliferator-activated receptors α and γ are activated by indomethacin and other non-steroidal anti-inflammatory drugs. J Biol Chem 272: 3406–3410, 1997
29. Lee SST, Pineau T, Drago J, Lee EJ, Owens JW, Kroetz DL, Fernandez-Salguero PM, Westphal H, Gonzalez FJ: Targeted disruption of the isoform of the peroxisome proliferator-activated receptor gene in mice results in abolishment of the pleiotropic effects of peroxisome proliferators. Mol Cell Biol 15: 3012–3022, 1995
30. Brun RP, Tontonoz P, Forman BM, Ellis R, Chen J, Evans RM, Spiegelman BM: Differential activation of adipogenesis by multiple PPAR isoforms. Genes & Dev 10: 974–984, 1996
31. Klyde BJ, Hirsch J: Increased cellular proliferation in adipose tissue of adult rats fed a high-fat diet. J Lip Res 20: 705–715, 1979
32. Shillabeer G, Lau DC: Regulation of new fat cell formation in rats: The role of dietary fats. J Lipid Res 35: 592–600, 1994
33. Amri EZ, Ailhaud G, Grimaldi PA: Fatty acids as signal transducing molecules: Involvment in the differentiation of preadipose to adipose cells. J Lipid Res 35: 930–937, 1994

Molecular and Cellular Biochemistry **192:** 69–75, 1999.

Novel classes of fatty acid and retinol binding protein from nematodes

Lindsay McDermott, Alan Cooper and Malcolm W. Kennedy

Division of Infection and Immunity, Institute of Biomedical and Life Sciences and Department of Chemistry, University of Glasgow, Glasgow, Scotland, UK

Abstract

Parasitic nematodes have recently been found to produce proteins which represent two new classes of fatty acid and retinoid binding protein. The first is the nematode polyprotein allergens/antigens (NPAs) which, as their name suggests, are synthesised as large polyproteins which are subsequently cleaved at regularly spaced sites to form multiple copies of a fatty acid binding protein of approximately 14.5 kDa. Binding studies using molecular environment-sensitive fluorescent ligands have shown that the binding site is highly unusual, producing blue-shifting in fluorescence to an unprecedented degree, suggesting a remarkably non-polar environment and isolation from solvent water. Computer-based structural predictions and biophysical observations have identified the NPAs as highly helical proteins which might form a four helix bundle, so constitute a new class of lipid binding protein from animals. The second class, like the NPAs, binds both fatty acids and retinol, but with a higher affinity for the latter. These are also highly helical but are structurally distinct from the NPAs. The biological function of these new classes of protein are discussed in the context of both the metabolic requirements of the parasites and the possible role of the proteins in control of the immune and inflammatory environment of the tissue sites parasitised. (Mol Cell Biochem **192**: 69–75, 1999)

Key words: nematodes, parasites, polyproteins, fatty acid binding proteins, retinol binding proteins

Introduction

As will be clear from the accompanying articles, lipid binding proteins are functionally and structurally diverse, but it is also becoming apparent that we have yet to appreciate this diversity to the fullest. Amongst the small, soluble proteins which function as lipid carrier proteins, the structure and function of the beta strand rich proteins of the FABP/P2/CRBP/CRABP and lipocalin families are understood at a considerable level of sophistication. Proteins of the former family are particularly well understood, and have been found across the animal phyla, from vertebrates to insects, and parasitic flatworms to nematodes [1–4]. It is becoming clear, however, that fatty acid and retinoid transport functions can be carried out by proteins which differ dramatically in structure from the beta-rich forms which appear to predominate in vertebrates, as exemplified by the helix-rich proteins of plants [5], and now by two different classes of proteins with fatty acid and retinoid-binding activities from nematodes.

This review sets out to summarise the essential and unusual characteristics of the newly described proteins from nematodes, concentrating mainly on the polyprotein antigens/allergens. The structure, function and biosynthesis of these are highly unusual and are under study in our laboratories using a variety of biophysical methods and site directed mutagenesis. A second class has only recently been discovered and shows that there is much yet to learn about the structures and biological functions of small lipid carrier proteins.

Nematode polyprotein allergens/antigens (NPAs)

The first NPA to be characterised was the ABA-1 protein allergen of the parasitic nematode *Ascaris*, which is abundant

Address for offprints: M.W. Kennedy, Division of Infection and Immunity, Institute of Biomedical and Life Sciences, Joseph Black Building, University of Glasgow, Glasgow, G12 8QQ, Scotland, UK

within, and is secreted by, the parasites, thereby potentially contributing to the pulmonary hypersensitivity reactions associated with the infection [6–8]. Interest in this allergen was further stimulated by the fact that there is strong genetic control of immune responses to ABA-1 and similar proteins from other parasites, such that certain individuals will not produce an immune response to it even in the face of extreme immune stimulation [9, 10]. With the cloning of cDNA encoding NPAs for a wide range of disease-causing nematode parasites of humans and domestic animals, including those which cause lymphatic filariasis in humans, highly unusual features of their biosynthesis as polyproteins began to emerge. But first we will deal with their fatty acid and retinoid binding functions, which themselves represent dramatic departures from established norms for lipid binding proteins.

Ligand binding characteristics

The ligand binding function of the NPAs has now been examined using either parasite-derived protein or recombinant protein produced in *E. coli*. With one exception (*Ascaris*), recombinant proteins have been essential to the analysis of NPA binding function because most parasites are not available in sufficient quantity to provide protein directly, usually because they infect only humans or are otherwise difficult to obtain.

The binding assays used have all been fluorescence-based, using molecular environment-sensitive artificial or naturally fluorescent ligands, and non-fluorescent competitors, to detect both binding itself and to provide information on the nature of the binding site. Particularly valuable in this respect has been the dansylated fatty acid 11-((5-dimethylaminonaphthalene-1-sulphonyl)amino)undecanoic acid (DAUDA), which has been extensively used for analysis of mammalian intestinal liver fatty acid binding protein (L-FABP) [11, 12]. This artificial ligand exhibits a dramatic increase in fluorescence upon binding to a protein and a blue shift in the wavelength of fluorescence emission which increases with increasing non-polarity of the binding site and isolation of the fluorophore from solvent water [13]. In the case of the NPAs, unprecedented fluorescence enhancement and blue shift were observed, with a dramatic increase in the level of DAUDA fluorescence emission over the fluorescence of the probe in water and a blue-shift greater than has been recorded either by ourselves or any other laboratory for proteins such as L-FABP and serum albumin (Fig. 1 and refs [11, 14]). Moreover, the degree of this blue-shift is equivalent to that which DAUDA undergoes in organic solvents such as cyclohexane, suggesting a highly unusual binding site. Fluorescence titration experiments provided dissociation constants of the order 8.8×10^{-8} M for DAUDA binding to ABA-1, which falls in the range reported for other small lipid carrier proteins [12], and the K_d for nonfluorescent natural ligands such as oleic acid, measured by competition with DAUDA, is of the same order of magnitude. Other potential hydrophobic ligands such as cholesterol, squalene and tryptophan are not observed to bind. See Table 1 for lists of ligands found to bind or not to NPAs.

Blue-shifts in fluorescence emission similar to that found with DAUDA were also found with a fluorescent fatty acid probe in which the dansyl fluorophore is attached at the α carbon (dansyl-DL-α-aminocaprylic acid; DACA), rather than to the hydrocarbon terminal, as in DAUDA [14]. This similarity in behaviour of the two probes probably indicates that the fatty acid ligand is held entirely within the binding site and isolated from solvent. This complete entry of ligands into a binding protein is also found in most other fatty acid binding proteins, presumably to protect them from degradation [1].

The NPAs were also found to bind retinol and retinoic acid, which is of potential importance given the association between *Ascaris* infection and vitamin A deficiency (although this remains controversial; [15]), and the causative agent of river blindness, *Onchocerca* volvulus, is known to accumulate retinoids to a level eight times the concentration in the tissue it inhabits [16, 17]. The reason for this accumulation is unknown, but possibilities range from their use by the parasite as antioxidants in self-defence against host immune

Table 1. Ligand binding by nematode polyprotein allergens/antigens

Binding	No binding
oleic acid	cholesterol
palmitic acid	tryptophan
stearic acid	caproic acid
retinoic acid	squalene
retinol	tocopherol
oleoyl-CoA	tocopherol acetate
bilirubin*	succinyl CoA
cis parinaric acid	2-methylbutyric acid
trans parinaric acid	2-methylvaleric acid
arachidonic acid	biliverdin
lysophosphatidic acid	mebendazole**
lysophosphatidyl ethanolamine	albendazole
lysophosphatidyl choline	thiabendazole
platelet activating factor	oxibendazole
lysoplatelet activating factor	piperazine
leukotrienes B_4, C_4, D_4, E_4*	tetramisole
	pyrantel
	DEC
	levamisole

2-methylbutyric acid and 2-methylvaleric acid are branched chain fatty acids produced as excretory products by certain nematodes. *Binding with these compounds was weak but detectable. **Mebendazole and the compounds below it in the list are widely used anti-nematode drugs. The binding studies were carried out either by displacement of a fluorescent ligand (DAUDA), or by detecting changes in the fluorescence of natural ligands (e.g. retinol, parinaric acid). Data from refs [14, 28, 29] and N. Heaney, A. Cooper and M.W. Kennedy, unpublished.

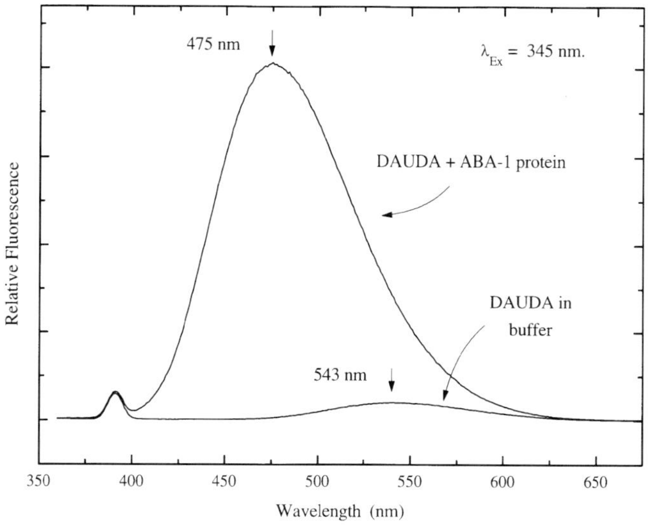

Fig. 1. Dramatic changes in fluorescence emission by the fatty acid analogue 11-((5dimethylaminonaphthalene-l-sulphonyl)amino)undecanoic acid (DAUDA) upon binding to parasite-derived ABA-1 protein of *Ascaris*. Note the large increase in fluorescence, but, more significantly, the dramatic blue shift in the peak wavelength of emission to a degree which is unsurpassed by any other fatty acid binding protein yet reported. The fluorescence emission spectrum of 1 μM DAUDA solution was taken in phosphate buffered saline and upon subsequent addition of ABA-1 protein.

mechanisms, or as essential requirements in morphogenesis of their developing juvenile forms.

The ligand specificity of the NPAs can therefore be summarised as a requirement for a simple aliphatic hydrocarbon tail with a charged carboxylate or hydroxyl head group. The binding site can nevertheless accommodate the relatively large fluorescent reporter groups used in probes such as DAUDA and the ionone ring of retinoids.

Biosynthesis as polyproteins

Before the binding function of the NPAs had been discovered, their immunological activity had led to the isolation of cDNA encoding them from several organisms, usually through the screening of cDNA expression libraries with serum from infected humans or animals or by PCR using degenerate primers based on the known encoding DNA sequences from related parasites [18–22]. This led to the initially perplexing finding that the mRNA and gene comprised tandemly repeated units, each of which encodes a single molecule of the NPA comprising, in the case of *Ascaris,* a single 14.4 kDa ABA-1 allergen molecule. It was Selkirk and colleagues who convincingly demonstrated that this was not an artifact of cloning, and in pulse chase experiments with isotopically-labelled methionine showed that the NPAs are synthesised as large polyproteins which are then progressively cleaved to form multiple copies of the NPA units [18, 19]. The intermediates produced during this process give a characteristic ladder of bands on SDS-PAGE [18, 21, 22], hence the alternative description of the NPAs as 'ladder' proteins. See Fig. 2 for a diagrammatic representation of the structure of the mRNA encoding the only NPA which has been fully sequenced, that of the cattle parasite *Dictyocaulus viviparus* [23].

The junctions between adjacent units of a NPA precursor polypeptide are marked by clusters of basic amino acids (in the case of ABA-1 this comprises four arginines), which are consensus cleavage sites for posttranslational processing enzymes of the subtilisin family of endoproteinases [24]. That cleavage occurs at this site was made clear from N-terminal amino acid sequence analysis of parasite-derived ABA-1 protein, and by the point at which repetition ceases in the polypeptide encoded by cDNA [7, 8, 20, 23]. Also, there appear to be no introns in the genomic DNA encoding the NPAs, so this and synthesis as a polyprotein provides the parasites with a means of producing large quantities of these particular proteins in a short period of time and in a metabolically highly efficient manner. Such a means of production is rarely seen for proteins, although certain hormones and structural proteins are known to be synthesised in a similar manner, but without the efficiency of the NPAs [25].

Sequence diversity within NPA arrays

The first NPAs for which genetic information became available had shown that there were few if any differences between the amino acid sequences of units within the polyprotein array [18–21]. It subsequently became increasingly clear that the units within a given array could be highly diverse, that close similarity between the units might be the exception rather than the rule, and that diversity might yet exist in those species in which it is currently considered to be limited [18]. In the one organism for which the sequence of a complete array is available, none of the units are identical and the sequence identity varies from 18% to 97% [23]. Finally, in *Caenorhabditis elegans* (the free-living nematode for which the sequence of the entire genome will shortly be available) the units of the array are similarly diverse. The adaptive significance of this diversity is not known, although DNA encoding an array of identical units might be susceptible to disruption by homologous recombination, so once multiplication of a single unit had occurred then there would be a selective advantage to diversification so long as functional and structural requirements were met.

Another feature of the emerging diversity was that the consensus proteinase cleavage sites were not always present between adjacent units, so it might either be that proteinases with different cleavage propensities are involved in pro-

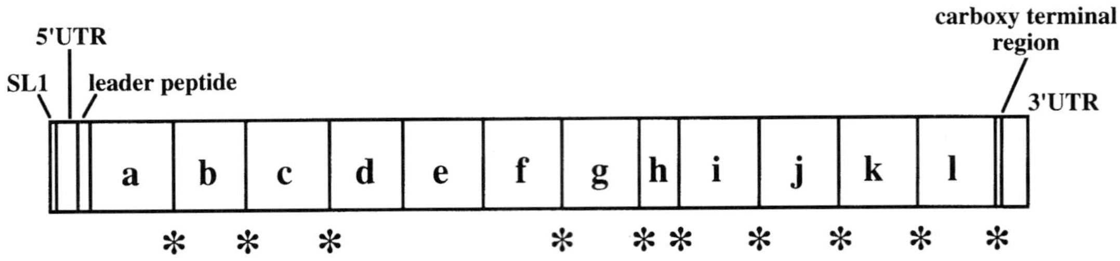

Fig. 2. Diagramatic representation of the mRNA encoding the nematode polyprotein allergen/antigen DvA-1 from *Dictyocaulus viviparus*, the pulmonary parasite of cattle. The 5′ and 3′ untranslated regions (UTR), the spliced leader sequence of nematode mRNAs (SL1) and the regions encoding the hydrophobic leader and C-terminal extension are indicated. The individual units encoding the approximately 15 kDa fatty acid binding polypeptides are labelled A through L, and the junctions having consensus K/R.X.K/R.K proteinase cleavage sites are indicated by asterisks (*). The mRNA is translated as a polyprotein and then processed by progressive cleavage at these sites to generate multiple copies of the fatty acid binding polypeptides. Recognisable cleavage sites do not appear to exist between certain of the units, which might explain the NPA molecules which appear to comprise two or three units. In DvA-1, none of the repeats are the same, varying in sequence identity from 18–97%, although each has the conserved tryptophan and cysteine fingerprint of all NPAs, with the single exception of the truncated unit e indicated in the diagram. See ref. [23] for further details.

cessing, or that some of the units function as parts of larger entities, and there is some evidence that units of twice and three times the expected size do occur [23, 26]. Sequence analysis has also indicated that the array originally derived from an ancient duplication event to produce the single modern NPA unit of approximately 14.4 kDa [20], and this is supported by the existence of a half unit in one array [23], and by sequence similarities between halves of different units in the array of another species [26].

Structure – predictions and observation

No crystal or NMR structure for any NPA is yet available, so the idea that they represent something unusual has emerged from computer-based predictions and direct biophysical observation. The computer predictions were greatly assisted by the construction of multiple sequence alignments using sequences from different species of parasite and for different members of the array of a single species. Using a range of programs, the PHD suite of programs in particular [14], it was predicted that a single NPA unit is extensively helical, with no evidence of beta structure. Moreover, the protein fell into four regions of predicted helix, so it might therefore form a four helix bundle similar to some invertebrate carrier proteins (e.g. myohemerythrin) and cytochrome b_{562} [27]. Circular dichroism observations confirmed the predominance of helix, and the percentage of residues involved agreed closely with that predicted [14].

If the NPAs were to form four helix bundles, then one of the expected characteristics would be high stability. When the stability of the proteins was investigated by CD combined with guanidine denaturation or by differential scanning calorimetery (DSC), then the NPAs (either natural or recombinant), were indeed found to be highly stable (Fig. 3). With DSC, this stability was particularly remarkable, requiring temperatures greater than 90°C for unfolding to occur at physiological pH, and the proteins were found to refold with high efficiency upon cooling (refs [14, 28] and Fig. 3). The measurements taken by DSC, and direct sizing by gel exclusion chromatography, also revealed the NPAs to form dimers [14, 28, 29], the interaction between the members of which possibly contributing to the high stability of the proteins, as is known to occur for subunits in other helical proteins [30].

A final structurally unusual feature of the NPAs was the position of their single tryptophan residue within the protein. Most lipid binding proteins, including the small β clam and β barrel structures and serum albumin, have a tryptophan with its side chain projecting into the binding cavity or embedded in the walls of the cavity [1, 31], and the fluorescence emission by such a residue is often found to alter upon ligand binding [32]. In the case of the Trp of the NPAs, the intrinsic fluorescence of the side chain peaks at a wavelength which is shorter than is the case for the majority of proteins, lipid-binding or not [14, 28, 29]. This means that it is isolated from solvent water to an unusual degree, and so unlikely to be in a binding pocket. The Trp seems therefore to be deeply buried within the structure of the protein, and unlikely to be involved in ligand binding, as was also inferred from the lack of a change in its fluorescence emission spectrum upon ligand binding [14, 28, 29] and by mutational analysis (see below).

Do the different members of the polyprotein array differ biochemically or structurally?

Among the potential consequences of the diversity in the amino acid sequences of members of a single NPA array is

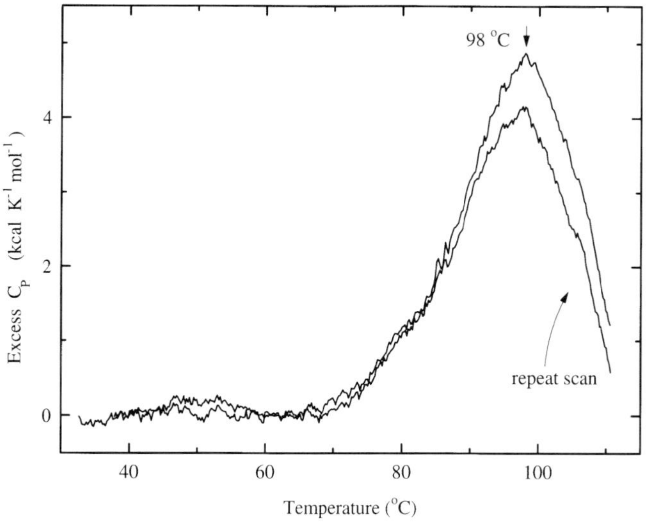

Fig. 3. Differential scanning calorimetry (DSC) data for DvA-1 illustrating the extreme thermal stability of these proteins. The DvA-1 protein used here is recombinant protein produced in *E. coli* bearing a plasmid encoding the DNA for only the L unit of the DvA-1 polyprotein. The unfolding of the protein was observed to occur at approximately 98°C, and re-folding was found to occur efficiently upon cooling ('repeat scan'). The cooperativity of this folding transition is consistent with the protein acting as a dimer. See ref. [23] for further details.

that each performs distinct biochemical functions. This has been examined in our unpublished work in which PCR was used to amplify DNA encoding two individual units from the NPA array of Ascaris which represent the extremes of the diversity known for that organism, in which sequence identity is only 49% (L. McDermott, J. Moore, A. Cooper and M.W. Kennedy, unpublished). When the two were compared for binding using the environment-sensitive fluorescent and natural ligands detailed above, they were found to be virtually indistinguishable. This similarity was also reflected in structural observations in terms of their circular dichroism spectrum, stability to guanidine denaturation, and the unusual fluorescence spectrum of the proteins' single tryptophan.

It remains possible that meaningful differences in the binding function between units of the array will be found, but current observations show no hint of this. It might therefore indeed be that diversity within the NPA array has evolved to increase the stability of the DNA encoding it, or that the arrays are so ancient that there has been gradual drift in amino acid residues between the units but with functional and structural conservation. If so, this leaves the question of why certain species of parasitic nematodes have several members of their arrays which are identical or virtually so, yet others in which none are the same [18–23, 26]. Also, there is as yet no biochemical function identified for the truncated and multiple NPA units which have been found, so there remains scope for multifunctionality.

Mutants

The multiple sequence alignments available for the NPAs identified several residues which are absolutely conserved, and we have been examining the function of these in ligand binding and in the structural integrity of the proteins using site-directed mutagenesis of the ABA-1 protein. The residues concerned are the single tryptophan (W16 ->R), and two conserved residues which occur in the presumptive external hydrophilic face of the first two helices (Q20 -> L and L42 -> R) which are currently hypothesised to be involved in subunit interaction [14]. All of the mutants exhibited reduced stability to denaturation by guanidine, particularly the L42R and W16R proteins, but only the L42R mutant had altered ligand binding activity, having entirely lost its binding capacity for all of the ligands detailed above (L. McDermott, A. Cooper and M.W. Kennedy, unpublished).

As summarised above, ABA-1 is observed to bind predominately hydrophobic ligands with aliphatic tails, and the only exception we have found to this is anilinonaphthalene sulfonic acid (ANS). This fluorescent compound is considered to bind to exposed hydrophobic regions of proteins, as detected by a dramatic increase in the probe's fluorescence emission, and so is used to detect unfolded states [33]. ANS was found to interact with wild type and all the mutants of ABA-1, and the probe was displaced by oleic acid in all but the L42R protein. This suggest that ANS enters the binding site and/or interacts with an exposed hydrophobic region at the mouth of the binding site of ABA-1, but in the case of the ligand binding defective L42R mutant the mutation has disrupted a critical feature of the binding site, but left an exposed hydrophobic surface or cavity.

A second new class?

To the NPAs as representatives of a new class of hydrophobic ligand binding protein has now been added another class of helix-rich proteins from nematodes which bind retinol, but with a lesser affinity for fatty acids. These are the Ov20 and Bm20 proteins of the highly pathogenic nematodes *O. volvulus* and *Brugia malayi,* respectively, and GpSec2, which is secreted by the plant parasitic nematode *Globodera pallida* ([34] and A. Prior, M.W. Kennedy, W. Robertson and J. Jones, unpublished results). These proteins are slightly larger (~ 20 kDa) than the NPAs, and are distinct in having neither the conserved residues of the NPAs nor are they synthesised as polyproteins [35]. It therefore seems that nematodes have evolved, or retained from their precursors, helical hydrophobic carrier protein counterparts to the beta-rich proteins of vertebrates, and it is quite possible that more remain to be discovered. This is not to say that nematodes do not produce beta barrel FABPs which are closely similar to those of vertebrates [4]. Why it is that nematodes have specialised in the use of the helical proteins will remain a matter of debate while their function in vivo and their particular suitability for nematode life styles are established, but it is clear that the beta barrels evolved before the divergence of the nematode and vertebrate lines. It is conceivable, however, that helix-rich FABPs also evolved before divergence, but that the vertebrates have elaborated them into more familiar proteins, possibly with only vaguely related or quite different functions, potential examples including serum albumin, or acyl CoA-binding protein which is of a similar size and helix content to the nematode proteins although it cannot bind free fatty acids [36].

In vivo *function*

The NPAs are extracellular proteins which are abundant in the pseudocoelomic fluid of nematodes, in which all the internal organs of nematodes are bathed, so is analogous to our blood. It is therefore likely that the NPAs act as carriers in the distribution of hydrophobic ligands within the parasite, for example in the distribution of nutrients from the gut to the other organs, or the transport of signalling molecules such as moulting hormones. The Ov20 protein is also abundant within its respective parasite, particularly in the body wall and the larvae developing within the uteri, and its function may be similar to the NPAs. Both types of protein probably have major roles in the metabolism of the parasites, and they might be involved in the transport or sequestration of hydrophobic anti-nematode drugs. The ligands they bind and presumably protect from degradation are of great importance to parasitic nematodes because they cannot synthesise complex lipids themselves and must acquire them from their host [37]. The major interest in both the NPAs and Ov20-type proteins with respect to infections is that they are secreted by the parasites and so could be involved in controlling the tissue environment or acting to counter immune defense reactions. If we look to their binding propensites for clues to such functions, then the binding, sequestration and/or delivery of lipid mediators are attractive possibilities. Among the ligands bound by ABA-1, for instance, are arachidonic acids and lysophospholipids, which are important inflammatory mediators themselves or are precursors of mediators [38, 39]. Retinoids are important in morphogenesis and gene activation [40], and retinol deficiency can compromise protective immune responses against nematodes [41], so binding proteins released by parasites could control both the local tissue and immunological environments. The helix-rich nematode proteins therefore merit attention not only for their intrinsic interest, but also for their potential as drug targets given that they appear to have no counterparts in vertebrates and their ligand binding sites clearly have unusual properties.

Acknowledgements

The work reported in this review was largely supported by grants to MWK and AC from the Wellcome Trust, and to Dr Janette Bradley of Salford University from the Edna McConnell Clark foundation and the STD programme of the European Economic Community. Calorimetric work (AC) is supported by the Biotechnology and Biological Sciences Research Council and the Engineering and Physical Sciences Research Council (UK).

References

1. Sacchettini JC, Gordon JI: Rat intestinal fatty acid binding protein. A model system for analyzing the forces that can bind fatty acids to proteins. J Biol Chem 268: 118399–18402, 1993
2. Haunerland NH, Jacobsen BL, Wesenberg G, Rayment I, Hoiden HM: Three-dimensional structure of the muscle fatty-acid-binding protein isolated from the Desert Locust *Schistocerca gregaria*. Biochemistry 33: 12378–12385, 1994
3. Becker MM, Kalinna BH, Waine GJ, McManus DP: Gene cloning, over-production and purification of a functionally active fatty acid-binding protein (Sj-FABP$_c$) from the human blood fluke *Schistosoma japonicum*. Gene 148: 321–325, 1994
4. Mei B, Kennedy MW, Beauchamp J, Komuniecki PR, Komuniecki R: Secretion of a novel, developmentally regulated fatty acid binding protein into the perivitelline fluid of the parasitic nematode, *Ascaris suum*. J Biol Chem 272: 9933–9941, 1997
5. Gomar J, Petit M-C, Sodano P, Sy D, Marion D, Kader J-C, Vovelle F, Ptak M: Solution structure and lipid binding of a nonspecific lipid transfer protein extracted from maize seeds. Protein Science 5: 565–577, 1996
6. Kennedy MW, Qureshi F: Stage-specific secreted antigens of the parasitic larval stages of the nematode *Ascaris*. Immunology 58: 515–522, 1986
7. Christie JF, Dunbar B, Davidson I, Kennedy MW: N-terminal amino acid sequence identity between a major allergen of *Ascaris lumbricoides* and *Ascaris suum*, and MHC-restricted IgE responses to it. Immunology 69: 596–602, 1990
8. McGibbon AM, Christie JIF, Kennedy MW, Lee TDG: Identification of the major *Ascaris* allergen and its purification to homogeneity by high-performance liquid chromatography. Mol Biochem Parasitol 39: 163–172, 1990
9. Tomlinson LA, Christie JF, Fraser EM, McLaughlin D, McIntosh AE, Kennedy MW: MHC restriction of the antibody repertoire to secretory antigens, and a major allergen, of the nematode parasite *Ascaris*. J Immunol 143: 2349–2356, 1989
10. Christie JF, Fraser EM, Kennedy MW: Comparison between the MHC-restricted antibody repertoires to *Ascaris* antigens elicited by adjuvant-assisted immunisation or infection. Parasite Immunol 14: 59–73, 1992
11. Wilkinson TCI, Wilton DC: Studies on fatty acid-binding proteins. The detection and quantification of the protein from rat liver using a fluorescent fatty acid analogue. Biochem. J 238: 419–424, 1986

12. Thumser AEA, Evans C, Worrall AF, Wilton DC: Effect on ligand binding of arginine mutations in recombinant rat liver fatty acid-binding protein. Biochem J 297: 103–107, 1994
13. Macgregor RB, Weber G: Estimation of the polarity of the protein interior by optical spectroscopy. Nature 319: 70–73, 1986
14. Kennedy MW, Brass A, McCruden AB, Price NC, Kelly SM, Cooper A: The ABA-1 allergen of the parasitic nematode *Ascaris suum*: Fatty acid and retinoid binding function and structural characterization. Biochemistry 34: 6700–6710, 1995
15. Nesheim MC: Ascariasis and human nutrition. In: D.W.T. Crompton, M.C. Nesheim, Z.S. Pawlowski (eds). Ascariasis and its Prevention and Control. Taylor & Francis, London, 1989, pp 87–107
16. Sani BP, Vaid A: Specific interaction of ivermectin with retinol-binding protein from filarial parasites. Biochem J 249: 929–932, 1988
17. Sani BP, Vaid A, Comley JCW, Montgomery JA: Novel retinoid-binding proteins from filarial parasites. Biochem J 232: 577–583, 1985
18. Tweedie S, Paxton WA, Ingram L, Maizels RM, McReynolds LA, Selkirk ME: Brugia pahangi and Brugia malayi – a surface-associated glycoprotein (gp15/400) is composed of multiple tandemly repeated units and processed from a 400-kda precursor. Exp Parasitol 76: 156–164, 1993
19. Paxton WA, Yazdanbakhsh M, Kurniawan A, Partono F, Maizels RM, Selkirk ME: Primary structure and IgE response to the repeat subunit of gp15/400 from human lymphatic filarial parasites. Infec Immun 61: 2827–2833, 1993
20. Spence HJ, Moore J, Brass A, Kennedy MW: A cDNA encoding repeating units of the ABA-1 allergen of Ascaris. Mol Biochem Parasitol 57: 339–344, 1993
21. Culpepper J, Grieve RB, Friedman L, Mika-Grieve M, Frank GR, Dale B: Molecular characterization of a *Dirofilaria immitis* cDNA encoding a highly immunoreactive antigen. Mol Biochem Parasitol 54: 51–62, 1992
22. Poole CB, Grandea AG, Maina CV, Jenkins RE, Selkirk ME, McReynolds LA: Cloning of a cuticular antigen that contains multiple tandem repeats from the filarial parasite *Dirofilaria immitis*. Proc Nat Acad Sci (USA) 89: 5986–5990, 1992
23. Britton C, Moore C, Gilleard JS, Kennedy MW: Extensive diversity in repeat unit sequences of the cDNA encoding the polyprotein antigen/allergen from the bovine lungworm *Dictyocaulus viviparus*. Mol Biochem Parasitol 72: 77–88, 1995
24. Barr PJ: Mammalian subtilisins: The long-sought dibasic processing endoproteases. Cell 66: 1–3, 1991
25. Rothnagel JA, Steinert PM: The structure of the gene for the mouse filaggrin and a comparison of the repeating units. J Biol Chem 2165: 1862–1865, 1990
26. Poole CB, Hornstra LJ, Benner JS, Fink JR, McReynolds LA: Carboxy terminal sequence divergence and processing of the polyprotein antigen from *dirofilaria-immitis*. Mol Biochem Parasitol 82: 51–65, 1996
27. Branden C, Tooze J: Protein Structure. Garland Publishing. New York and London, 1991
28. Kennedy MW, Britton C, Price NC, Kelly SM, Cooper A: The DvA-1 polyprotein of the parasitic nematode *Dictyocaulus viviparus*: A small helix-rich lipid-binding protein. J Biol Chem 270: 19277–19281, 1995
29. Kennedy MW, Allen JE, Wright AS, McCruden AB, Cooper A: The gp15/400 polyprotein antigen of Brugia malayi binds fatty acid and retinoids. Mol Biochem Parasitol 71: 41–50, 1995
30. Steif C, Weber P, Hinz H-J, Flossdorf J, Cesareni G, Kokkinidis M: Subunit interactions provide a significant contribution to the stability of the dimeric 4-alpha-helical-bundle protein ROP. Biochemistry 32: 3867–3876, 1993
31. He XM, Carter DC: Atomic structure of human serum albumin. Nature 358: 209–215, 1992
32. Frapin D, Dufour E, Haertle T: Probing the fatty acid binding site of beta-lactoglobulins. J Prot Chem 12: 443–449, 1993
33. Varley PG: Fluorescence spectroscopy. Meth Mol Biol 22: 203–218, 1994
34. Kennedy MW, Garside LH, Goodrick LE. McDermott L, Brass A, Price NC, Kelly SM, Cooper A, Bradley JE: The Ov20 protein of the parasitic nematode *Onchocerca volvulus*. A structurally novel class of small helix-rich retinol-binding protein. J Biol Chem 272: 29442–29448, 1997
35. Tree TIM, Gillespie AJ, Shepley KJ, Blaxter ML, Tuan RS, Bradley JE: Characterisation of an immunodominant glycoprotein antigen of *Onchocerca volvulus* with homologues in other filarial nematodes and *Caenorhabditis elegans*. Mol Biochem Parasitol 69: 185–195, 1995
36. Kragelund BB, Andersen KV, Madsen JC, Knudsen J, Poulsen FM: 3-Dimensional structure of the complex between acyl coenzyme A binding-protein and palmitoyl coenzyme A. J Mol Biol 230: 1260–1277, 1993
37. Barrett J: Biochemistry of Parasitic Helminths. University Park Press, Baltimore, 1981
38. Murphy RC, Fitzpatrick FA (eds).: Arachidonate Related Lipid Mediators. Meth Enzymol 187. Academic Press Inc., San Diego, New York, London, 1990
39. Gerrard JM: Prostaglandins and Leukotrienes: Blood and Vascular Cell Function. Marcel Dekker Inc., New York and Basel, 1985
40. Leid M. Kastner P, Chambon P: Multiplicity generates diversity in the retinoic acid signalling pathways. Trends Biochem Res 17: 427–433, 1992
41. Carman JA, Pond L. Nashold F, Wassom DL, Hayes CE: Immunity to *Trichinella spiralis* infection in vitamin-A deficient mice. J Exp Med 175: 111–120, 1992

Molecular and Cellular Biochemistry **192:** 77–85, 1999.

Fatty acid interactions with native and mutant fatty acid binding proteins

Gary V. Richieri, Ronald T. Ogata and Alan M. Kleinfeld
Medical Biology Institute, La Jolla, CA, USA

Abstract

The interactions of long chain fatty acids (FA) with wild type (WT) fatty acid binding proteins (FABP) and engineered FABP mutants have been monitored to determine the equilibrium binding constants as well as the rate constants for binding and dissociation. These measurements have been done using the fluorescent probes, ADIFAB and ADIFAB2, that allow the determination of the free fatty acid (FFA) concentration in the reaction of FA with proteins and membranes. The results of these studies indicate that for WT proteins from adipocyte, heart, intestine, and liver, K_d values are in the nM range and affinities decrease with increasing aqueous solubility of the FA. Binding affinities for heart and liver are generally greater than those for adipocyte and intestine. Moreover, measurements of the rate constants indicate that binding equilibrium at 37°C is achieved within seconds for all FA and FABPs. These results, together with the level of serum (unbound) FFA, suggests a buffering action of FABPs that helps to maintain the intracellular concentration of FFA so that the flux of FFA between serum and cells occurs down a concentration gradient. Measurements of the temperature dependence of binding reveal that the free energy is predominately enthalpic and that the enthalpy of the reaction results from FA-FABP interactions within the binding cavity. The nature of these interactions were investigated by determining the thermodynamics of binding to engineered point mutants of the intestinal FABP. These measurements showed that binding affinities did not report accurately the changes in protein-FA interactions because changes in the binding entropy and enthalpy tend to compensate. For example, an alanine substitution for arginine 106 yields a 30 fold *increase* in binding affinity, because the loss in enthalpy due to the elimination of the favorable interaction between the FA carboxylate and Arg^{106}, is more than compensated for by an increase in entropy. Thus understanding the effects of amino acid replacements on FA-FABP interactions requires measurements of enthalpy and entropy, in addition to affinity. (Mol Cell Biochem **192**: 77–85, 1999)

Key words: fatty acid binding protein, fatty acid binding affinities

Introduction

Fatty acid binding proteins (FABP)[3] are 14–15 kDa proteins that are found in a variety of tissues and may be involved in fatty acid (FA) metabolism [1–8]. X-ray crystallography and NMR studies have determined the structures of the apo and holo (with FA bound) forms of several of these proteins [9–17]. These studies reveal that the FA binds in a cavity within the protein and that the bound FA shares this cavity with a large number of ordered and disordered water molecules. The conformation of the FA within the cavity differs among the FABPs and can also be different for the different FA binding within the same FABP. These results suggest that a complex set of interactions govern FA binding to these proteins.

To investigate these interactions we have developed a method that allows the determination of the concentration of the FFA in the interaction, $FFA + FABP_{apo} \Leftrightarrow FABP_{holo}$ ([12, 18–20] and Richieri *et al.* in this same issue). This method uses the response of the fluorescent probes of FFA, ADIFAB and ADIFAB2, which are fluorescent derivatives of the FABP from rat intestine (I-FABP). These probes undergo a red shift in fluorescence when binding FA and therefore, FFA can be determined simply by measuring the ratio of the probe's holo to apo intensities. Because these probes monitor [FFA], they can be used to measure the interaction of FA with any protein or membrane, without consideration for the properties of the protein or membrane. We have therefore used these probes to determine both the equilibrium and kinetic parameters of

Address for offprints: A.M. Kleinfeld, Medical Biology Institute, 11077 North Torrey Pines Road, La Jolla, CA, USA

the interaction of FA with wild type and mutant FABPs [12, 19, 20].

Materials and methods

Details of the methods used in these investigations have been described previously [12, 18–21] and in the accompanying paper (Richieri *et al.* this issue). Briefly, all measurements were done using the sodium salts of the FA in a buffer consisting of 10 mM HEPES, 150 mM NaCl, 5 mM KCl, and 1 mM $NaHPO_4$, at pH 7.4. All WT FABPs were recombinant proteins that were expressed in the BL21 (DE3)/pET11 host/vector expression system and the proteins were purified as described previously [12, 18]. Protein for the I-FABP mutants, except E51A and F93A, were isolated from cell lysates. E51A and F93A were expressed as inclusion bodies and were solubilized by denaturation in 4M GndHCL followed by renaturation as described [22]. ADIFAB, which is available from Molecular Probes, Eugene, OR, was prepared as described [18] and ADIFAB2, the acrylodan derivatized L72A mutant of I-FABP, as in [20].

Measurements of equilibrium binding were done by titrating a solution containing FABP and ADIFAB with FA, and at each step the ADIFAB fluorescence was used to determine the concentration of FFA. These binding isotherms were determined for temperatures between 10 and 45°C. Equilibrium dissociation constants, K_d, were determined at each temperature and van't Hoff plots were used to obtain ΔH^o values. ΔG^o values were evaluated from the K_ds measured at 25°C and -$T\Delta S^o$ was calculated as $\Delta G^o - \Delta H^o$ [12, 19, 21]. Measurements of the rate constants for dissociation and binding were determined by stopped-flow fluorescence. First FA and ADIFAB and/or ADIFAB2 in one syringe were mixed with fatty acid free BSA in the second, to obtain the rate constants for ADIFAB and then dissociation of FA from FABP was measured by mixing a solution of FA and FABP in one syringe with ADIFAB in the second syringe [20, 21]. Differences in the enthalpy ($\Delta\Delta H^o$), entropy ($T\Delta\Delta S^o$), and free energy ($\Delta\Delta G^o$) of FA binding between each mutant and the WT protein were computed as $\Delta G^o_{mut} - \Delta G^o_{WT}$, $\Delta H^o_{mut} - \Delta H^o_{WT}$, and $T\Delta S^o_{mut} - T\Delta S^o_{WT}$, respectively. Values for ΔG^o, ΔH^o, and $T\Delta S^o$ were determined at 25°C for each of the FAs.

Results and discussions

Equilibrium binding of FA and WT FABPs

Figure 1 summarizes the equilibrium binding measurements done at 37°C for ADIFAB and WT FABPs from adipocyte, heart, intestine, and liver binding to the 18 carbon series of FA. These results show that K_ds are in the nM range and that there is considerable variation in affinity with both tissue type of FABP and FA molecular species. With the exception of the adipocyte FABP, affinities decrease exponentially, with double bond number for the 18 carbon series of FA, or more generally, with increasing aqueous solubility of the FA [12, 18]. For example, the affinity for binding to I-FABP of stearate, with no double bonds, is 60 fold greater than for linolenate, with 3 double bonds. These results emphasize that unless the interactions within the binding site can overcome the free energy cost of solubility, binding affinities will generally be inversely related to the ligand solubility. This exponential dependence on ligand solubility is reminiscent of the partition of such ligands between water and an isotropic non-polar solvent [23], suggesting that FABP binding cavities behave to some extent like a non-polar solvent, albeit with an unusual affinity for water [7, 19].

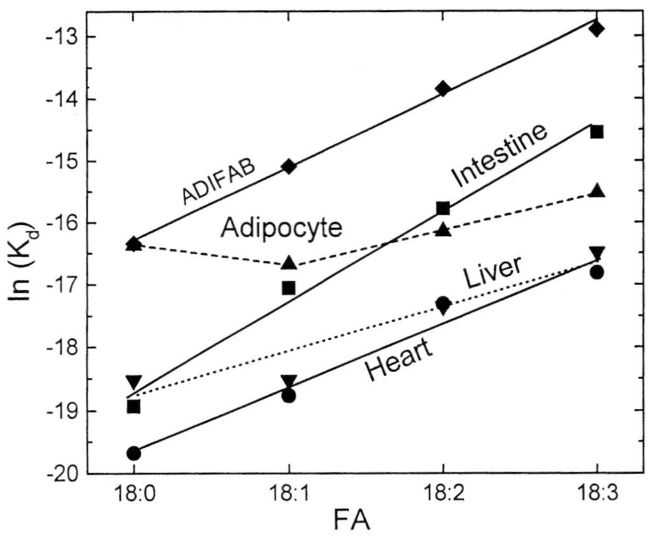

Fig. 1. Binding of the series of 18 carbon FA to FABPs from different tissues. Dissociation constants for recombinant FABPs from adipocyte, heart, intestine, and liver, as well as ADIFAB, the acrylodan derivative of the rat intestinal FABP from reference [12]. Measurements were done at 37°C with stearate (18:0), oleate (18:1), linoleate (18:2), and linoleate (18:3).

Binding of palmitate and oleate at 37°C provides further insight about the tissue and species dependence of equilibrium binding (Fig. 2). These results reveal that although K_d values range from about 4–80 nM for different tissues, binding to FABPs from the same tissue but different species are similar, supporting the notion that the FA/FABP interaction is important because it is conserved in different species. The results also reveal that, in these FABPs from the major FA metabolizing tissues, adipocyte displays the smallest and heart the largest binding affinity. These affinity differences, as discussed below, may play a role in determining the direction of intercellular FA trafficking. Finally, all the FABPs bind a single FA except liver, which has two sites [19].

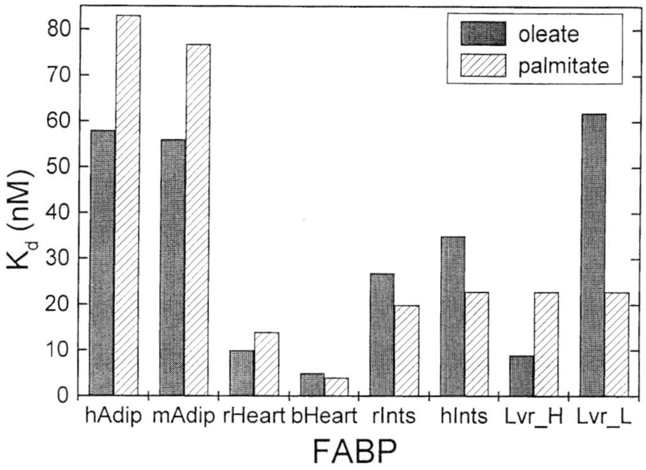

Fig. 2. Binding constants for oleate and palmitate binding to FABPs from different tissues and species. Measurements were done at 37°C and the data is from ([12] and the present study). The abbreviations used for the FABPs are: h & m Adip; human and murine adipocyte, r & b Heart; rat and bovine heart, r & h intestine; rat and human intestine, and Lvr_H and Lvr_L are respectively the high and low affinity binding sites of the rat liver FABP.

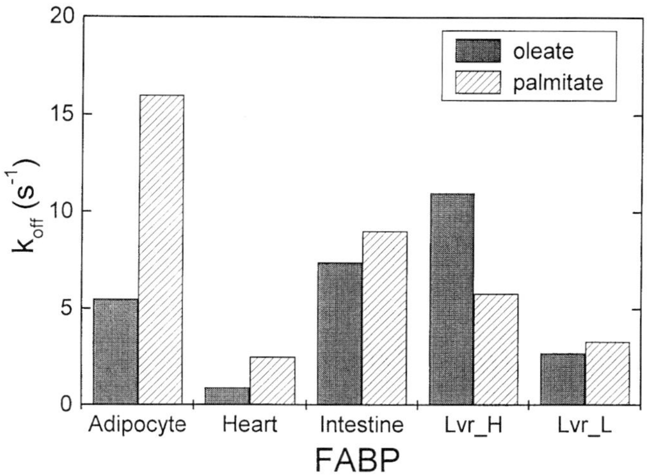

Fig. 3. Rate constants for FA dissociation from WT FABPs. Measurements were done at 37°C using recombinant FABPs as described in Figs 1 and 2. Results are from [20].

These 2 sites are quite different, and reveal at 37°C, about 10 fold differences in affinity for unsaturated but equal affinities for the long chain saturated FA [21]. The recently solved crystal structure of the liver FABP complexed with oleate is consistent with theses results, revealing that the 2 oleates bound to the protein are in quite different, approximately orthogonal, conformations [15].

FA binding kinetics for WT FABPs

As shown in Fig. 3 the kinetics of FA interactions with FABPs is also a sensitive function of tissue type and the molecular species of FA. These results show that at 37°C dissociation rate constants vary from about 1 s^{-1} for oleate dissociating from rat heart FABP to about 15 s^{-1} for palmitate leaving adipocyte FABP. Together with the on-rate constants calculated from the measured k_{off} and K_ds values, these results indicate that equilibrium occurs within about 1sec at 37°C in the binding of these FABPs and all FA [20, 21]. (Rates for the poly-unsaturated FA are faster than for palmitate and oleate [20, 21]). The variation of the dissociation rate constants of Fig. 3 with the tissue origin of the FABP might be expected from the variation of the equilibrium constants, heart has the highest affinity and the slowest rate of dissociation and the converse is true for adipocyte. For liver FABP, however, the rate of dissociation from the high affinity site is significantly greater than that for the low affinity site. Thus the greater affinity of the high affinity site is due to a more than 10 fold greater on-rate constant than for the binding of the second FA. This raises the possibility of an interaction between the two FAs within L-FABP binding cavity, consistent with the crystallographic results of Thompson *et al.* [15].

Equilibrium and kinetic characteristics of WT FABPs suggest a tissue specific role in buffering intracellular FFA levels

These equilibrium and kinetic characteristics suggest that by buffering intracellular FFA levels, FABPs may play a role in directing intercellular FA trafficking. If we assume that FA trafficking between serum and cells is governed by the FFA concentration difference between cytosol and serum, and that the direction of FFA flux is down its concentration gradient, then intracellular FFA levels should be higher or lower for cells that export or import FFA, respectively. Measurements of healthy individuals indicate that serum FFA concentrations are about 7 nM [24]. By arranging the K_ds for oleate binding to FABPs from the different tissues relative to the 7 nM serum FFA levels we see that for adipocyte and intestine, cells that transport FA from the cytosol to serum, K_ds are significantly higher than the serum FFA concentration (Fig. 4). Conversely, the liver and heart, which import large quantities of FA, have K_ds that are on the order of serum FFA levels. Thus, if intracellular FFA levels are similar to the Kds of their respective FABPs, an intracellular buffering role for FABPs is consistent with the measured binding constants. What is needed to establish this role more firmly, are measurements of the actual intracellular FFA levels and studies to do this, using fluorescence ratio microscopy of cells microinjected with ADIFAB, are currently in progress.

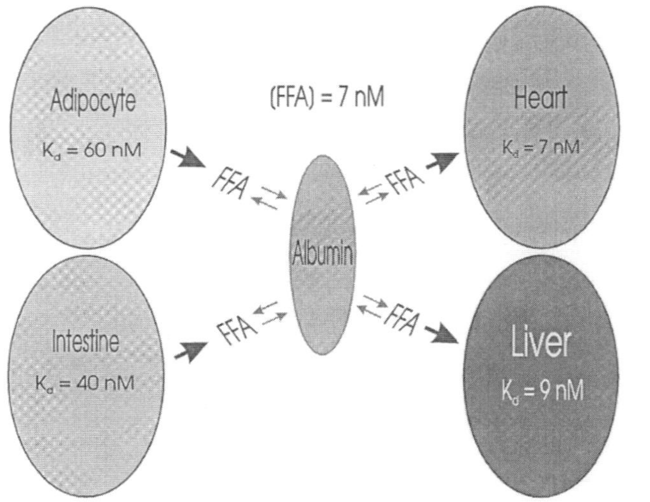

Fig. 4. The potential intracellular FFA buffering role for FABP. Values for Kds are from [12] and for serum FFA levels from [24]. These K_d values are for oleate at 37°C.

Binding thermodynamics of the WT FABPs

To understand better the molecular interactions that give rise to the binding constants, the free energy, enthalpy and entropy changes were determined for long chain FA binding to the WT FABPs (Fig. 5A,B). These results reveal that, with the exception of the low affinity site of L-FABP, relatively small (≤ 2 kcal/mol) |TΔS°| values contribute to the free energy. These entropic contributions tend to become more favorable with decreasing FA solubility, as predicted by the hydrophobic effect [23]. The variation of enthalpy with FA solubility also reveals a pattern that may reflect the homology groupings [6] of the FABPs. Thus |ΔH°| decreases sharply from about 12–6 kcal/mol for adipocyte and heart while remaining virtually constant at –11 kcal/mol for intestine, and liver shows a variation that is different from both these groups. Moreover, in contrast to the other FABPS, liver FABP reveals large values of heat capacity differences [21]. That these differences in patterns are not serendipitous is supported by the similarity of the thermodynamic parameters for adipocyte and heart although their affinities differences are the largest of all the WT proteins. This happens because the binding affinity differences are relatively small (|ΔΔG°| values < 1.5 kcal/mol) compared to |ΔH°|. These results suggest that each of the homology groups have distinctive structural elements which dictate specific thermodynamic features that, in turn, may have important functional consequences.

Another striking aspect of these thermodynamic studies, is that the free energy of FA binding to these FABP is a net enthalpic process [19, 21, 25–28]. This enthalpy domination is surprising, given that the binding cavity is suggestive of a non-polar solvent and that water-solvent partition for small hydrophobic molecules is generally entropically driven [19, 23, 29]. The thermodynamic parameters of Fig. 5 are for the combined processes of FA desolvation and insertion into the binding cavity of the protein. Desolvation

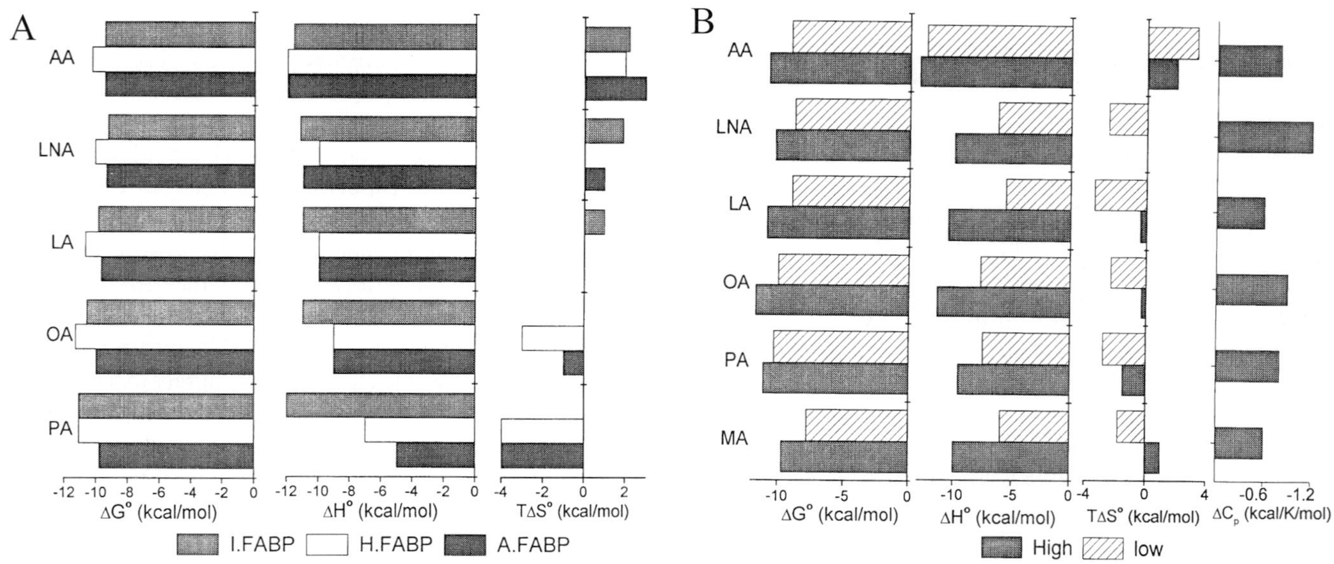

Fig. 5. Thermodynamic parameters for FA binding to WT FABPs. Binding enthalpies (ΔH°) were determined from van't Hoff analyses of the temperature dependence of K_d and the free energies (ΔG°), and entropies (ΔS°) were determined at 25°C using ΔG° = ΔH° - TΔS°, as described [19, 21]. (A) Recombinant FABPs from adipocyte, heart, and intestine binding to palmitate (PA), oleate (OA), linoleate (LA), linolenate (LA), and arachidonate (AA). Results are from [19]. (B) Recombinant rat liver FABP binding to the 5 FA in A) plus myristate (MA). The last column in this figure shows the heat capacity changes for the high affinity binding site. Results are from [21].

thermodynamic parameters were determined separately, by measuring FA partition between water and membrane, and these results reveal a favorable free energy that is about –10 kcal/mol but is virtually all entropic [19]. These results suggest that long chain FA are driven out of the water by the favorable entropic forces of the hydrophobic effect and then enter the FABP cavity where interactions within the cavity contribute a favorable enthalpy of about –10 kcal/mol. However, these interactions in the cavity impose additional constraints on the FA and the amino acid side chains so that this favorable enthalpy is compensated for by unfavorable entropic interactions (about +10 kcal), resulting in no net entropy for the reaction. Clearly, interactions between FA and the amino acid residues within the cavity are not simply those corresponding to an isotropic non-polar solvent.

Equilibrium binding of FA to site specific mutants of I-FABP reveal large changes in affinities for single alanine substitutions

In an effort to understand the interactions between FA and FABP within the cavity, we have generated site specific

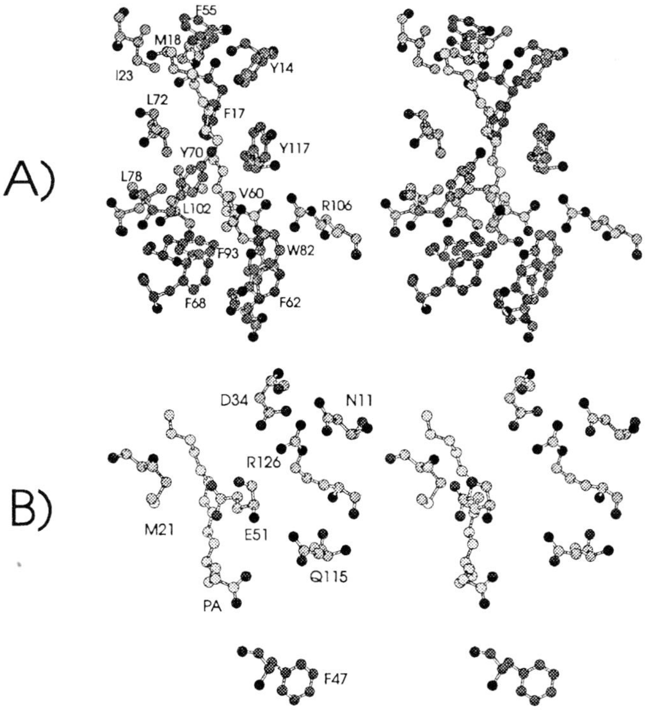

Fig. 6. Amino acid residues of rat intestinal FABP which were subjected to site-specific mutation. Coordinates for these structures are from the X-ray crystallography results of Sacchettini *et al.* [30] for rat intestinal FABP complexed with palmitate. (A) The 16 residues within the binding cavity for which Ala and Gly (R106) were substituted. (B) The 7 more distal residues that were investigated. This figure is taken from [22].

mutants of the intestinal FABP and have measured the interaction of FA with each of these mutants [22]. Mutations were made at sites that line the binding cavity, as revealed by the crystal structure (Fig. 6A). We substituted alanine for 16 of the 19 residues that define the cavity as well as Gln at residue 106. We also substituted Ala for 7 more distal amino acid residues as shown in Fig. 6B. For each substitution FA binding was measured as a function of temperature for up to 6 different FA. Dissociation constants (K_d) determined at 37°C for binding of 6 long chain FA to the wild type protein and to each of the 24 mutants were found to range between 0.5 and 4500 nM (Table 1 and Fig. 7). Single Ala substitutions in I-FABP generate proteins that bind with affinities ranging from 30 fold higher (L72, R106) to almost 30 fold lower (M18, F68) than the wild type (WT) protein. These results also show that for each mutant the binding affinity decreases with increasing solubility of the FA; for all mutants K_d values vary according to SA<OA<LA<LNA as observed previously for the WT proteins [12, 18, 19, 22].

To better illustrate how these substitutions affect each mutant relative to the WT proteins, we have expressed the

Table 1. Dissociation constants (K_d) for fatty acid binding to wild type and site specific mutants of rat intestinal fatty acid binding protein.[a]

FABP[b]	PA[c]	SA	OA	LA	LNA	AA
WT	20	6	27	90	367	200
L72A	1	0.5	2	8	8	12
R106A	4	1	2	4	10	8
L102A	7	1	3	12	26	27
Y117A	5	2	4	10	35	20
Q115A	13	6	19	49	127	60
E51A	14	6	23	83	230	244
Y14A	14	8	24	45	121	147
W82A	38	25	61	88	155	100
F93A	19	16	55	83	100	66
Y70A	30	16	44	177	611	322
N11A	32	10	47	151	478	383
L78A	27	16	43	243	620	572
M21A	53	39	94	227	408	424
V60A	45	13	60	182	437	624
F17A	56	27	73	211	461	205
F47A	82	21	80	404	1168	855
F62A	89	27	135	452	1565	1162
I23A	94	32	186	509	1109	589
D34A	70	65	247	493	1330	1590
F55A	104	54	248	478	1376	1515
R126A	148	148	352	520	1250	1550
M18A	149	93	366	748	1640	2300
F68A	510	365	878	2183	4500	1890

[a] K_d values were measured at 37°C are given in nM. Measurement standard deviation are approximately 10%. This table is from [22]. [b] FABPs are designated as: WT is the wild type I-FABP and single letter designations are used to indicate the WT residue, amino acid sequence position, and the substituted amino acid residue for each mutant. [c] Abbreviations for the fatty acids are: PA, palmitate (16:0); SA, stearate (18:0); OA, oleate (18:1); LA, linoleate (18:2), LNA, linolenate (18:3), AA, arachidonate (20:4).

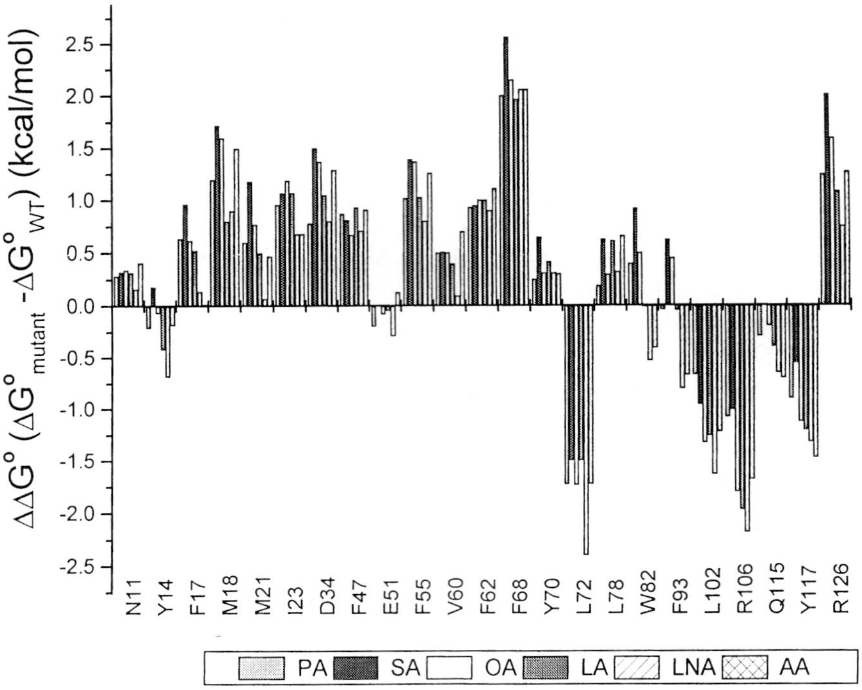

Fig. 7. Differences in free energy changes between alanine mutants and WT I-FABP. These are results obtained at 37°C for single alanine substitutions at sites that are inside the binding cavity of rat intestinal FABP. Results are from [22].

affinity measurements in terms of the free energy differences for binding between each mutant (ΔΔG°) and the WT protein in Fig. 7. We see that these single Ala mutants result in ΔΔG° values that range from about –2 kcal/mol (increased affinity) to about +2 kcal/mol (decreased affinity). If we interpret these results as indicating the roles played by each of the residues in the interaction with the ligand we might, for example, conclude that L72, Y117, L102, and R106 contribute large repulsive and R126, F68, M18 large attractive, interactions in the WT protein because the mutants have larger and smaller affinities, respectively. We would also conclude that residues such as Y14 and L78 do not interact significantly with the ligand in the WT protein because ΔΔG° is consistent with no change for these substitutions. Although this kind of analysis is often used to interpret the effects of mutation on ligand binding or enzyme activity, it fails to account for the concomitant changes in enthalpy and entropy that also occur for each amino acid residue substitution.

Enthalpy-entropy compensation plays a critical role in understanding the interactions between FA and FABP

In Fig. 8 results for the differences in enthalpy (ΔΔH°), entropy (TΔΔS°), and free energy (ΔΔG°) for each mutant, determined from the temperature dependence of the K_d values, are arranged so that the ΔΔG° values for linoleate increase monotonically [22]. These results demonstrate that changes in affinity, equivalent to ΔΔG° values, are not correlated with ΔΔH° and TΔΔS° or, equivalently, amino acid substitution induced changes in the underlying molecular interactions are not correlated with affinity changes. This lack of correlation results because the enthalpy and entropy changes tend to compensate in these binding reactions. As a consequence, the mutation-induced changes in binding enthalpy and entropy are almost always larger than ΔΔG°, so that relatively small mismatches in the enthalpy/entropy compensation can result in highly significant changes in binding affinity.

An important aspect of these results is that changes in affinity are not uniquely determined by the changes in enthalpy and entropy. For example, although Ala substitutions for L72, R106, L102, and Y117 all result in substantial increases in affinity, the molecular interactions that generate these increases are different in each case. Similarly, M18A and F68A both have substantially lower affinities than the WT protein but this is achieved by large decreases in |ΔH°| and smaller increases in TΔS° for M18A and relatively small additive enthalpy and entropy changes for F68. Virtually no changes in affinity occurs for Y14 and L78 but this is due to large but almost exactly compensating changes in enthalpy

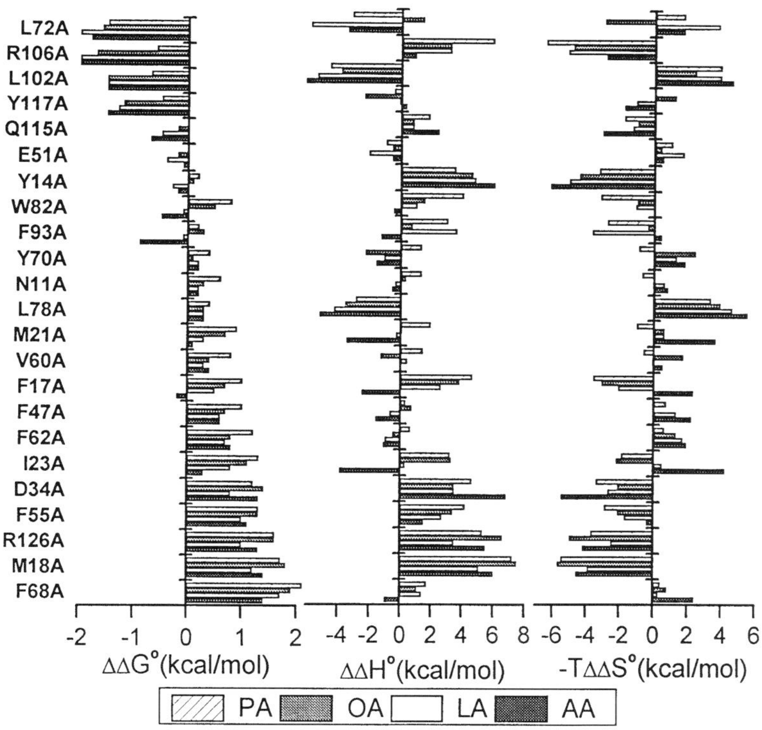

Fig. 8. Thermodynamic parameter differences between alanine mutants and WT I-FABP. Differences in alanine mutants and WT I-FABP free energies ($\Delta\Delta G^o$), enthalpies ($\Delta\Delta H^o$), and entropies ($-T\Delta\Delta S^o$) are organized so that the $\Delta\Delta G^o$ for linoleate increases monotonically from negative to positive values. The $\Delta\Delta G^o$ and $-T\Delta\Delta S^o$ values were calculated from the affinities measured at 25°C. This figure is taken from [22].

and entropy, but of opposite signs for the two residues, rather than a lack of interaction between the FA and the WT residue. These large interactions are consistent with the crystal structure which indicates close contacts between Y14 and L78 and the FA; the lack of a change in affinity and thermodynamic parameters for N11 is consistent with its more than 10 Å separation from any part of the FA [30]. Thus, enthalpy and entropy measurements provide greater insight about FA-FABP interactions than only affinity measurements.

Results for Arg[106] represent a particularly striking example of the value of the complete thermodynamic measurements. The crystal structure strongly suggests that for I-FABP the negative carboxylate of the FA and the positively charged Arg[106] should provide an important and attractive contribution to the free energy of FA binding [30]. So how does one explain the surprising result that an Ala mutation, which should eliminate this attractive interaction, results in a protein which binds FA with more than 30 fold *greater* affinity than the WT protein? The answer lies in the enthalpy-entropy compensation that accompanies amino acid substitutions, as indicated in Fig. 9. These results show that changes in affinities upon substitution of either Gln or Ala are apparently inconsistent with the reduction in binding expected from the elimination of an attractive interaction.

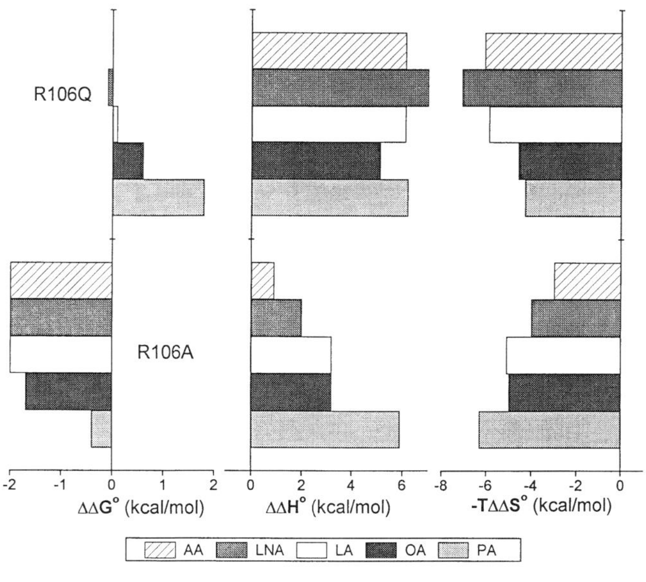

Fig. 9. Thermodynamic parameter differences for Ala and Gln mutants of Arg[106]. Thermodynamic parameter differences were determined for Ala and Gln mutants of R106 of I-FABP at 25°C for binding of the 4 FA of Fig. 8 plus linolenate. The disparity between previous calorimetry measurements showing a 20 fold reduction in oleate binding to the R106Q mutant [25], as compared to the 3 fold reduction observed in the present study, is likely result of the inability of calorimetry to measure accurately the binding affinities of FA/FABP complexes [28, 31]. This figure is taken from [22].

However, both the Gln and Ala mutants result in substantial reductions in the enthalpy of interaction, as would be expected for the loss of the favorable electrostatic interaction. At the same time, these losses are either exactly (Gln) or more than (Ala) compensated for by entropy increases. Thus, the Ala mutation results in an up to 30 fold increase in affinity because the entropy of this reaction is much greater than for the WT protein. These results, which imply higher entropies for the mutants than for the WT protein, are consistent with the crystal structure. The structure of R106Q with bound oleate shows that the FA has virtually the same conformation as in the WT protein but, especially at the carboxylate end, shows much greater disorder than in the WT I-FABP [17]. At least qualitatively, therefore, the structural and binding results are consistent and these results suggest that measurements of the thermodynamics are essential for understanding the connection between structure and ligand interactions.

Acknowledgements

This work was supported by grant GM46931 from the National Institute of General Medical Sciences of the National Institutes of Health. We thank Dr. James Sacchettini for the cDNA clone of the human intestinal FABP.

References

1. Ockner RK, Manning JA, Poppenhausen RB, Ho WKL: A binding protein for fatty acids in cytosol of intestinal mucosa, liver, myocardium, and other tissues. Science 177: 56–58, 1972
2. Mishkin S, Stein L, Fleischner G, Gatmaitan Z, Arias IM: Z protein in hepatic uptake and esterification of long-chain fatty acids. Am J Physiol 228: 1634–1640, 1975
3. Bass NM: The cellular fatty acid binding proteins: Aspects of structure, regulation, and function. Intern Rev Cytol 111: 143–184, 1988
4. Veerkamp JH, Peeters RA, Maatman RGHJ: Structural and functional features of different types of cytoplasmic fatty acid-binding proteins. Biochim Biophys Acta 1081: 1–24, 1991
5. Veerkamp J, Maatman R, Prinsen C: Fatty acid-binding proteins: Structural and functional diversity. Biochem Soc Trans 20: 801–805, 1992
6. Banaszak L, Winter N, Xu Z, Bernlohr DA, Cowan S, Jones TA: Lipid-binding proteins; A family of fatty acid and retinoid transport proteins. Adv Protein Chem 45: 89–151, 1994
7. Sacchettini JC, Gordon JI: Rat intestinal fatty acid binding protein. J Biol Chem 268: 18399–18402, 1993
8. Haunerland NH, Jacobson BL, Wesenberg G, Rayment I, Holden HM: Three-dimensional structure of the muscle fatty acid binding protein isolated from the desert locust *Schistocerca gregaria*. Biochemistry 33: 12378–12385, 1994
9. Sacchettini JC, Scapin G, Deshmukh G, Gordon JI: Refinement of the structure of *Escherichia coli*-derived rat intestinal fatty acid binding protein with bound oleate to 1.75-A resolution. Correlation with the structures of the apoprotein and the protein with bound palmitate. J Biol Chem 267: 23534–23545, 1992
10. Zanotti G, Scapin G, Spadon P, Veerkamp JH, Sacchettini JC: Three-dimensional structure of recombinant human muscle fatty acid-binding protein. J Biol Chem 267: 18541–18550, 1992
11. Xu Z, Bernlohr DA, Banaszak LJ: The adipocyte lipid-binding protein at 1.6-Å resolution. J Biol Chem 268: 7874–7884, 1993
12. Richieri GV, Ogata RT, Kleinfeld AM: Equilibrium constants for the binding of fatty acids with fatty acid binding proteins from intestine, heart, adipose, and liver; measured with the fluorescence probe ADIFAB. J Biol Chem 269: 23918–23930, 1994
13. Cowan SW, Newcomer ME, Jones TA: Crystallographic studies on a family of cellular lipophilic transport proteins. Refinement of P2 myelin protein and the structure determination and refinement of cellular retinol-binding protein in complex with all-trans-retinol. J Mol Biol 230: 1225–1246, 1993
14. Young ACM, Scapin G, Kromminga A, Patel SB, Veerkamp JH, Sacchettini JC: Structural studies on human muscle fatty acid binding protein at 1.4 Å resolution: Binding interactions with three C18 fatty acids. Structure 2: 523–534, 1994
15. Thompson J, Winter N, Terwey D, Bratt J, Banaszak L: The crystal structure of the liver fatty acid binding protein. A complex with two bound oleates. (in press) 1997
16. LaLonde JM, Bernlohr DA, Banaszak LJ: X-ray crystallographic structures of adipocyte lipid-binding protein complexed with palmitate and hexadecanesulfonic acid. Properties of cavity binding sites. Biochemistry 33: 4885–4895, 1994
17. Eads J, Sacchettini JC, Kromminga A, Gordon JI: *Eschericia coli*-derived rat intestinal fatty acid binding protein with bound myristate at 1.5 Å resolution and I-FABP Arg106-Gln with bound oleate at 1.74 ′ Resolution. J Biol Chem 268: 26375–26385, 1993
18. Richieri GV, Ogata RT, Kleinfeld AM: A fluorescently labeled intestinal fatty acid binding protein. Interactions with fatty acids and its use in monitoring free fatty-acids. J Biol Chem 267: 23495–23501, 1992
19. Richieri GV, Ogata RT, Kleinfeld AM: Thermodynamics of fatty acid binding to fatty acid-binding proteins and fatty acid partition between water and membranes measured using the fluorescent probe ADIFAB. J Biol Chem 270: 15076–15084, 1995
20. Richieri GV, Ogata RT, Kleinfeld AM: Kinetics of fatty acid interactions with fatty acid binding proteins from adipocyte, heart, and intestine. J Biol Chem 271: 11291–11300, 1996
21. Richieri GV, Ogata RT, Kleinfeld AM: Thermodynamics and kinetic properties of fatty acid interactions with rat liver fatty acid binding protein. J Biol Chem 271: 31068–31075, 1996
22. Richieri GV, Low PJ, Ogata RT, Kleinfeld AM: Mutants of rat intestinal fatty acid binding protein illustrate the critical role played by enthalpy-entropy compensation in ligand binding. J Biol Chem 272: 16737–16740, 1997
23. Tanford C: The hydrophobic effect and the organization of living matter. Science 200: 1012–1018, 1978
24. Richieri GV, Kleinfeld AM: Unbound free fatty acid levels in human serum. J Lipid Res 36: 229–240, 1995
25. Jakoby MG, Miller KR, Toner JJ, Bauman A, Cheng L, Li E, Cistola DP: Ligand-protein electrostatic interactions govern the specificity of retinol- and fatty acid-binding proteins. Biochemistry 32: 872–878, 1993
26. LaLonde JM, Levenson MA, Roe JJ, Bernlohr DA, Banaszak LJ: Adipocyte lipid-binding protein complexed with arachidonic acid: Titration calorimetry and X-ray crystallographic studies. J Biol Chem 269: 25339–25347, 1994
27. Miller KR, Cistola DP: Titration calorimetry as a binding assay for lipid-binding proteins. Mol Cell Biochem 123: 29–37, 1993
28. Kurian E, Kirk WR, Prendergast FG: Affinity of fatty acid for rRat intestinal fatty acid binding protein: Further examination. Biochemistry 35: 3865–3874, 1996

29. Wenk MR, Alt T, Seelig A, Seelig J: Octyl-Beta-D-glucopyranoside partitioning into lipid bilayers: Thermodynamics of binding and structural changes of the bilayer. Biophys J 72: 1719–1731, 1997
30. Sacchettini JC, Gordon JI, Banaszak LJ: Crystal structure of rat intestinal fatty acid binding protein. Refinement and analysis of the Escherichia coli derived protein with bound palmitate. J Mol Biol 208: 327–339, 1989
31. Cistola DP, Kim K, Rogl H, Frieden C: Fatty acid interactions with a helix-less variant of intestinal fatty acid-binding protein. Biochemistry 35: 7559–7565, 1996

Molecular and Cellular Biochemistry **192:** 87–94, 1999.

The measurement of free fatty acid[1] concentration with the fluorescent probe ADIFAB: A practical guide for the use of the ADIFAB probe

G.V. Richieri, R.T. Ogata and A.M. Kleinfeld

Medical Biology Institute, CA, USA

Abstract

The aqueous phase monomers of fatty acids (FFA) appear in many steps of fat metabolism. Understanding metabolism requires that accurate measurements of FFA levels be determined in enzyme-mediated as well as in membrane and protein binding reactions. Measuring long chain FFA levels with sufficient sensitivity and temporal resolution is now possible using fluorescent probes constructed by ligating fluorescent groups and fatty acid binding proteins. In this paper we provide a practical description of the use of ADIFAB, the acrylodan labeled intestinal fatty acid binding protein. We describe with specific examples how ADIFAB can be used to determine, (1) FFA concentrations in aqueous solutions, (2) binding affinities of fatty acid binding proteins, (3) membrane/water partition coefficients, (4) lipase activities, and (5) serum levels of FFA. (Mol Cell Biochem **192**: 87–94, 1999)

Key words: free fatty acids, ADIFAB, fluorescence, FABP

Introduction

Recently, a method has been developed to measure the concentration of fatty acids (FA) in aqueous solution (1). Here we describe this methodology in detail, and illustrate the use of this technique with several examples. To measure accurately free fatty acids (FFA)[1] concentrations it is necessary to probe the aqueous solution with a high degree of specificity and sensitivity. To accomplish this we constructed the fluorescent probe ADIFAB, which is an acrylodan labeled intestinal fatty acid binding protein. ADIFAB obtains its specificity from the nature of the FABP binding site and its sensitivity from the use of fluorescence. The fluorescent characteristics of ADIFAB are such that its emission wavelength changes upon binding a fatty acid from 432 mn in the unbound state, to 505 nm with fatty acid bound (see Fig. 1). Thus in a solution containing FA and ADIFAB the ratio of fluorescence intensity emitted at 505 to that at 432 mn reflects the fraction of bound ADIFAB and together with the binding constant, this ratio allows one to determine the concentration of free fatty acid.

Materials and methods

ADIFAB preparation

ADIFAB (MW = 15000) can be purchased from Molecular Probes (Eugene OR) as 200 μg of lyophilized powder. For stability it is best to store ADIFAB at a high concentration. Thus bring up the 200 μg of powder in 130 μl of *storage* buffer consisting of 50 mM TRIS, 1 mM EDTA, 0.5 mM PMSF, and 0.05% Na Azide at pH = 8.0. This will result in a stock of approximately 100 μM ADIFAB. This low salt, high pH buffer is used only for *storage*. For the measure-

[1]The term free fatty acid (FFA), sometimes referred to as unbound FA, is used to designate the monomer of either the acid, the salt, or the ionized molecule in the aqueous phase. At pH 7.4 and for concentrations below the FA solubility limit (where the studies described in this paper were done) most of the molecules are ionized, because fatty acids have an apparent pKa of 4.8 (15).

Address for offprints: A.M. Kleinfeld, Medical Biology Institute, 11077 North Torrey Pines Road, La Jolla, CA 92037, USA

ments described here and done previously, the buffer contained > 10 mM NaCl and had a pH of 7.4.

Fatty acid stock preparation

ADIFAB has been found to react with organic solvents such as ethanol and methanol (although less so with DMSO), with even small amounts reducing the peak at 432 and thereby giving the appearance of bound fatty acid [1]. Thus we recommend against using FA dissolved in a vehicle such as ethanol when doing measurements with ADIFAB. We prepare aqueous solutions of the FA and use the sodium salt (or soap) rather than the acid to prepare these solutions because the salt appears to dissolve more rapidly. Primary stocks of these salts, which in solution are anions but for convenience we denote as fatty acids, should be prepared in water containing 50 μM butylated hydroxytoluene at concentrations of 20–50 mM at pH > 9, and at temperatures above the melting point of the fatty acid. Lower stock concentrations are made by dilutions of the primary stock with water. The fatty acid stocks are kept under argon at –20°C.

Instrumentation

Upon excitation at 386 nm, ADIFAB fluoresces at 432 nm in the absence of fatty acid and at 505 nm in the presence of fatty acids. Two fluorometer configurations have been used to measure ADIFAB fluorescence. In the first, a single emission monochromator is used to measure, alternatively, intensities at 505 nm and at 432 nm; and in the second the two intensities are measured simultaneously using two photomultiplier tubes in a T-format with appropriate 432 and 505 nm bandpass filters.

Single detector
The most efficient way to measure the ratio of intensities at 505 and 432 nm using a single detector is to measure alternately individual intensities at 505 and 432 nm, which can be done automatically with most modern fluorometers. For example, the SLM series of fluorometers can be programmed to (1) measure the 432 and 505 intensities of a blank sample (complete sample without ADIFAB), (2) measure *n* replicates of the intensities of the ADIFAB-containing sample alternately at 432 and 505 nm, (3) subtract the respective blank intensities from each sample intensity, (4) calculate each of the *n* ratio values, and (5) calculate the average and standard deviation. Typically, each intensity is collected for about 4 seconds, and therefore 10 replicates can be measured in about 2 minutes.

ADIFAB fluorescence can also be monitored as a function of time with this single detector configuration. Using this arrangement, temporal resolution is limited by the time required for the monochromator to switch between the 432 and 505 wavelengths. Considerably faster changes can be resolved by using the T format as described below.

T-Format
Using two photomultiplier tubes placed orthogonal to the incident light beam, on either side of the sample cuvette (T format) with bandpass filters of 505 and 432 nm, allows virtually continuous monitoring of ADIFAB fluorescence, limited only by the detector response time. Typically this arrangement would be used in conjunction with a stopped-flow device for rapid mixing of reactants [2].

Determining free fatty acid concentrations

Full emission scans of ADIFAB with and without fatty acid are shown in Fig. 1. The binding of fatty acid causes the emission peak at 432 to decrease and the emission peak at 505 to increase. Thus the ratio of 505 to 432 (R value) is indicative of the amount of fatty acid present. ADIFAB has been calibrated with 12 fatty acids and the dissociation constants (K_d), at 37°C, from this determination are listed in Table 1. The buffer used to measure these K_ds in Table 1 consisted of 20 mM HEPES, 150 mM NaCl, 5 mM KCl, 1 mM Na_2HPO_4, at pH = 7.4.

The basic procedure for measuring the free fatty acid concentration is as follows: First, the value of the ADIFAB ratio is measured without FA present (Ro). To do this measure the fluorescence intensities at 505 and 432 nm in a cuvette with buffer only (without ADIFAB). These are the blank intensities. Then add 0.2 μM ADIFAB to the cuvette and again measure the intensities at 505 and 432 nm. The Ro value is equal to:

$$Ro = \frac{I^{o}_{505} - I^{blank}_{505}}{I^{o}_{432} - I^{blank}_{432}}$$

Second, add to the cuvette a small amount of fatty acid from a stock solution. Once again measure the intensity at 505 and 432 nm. An R value is then calculated as:

$$R = \frac{I_{505} - I^{blank}_{505}}{I_{432} - I^{blank}_{432}}$$

Third, these Ro and R values are then used in Eq. (1), along with the appropriate K_d value from Table 1 to determine the concentration of the free fatty acid according to [1].

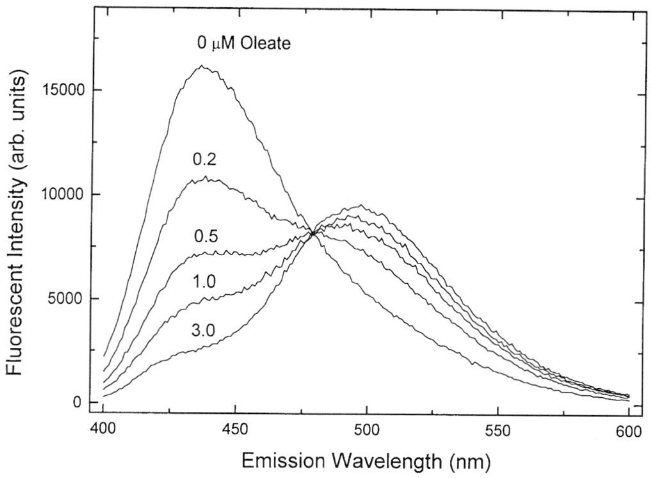

Fig. 1. Emission spectra of ADIFAB titrated with increasing amounts of oleate. An SLM 8100C fluorometer set to photon counting was used for these scans with the excitation wavelength at 386 nm, and excitation and emission slits set to 4 nm. The ADIFAB concentration was 0.2 μM in HEPES buffer containing 20 mM HEPES, 150 mM NaCl, 5 mM KCl, 1 mM Na_2HPO_4, pH 7.4, at 37°C. A background scan of buffer alone was subtracted from each ADIFAB scan.

$$[FFA] = K_d \bullet 19.5 \bullet \frac{(R - R_o)}{(11.5 - R)} \quad (1)$$

The amount of fatty acid that has bound to the ADIFAB probe can also be calculated as:

Table 1. Fatty acid properties:[a] ADIFAB dissociation constants partition coefficients, melting points, and solubilities

Fatty Acid	K_d (μM)[b]	K_p(× 10^{-4})[c]	MP(°C)[d]	Sol (μM)
Laurate (12.0)	14.00		44	>400
Myristate (14:0)	3.40		54	50
Palmitate (16:0)	0.34	46	63	4
Palmitoleate (16:1) 9 cis	2.90		1	40
Stearate (18:0)	0.08	220	70	<1
Oleate (18:1) 9 cis	0.28	38	16	6
Elaidate (18:1) 9 trans	0.22		44	6
Petroselinate (18:1) 6 cis	0.25		53	6
Vaccinate (18: 1) 11 cis	0.27		12	6
Linoleate (182) 9,12 cis	0.94	9	–7	14
Linolenate (183) 9,12,15 cis	2.50	3	–13	50
Arachidonate (20:4) 5,8,11,14 cis	1.62	8		22

[a]Measurements of K_d, K_p and solubility done in standard HEPES buffer: 20 mM HEPES, 150 mM NaCl, 5 mM KCl, 1 mM Na_2HPO_4, pH 7.4, at 37°C [b]Dissociation constants from 1, 3, 6, 12, and unpublished data. [c]Average value K_p for egg PC and egg PC + cholesterol vesicles from [9] [d]Melting points from the supplier. [e]Limiting concentrations above which FA aggregates. These values were determined from the peak R value for ADIFAB in a calcium free medium as described ([1], and unpublished data).

$$[ADIFAB_{bound}] = \frac{[ADIFABtotal] \bullet 19.5 \bullet (R - R_o)}{11.5 - R + 19.5 \bullet (R - R_o)} \quad (2)$$

Changes in buffer conditions, such as pH, temperature, or ionic strength, will alter the affinity (K_d) of FA for ADIFAB [3, 4]. Fatty acid binding to ADIFAB is extremely sensitive to pH, with affinities increasing with decreasing pH. For example, at 37°C, 150 mM NaCl, the K_d for oleate increases from 0. 18 μM at pH 7.0 to 0.52 μM at pH 7.8. The affinities of FA to ADIFAB are also sensitive to changes in temperature. The affinities increase (lower K_d) with decreasing temperature, as indicated in Table 2 [5]. The ionic strength of the buffer also effects the K_d, affinities increase with decreasing ionic strength. For example, at 37°C, and pH 7.4, K_d of oleate is 0.51 μM in 490 mM NaCl, while at 30 mM NaCl the K_d is 0.055 μM. The ADIFAB response at salt concentrations < 10 mM has not been calibrated. At these low salt concentrations the Ro value increases sharply. Thus for use of ADIFAB with a fatty acid not listed in Table 1, with a pH significantly different than 7.4, or with very low salt concentrations, specific binding characteristics must be determined, as described in the Appendix.

Results

In the following section we present five examples in which ADIFAB is used to determine the concentration of FFA. In these experiments we show how to determine: (1) the FFA concentration in an aqueous solution, (2) the fatty acid binding affinity of a protein, (3) the FA membrane partition coefficient, (4) phospholipase A2 activity, and (5) the FFA concentration in serum. Unless otherwise stated, the buffer used for all of the experiments that follow consisted of 20 mM HEPES, 150 mM NaCl, 5 mM KCl, 1 mM Na_2HPO_4, at pH = 7.4, and 37°C.

Determining the concentration of fatty acid in an aqueous solution

ADIFAB can be used to measure the concentration of FFA in a simple aqueous solution. To do this an aliquot of the fatty acid stock solution is added to a cuvette containing ADIFAB. Using Eqs. 1 and 2, the free fatty acid level and that bound to ADIFAB are calculated. Add to this the amount bound to the walls (see Appendix) and the concentration of the solution can be determined.

Example

ADIFAB was first added to a cuvette containing 1. 5 ml buffer, so that the final ADIFAB concentration was about 0.2 μM. The

Table 2. Temperature dependence of the K_ds for fatty acid binding to ADIFAB.[a]

Temperature (°C)	Laurate	Myristate	Palmitate	Oleate	Linoleate	Linolenate	Arachidonate
5	12.3	2.54	0.20	0.17	0.48	1.31	0.72
10	12.5	2.67	0.22	0.19	0.54	1.47	0.82
15	12.8	2.80	0.24	0.20	0.60	1.63	0.94
20	13.1	2.93	0.26	0.22	0.67	1.81	1.07
25	13.4	3.07	0.28	0.23	0.74	2.00	1.22
30	13.6	3.21	0.31	0.25	0.82	2.21	1.38
35	13.8	3.30	0.33	0.27	0.90	2.42	1.55
37	14.0	3.40	0.34	0.28	0.94	2.51	1.62
40	14.2	3.48	0.36	0.29	0.99	2.65	1.73
45	14.4	3.62	0.39	0.31	1.09	2.90	1.94
50	14.7	3.77	0.41	0.33	1.18	3.15	2.16

[a]Values were measured in standard HEPES buffer and are in units of µM.

Ro value (the intensity ratio of 505/432) was then measured and found to be 0.300. Next, 0.005 ml of the oleate stock solution was then added to the cuvette. The R value was then measured and found to be 0.420. Using eq. (1), eq. (2), and a K_d of 0.28 (Table 1), the FFA concentration is 59 nM and the amount of FA bound to ADIFAB ($ADIFAB_{bound}$) is 35 nM. About 20% of the sodium oleate binds to the cuvette walls under these conditions (Table 5) so the total amount of FA added to the cuvette was (59 + 35)/0.80 = 118 nm. Thus the stock concentration was 118 nM • 1.5 ml/0.005 ml = 35.4 µM.

Determining the fatty acid binding affinity of a protein

ADIFAB can be used to determine FA binding constants to any unlabeled protein by monitoring FFA in the presence of FA and a FA binding protein [6, 7]. To do this, a solution containing ADIFAB and the protein of interest is titrated with fatty acid and at each step in the titration the R value is measured. The amount of fatty acid bound to the protein, after each titration, is then calculated as [6]:

$$[FA]_{bound} = [FA]_{total} - [FFA] - [ADIFAB]_{bound} \quad (3)$$

In the case of single-affinity binding sites, the titration data can be analyzed by the Scatchard method as:

$$\frac{[FA]_{bound}/[Protein]_{total}}{[FFA]} = \frac{1}{Kd'} \frac{[FA]_{bound}}{[Protein]_{total}} + \frac{n}{Kd'} \quad (4)$$

where n is the number of fatty acid binding sites per protein monomer, and K_d' is the binding affinity of the fatty acid to the protein. Plotting the data as $FA_{bound}/Protein_{total}$ vs. $FA_{bound}/Protein_{total}/FFA$ yields a straight line with the slope of the line equal to $-1/K_d'$, and the x-axis intercept equal to n. For multiple binding sites of different affinities such a plot is nonlinear as discussed in [6, 8].

Example

In order to determine the affinity of sodium oleate to murine adipocyte fatty acid binding protein (mAFABP), 4. 0 µM of AFABP and 0. 2 µMADIFAB were added together in a cuvette containing 1.5 ml buffer. The Ro value was measured and gave a value of 0. 2671. Sodium oleate was added so that the OA concentration in the cuvette was 0. 53 µM, and after waiting 10 min for equilibrium, the R value was measured giving a value of 0.2855. Additional aliquots of sodium oleate were made and after each an R value was measured The results and analysis of these measurements are listed in Table 3. Generating a Scatchard plot from these data (column F) vs. (column G) resulted in a straight line (Fig. 2). The x-axis intercept is equal to 1.01, so the stoichiomety is 1:1. Thus binding to mAFABP yield a single site with a binding affinity of 62 nM.

Determination of fatty acid - membrane partition coefficients

The partition coefficient of a fatty acid between membranes and aqueous solution can be determined with ADFAB as described [9–11]. To do this, ADIFAB is added to a cuvette containing a membrane whose lipid concentration is known, and Ro is measured. This solution is then titrated with fatty acid, and after each aliquot, R is measured. The coefficient describing partition of FA between membrane and aqueous phase is defined as $K_p = [FA]_m/[FFA]$, where $[FA]_m$ is the concentration of FA in the membrane [9]. K_p can be expressed in terms of quantities measured directly as:

$$K_p = \frac{\frac{V_a}{V_m} \bullet ([FA]_{total} - [FFA])}{[FFA]} \quad (5)$$

Table 3: Analysis of binding measurements using ADIFAB to determine the affinity of sodium oleate and murine adipocyte FABP[a]

[A] FA Added (μM)	[B] Measured R Value	[C] $[FFA]^b$	[D] $ADIFAB_{bnd}{}^c$	[E] $FA_{bnd}{}^d$	[F] $\frac{FA_{bnd}{}^c}{FABP_{total}}$	[G] $\frac{FA_{bound}{}^f / FABP_{total}}{FFA}$
0	0.2671	0				
0.53	0.2855	8.96E-03	6.20E-03	0.515	0.129	14.368
1.06	0.3113	2.16E-02	1.43E-02	1.024	0.256	11.870
1.57	0.3424	3.68E-02	2.33E-02	1.510	0.377	10.244
2.09	0.3865	5.87E-02	3.46E-02	1.997	0.499	8.510
2.6	0.4612	9.60E-02	5.11E-02	2.453	0.613	6.387
3.1	0.5846	1.59E-01	7.24E-02	2.869	0.717	4.516
3.6	0.7977	2.71E-01	9.83E-02	3.231	0.808	2.984
4.09	1.09	4.32E-01	1.21E-01	3.537	0.884	2.049
4.58	1.477	6.59E-01	1.40E-01	3.781	0.945	1.434

[a]Measurements were done using sodium oleate and 4 μM A-FABP in standard HEPES buffer, at 37°C; [b]FFA was calculated using eq. (1) and R values from column B; [c]FA binding to ADIFAB was calculated with equation (2) and R values from column B; [d]FA bound to FABP was calculated according to eq. (3) as column A - column C - column D; [e]Column E / the total FABP concentration (bound/total. the abscissa for a Scatchard plot); [f]Column F / Column C (bound/total/free, the ordinate of the Scatchard plot).

where $[FA]_{total}$ is the added FA concentration, [FFA] is determined using ADIFAB and Eq. 1, and V_a and V_m are the volumes of the aqueous and membrane phase, respectively. (The ratio V_m/V_a can be determined from the lipid concentration as [9]: $V_m/V_a = 10^{-3}$ for 1 mM phospholipid). With sufficient membrane present no correction for wall binding is necessary because for typical conditions > 95% of the fatty acid will be bound to the membrane; very little will be free, bound to ADIFAB, or bound to the walls.

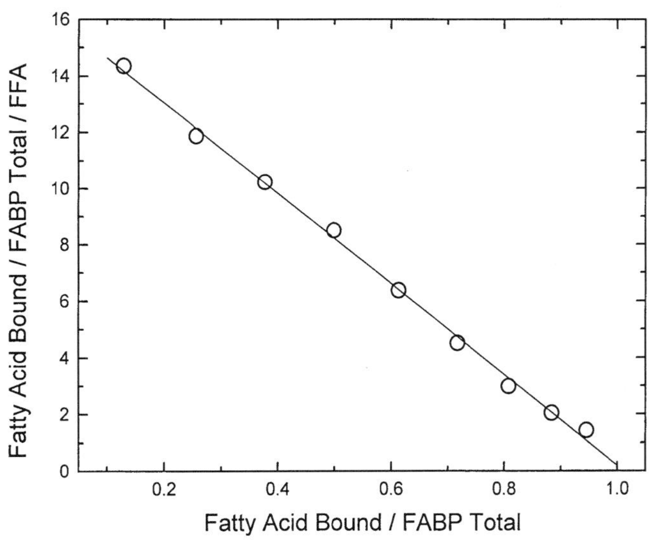

Fig. 2. Scatchard plot of oleate binding to murine adipocyte FABP (A-FABP), in standard HEPES buffer, pH 7.4 at 37°C. Sodium oleate was added to a 4 μM A-FABP solution, with the FFA level determined from the R values after each FA addition. The values are listed in Table 3. The data plotted here are from columns F and G from Table 3. The straight line represents the linear regression of the data. The slope of this line (–16.06 ± 0.31, r = 0.998) is equal to $-1/K_d$. Thus the $K_d' = 62$ nM. The x-axis intercept is equal to the number of FA binding sites on the protein, here equal to 1.0.

Example

In a cuvette with 2 ml of buffer, at 37°C, containing 100 μM egg phosphatidylcholine vesicles and 0.2 μM ADIFAB, the Ro is measured and found to be 0.260. After adding specific aliquots of sodium palmitate, and waiting a few minutes after each titration for equilibrium, the R value was measured. Using $V_a/V_m = 10000$, the K_p value calculated using Eq. 5 was found to be 373000. A complete set of R, FFA, and K_p values for a palmitate titration of EPC vesicles is listed in Table 4

Determination of phospholipase A_2 (PLA_2) activity

Here we wish to determine the time course for PLA_2 hydrolysis of membrane phospholipid [3]. For simplicity we

Table 4. Partition coefficient for palmitate in egg PC vesicles[a]

FA added (μm)	Measured R value	FFA (nM)	K_p
0	0.260	0	
3	0.335	82	356000
5	0.427	133	360000
7	0.516	185	368000
10	0.606	252	387000
15	0.718	375	390000
20	0.920	522	373000
25	1.150	651	374000
			Average = $3.7 \pm 1 \times 10^5$

[a]Measurments done using 100 μM vesicles in standard HEPES buffer, at 37°C.

consider lipid membranes composed of a single type of lipid in this example. Hydrolysis from complex membranes can also be determined as described in [3]. The absolute enzyme activity will be determined from the rate of change in the R value. Before PLA_2 addition, Ro is determined from the R values obtained with membrane plus ADIFAB alone. After about a minute of monitoring the Ro value, a specific amount of PLA_2 is added to the cuvette and the R value begins to increase, reflecting liberation of FA by PLA_2 hydrolysis (Fig. 3A). The total amount of fatty acid hydrolyzed is much greater than indicated by the R value change because a large fraction of the liberated FA partitions into the membrane. (Note: Such conditions do not apply to water soluble substrates as described by [4].) To obtain the total rate of hydrolysis the amount of fatty acid that remains membrane bound must be accounted for, and this can be done as:

$$[FA]_{total} = (K_p \bullet \frac{V_m}{V_a} \bullet [FFA]) + [FFA] \qquad (6)$$

Equation 6 is simply Eq. 5 solved for $[FA]_{total}$, with values of K_p, V_a, and V_m obtained as described above (see Eq. 5). From the total amount of fatty acid produced per unit time, a rate of hydrolysis can be calculated.

Example

To a cuvette at 37°C containing 1.5 ml of buffer containing 1 mM calcium and 1 mM magnesium was added 100 µm dioleoylphosphatidylcholine lipid vesicles (made by the extrusion method [3, 9]) and 0.2 µM ADIFAB. The cuvette was placed in the fluorometer and the R (Ro) value was monitored with time. After approximately 1 min, 0.5 ng of a PLA_2 (Bee Venom PLA_2 from Boehringer Mannheim # 1243063) was added to the cuvette with mixing, and the R values were then monitored for an additional 14 min (see Fig. 3A). The R values as shown in Fig. 3B were converted to total FA produced (or equivalently phospholipid hydrolyzed) with eq. 6, using a K_d of 0.28 (the FA released was oleate). The rate of phospholipid hydrolyzed was determined from the slope of the line in the Fig. 3B as 228 nM/min. Because the sample volume was 1.5 ml, the absolute amount hydrolyzed was 342 pmol/min. The amount of PLA_2 added was 0. 5 ng, thus the enzymatic activity was determined to be 685 µmol/min/mg PLA_2.

Determination of serum levels of FFA

The concentration of FFA in serum can be measured with ADIFAB [12, 13]. (In these studies FFA was referred to as unbound FFA.) For these measurements Ro is determined

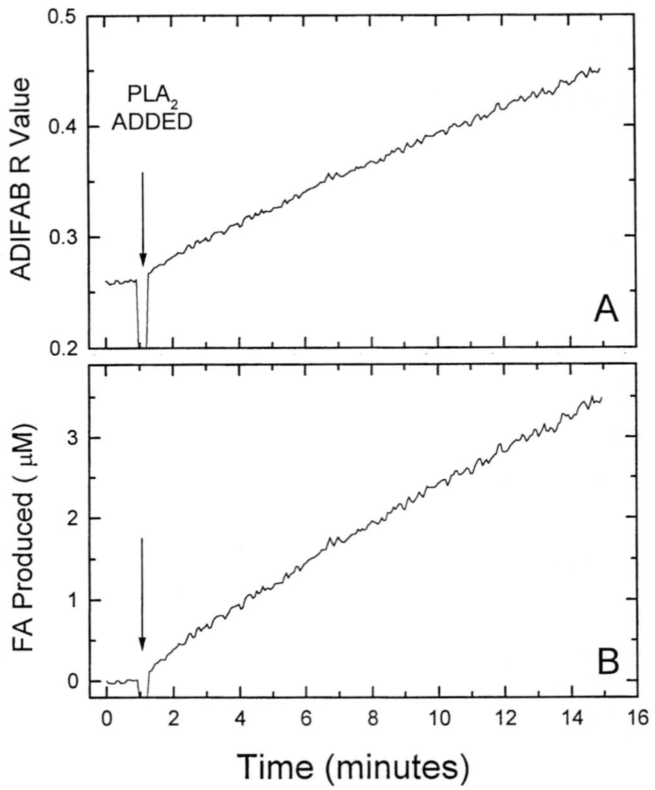

Fig. 3. (A) Time dependent change in ADIFAB R value in response to PLA_2 hydration. Bee Venom PLA_2 (333 µg/ml) was added to and mixed with 100 µM dioleoylphosphotidylcholine lipid vesicles, in standard HEPES buffer, pH 7.4 at 37°C, with 0.2 µM ADIFAB at approximately 1 min after start of scan. (B) Using both Eqs 1 and 6 the R values measured in (A) were converted to the total amount of FA produced upon addition of PLA_2.

by measuring R in the presence of 6 µM fatty acid free bovine serum albumin (BSA). The serum samples are prepared by a 100 fold dilution of 100% serum in buffer, yielding an albumin concentration of about 6 µM, about the same as used to determine Ro. This dilution does not effect [FFA] since the [FFA] level is buffered by the $[FA]_{total}$/albumin ratio [14]. Since the difference between R and Ro is very small (approximately 0.01 for serum from a healthy donor), an accurate measurement of both R and Ro is important. Thus, at least 10 measurements of R and Ro should be averaged. As mentioned in the Materials and methods section this can be done automatically on most modern fluorometers. To determine the actual concentration of free fatty acid in serum a weighted average of the K_ds for the different fatty acids in serum is used, and this value has been shown to be 0.44 µM [12]. Because a small increase in R with time occurs in the presence of albumin [12, 14], the actual serum sample and Ro should be measured at similar times after adding ADIFAB (typically after 10–20 min.)

Example

Start with 4 cuvettes containing 1.5 ml of buffer at 37°C, in which # 1) has 6 µM BSA, # 2) has 6 µM BSA + 0.5 µM ADIFAB, # 3) has 1/100 dilution serum, and # 4) has 1/100 dilution serum + 0.5 µM ADIFAB. After 20 min the R value of the BSA containing cuvette (# 2) was measured using cuvette # 1 as the blank, these measurements yield an Ro = 0.2742. Next the serum containing cuvette (#4) was measured using cuvette # 3 as the blank, to get R = 0.2835. Using Eq. 1 and K_d = 0.44, the amount of serum FFA is 7.1 nM. (Average value for human serum from healthy donors is about 7 nM [12]). (NOTE: If many samples are to be measured, a good approach is to do 2 Ro measurements, then 8–10 serum samples, 2 Ro, 8–10 serum samples, etc.). It is important that the serum samples contain little or no hemoglobin because hemoglobin will preferentially absorb the 432 nm fluorescence intensity which gives the appearance of higher than actual FFA levels. Although an inner filter correction can be applied, for the very small R value differences encountered in serum measurements, this correction will generate large uncertainties in the estimated FFA levels.

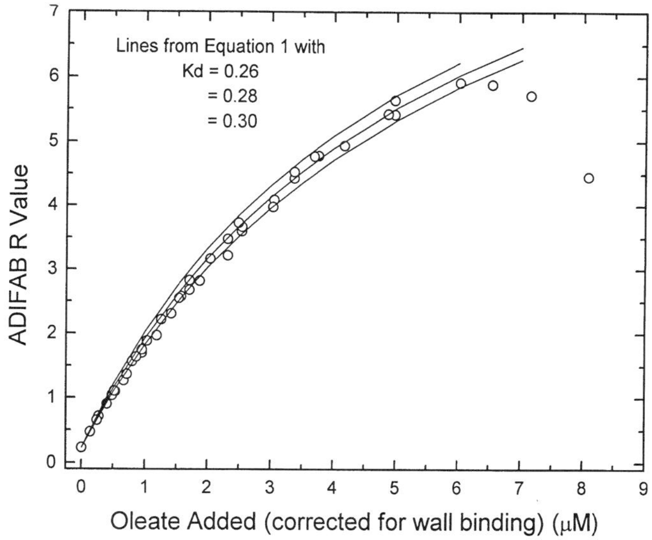

Fig. 4. Determination of binding affinity of sodium oleate to ADIFAB. Sodium oleate was added to a cuvette containing standard HEPES buffer at pH 7.4, 37°C, and 0.2 µM ADIFAB. After each titration the ADIFAB R value was measured. The amount of sodium oleate added to the cuvette was corrected for the amount bound to the walls of the cuvette (Table 5). The lines are generated from eq. 1 using different values of K_d.

Appendix

This appendix describes (1) the method to determine the K_d for FA not listed in Table 1, and (2) how to determine FA binding to the cuvette walls.

Determination of K_d

To calibrate ADIFAB for a particular fatty acid, titrate known amounts of fatty acid into a cuvette with ADIFAB, measuring the R value after each titration. Once the amount of fatty acid binding to the cuvette walls has been subtracted out (see next section), one can then fit these titration curves with different K_ds generated from Eq. 1 (see Fig. 4).

Figure 4 shows the titration of ADIFAB with oleate plotted as symbols and three lines generated from Eq. 1 with different K_d values. The decrease in R value with increasing amounts of fatty acid is due to aggregate formation [1]. Here we can see that for oleate the best fit occurs with a K_d of 0.28 ± 0.02 µM.

FA wall binding

Before ADIFAB can be calibrated for a particular fatty acid, the amount of fatty acid that binds to the walls of the cuvette must be determined. If fatty acid binds to the cuvette wall then the concentration of fatty acid in equilibrium with ADIFAB will be reduced. To determine the amount of wall binding - the change in R value is measured upon transferring a sample containing fatty acid from one cuvette to another. First the Ro value is determined in a sample containing ADIFAB in buffer without FA. Then a small amount of fatty acid is added and after waiting 5–10 min R is measured. Transfer the contents of this cuvette to a second cuvette, again wait for equilibrium, and measure R′. From the difference between R and R′ the fraction bound (BF) to the walls can be determined as: BF = (R–R′)/(R–Ro). The amount

Table 5. Degree of wall binding for fatty acids[a]

	Temperature (°C)			
Fatty Acid	10	30	37	50
Laurate (12:0)			1.5	
Myristate (14:0)			8.5	
Palmitate (16:0)	7	20	22.3	30
Palmitoleate (16:1) 9 cis			8.1	
Stearate (18:0)			>50	
Oleate (18:1) 9 cis	18	20	21	29
Elaidate (18:1) 9 trans			21	
Petroselinate (18:1) 6 cis			21	
Vaccinate (18: 1) 11 cis			21	
Linoleate (182) 9,12 cis	15	15	15	15
Linolenate (183) 9,12,15 cis	3	2	2	3
Arachidonate (20:4) 5,8,11,14 cis 1	14	14	14	14

[a]Amount of FA binding to the walls of glass cuvettes as a percent of the added FA concentration. Measurements done in standard HEPES buffer.

bound is different for different fatty acids, temperature, and cuvette material but is fairly constant over a large FA concentration range (0–4 μM). The percent binding to the walls of glass cuvettes for some fatty acids are presented in Table 5.

Acknowledgements

This work was supported by grant GM46931 from the National Institute of General Medical Sciences of the National Institutes of Health.

References

1. Richieri GV, Ogata RT, Kleinfeld AM: A fluorescently labeled intestinal fatty acid binding protein; Interactions with fatty acids and its use in monitoring free fatty-acids. J Biol Chem 267: 23495–23501, 1992
2. Richieri GV, Ogata RT, Kleinfeld AM: Kinetics of fatty acid interactions with fatty acid binding proteins from adipocyte, heart, and intestine. J Biol Chem 271: 11291–11300, 1996
3. Richieri G, Kleinfeld AM: Continuous measurement of phospholipase A2 activity using the fluorescence probe ADIFAB. Analyt Biochem 229: 256–263, 1995
4. She HS, Garsetti DE, Steiner MR, Egan RW, Clark RA: The substrate specificities of four different lysophospholipases as determined by a novel fluorescence assay. Biochem J 298: 23–29, 1994
5. Richieri GV, Ogata RT, Kleinfeld AM: Thermodynamics of fatty acid binding to fatty acid-binding proteins and fatty acid partition between water and membranes measured using the fluorescent probe ADIFAB. J Biol Chem 270: 15076–15084, 1995
6. Richieri GV, Ogata RT, Kleinfeld AM: Equilibrium constants for the binding of fatty acids with fatty acid binding proteins from intestine, heart, adipose, and liver; measured with the fluorescence probe ADEFAB. J Biol Chem 269: 23918–23930, 1994
7. Herr FM, Aronson J, Storch J: Role of portal lysine residues in electrostatic interactions between heart fatty acid binding protein and phospholipid membranes. Biochemistry 35: 1296–1303, 1996
8. Richieri GV, Ogata RT, Kleinfeld AM: Thermodynamics and kinetic properties of fatty acid interactions with rat liver fatty acid binding protein. J Biol Chem 271: 31068–31075, 1996
9. Anel A, Richieri GV, Kleinfeld AM: Membrane partition of fatty acids and inhibition of T cell function. Biochemistry 32: 530–536, 1993
10. Langner M, Isac T, Hui SW: Interaction of free fatty acids with phospholipid bilayers. Biochim Biophys Acta 1236: 73–80, 1995
11. Bent ED, Bell JD: Quantification of the interactions among fatty acid, lysophosphatidylcholine, calcium, dimyristoylphoshatidylcholine vesicles, and phospholipase A2. Biochim Biophys Acta 1995: 349–360, 1995
12. Richieri GV, Kleinfeld AM: Unbound free fatty acid levels in human serum. J Lipid Res 36: 229–240, 1995
13. Kleinfeld AM, Prothro D, Brown D, Davis RC, Richieri GV, DeMaria A: Increases in serum unbound free fatty acid levels following coronary angioplasty. Am J Cardiol 78: 1350–1354, 1996
14. Richieri GV, Anel A, Kleinfeld AM: Interactions of long chain fatty acids and albumin: Determination of free fatty acid levels using the fluorescent probe ADIFAB. Biochemistry 32: 7574–7580, 1993
15. Small DM: The physical chemistry of lipids, from alkanes to phospholipids. New York: Plenum Press, 1986

Molecular and Cellular Biochemistry **192:** 95–103, 1999.

Role of acylCoA binding protein in acylCoA transport, metabolism and cell signaling

Jens Knudsen, Mette Valentin Jensen, Jan Krogh Hansen, Nils J. Færgeman, Thomas B.F. Neergaard and Barbara Gaigg
Biokemisk Institut, Odense Universitet, Odense M, Denmark

Abstract

Long chain acylCoA esters (LCAs) act both as substrates and intermediates in intermediary metabolism and as regulators in various intracellular functions. AcylCoA binding protein (ACBP) binds LCAs with high affinity and is believed to play an important role in intracellular acylCoA transport and pool formation and therefore also for the function of LCAs as metabolites and regulators of cellular functions [1]. The major factors controlling the free concentration of cytosol long chain acylCoA ester (LCA) include ACBP [2], sterol carrier protein 2 (SCP2) [3] and fatty acid binding protein (FABP) [4]. Additional factors affecting the concentration of free LCA include feed back inhibition of the acylCoA synthetase [5], binding to acylCoA receptors (LCA-regulated molecules and enzymes), binding to membranes and the activity of acylCoA hydrolases [6]. (Mol Cell Biochem **192**: 95–103, 1999)

Key words: acylCoA, acylCoA binding protein, ACBP, transport, regulation and cell signaling

AcylCoA binding protein

AcylCoA binding protein (ACBP) was originally discovered as an impurity in a FABP preparation [7]. ACBP is a widely distributed 10 kDa cytosolic protein found in all eukaryotes tested from yeast and plants to reptiles, birds and man [8] and has been identified in liver, adipose tissue, kidney, heart, brain, intestine, skeletal muscle, mammary gland [9], erythrocytes [10] duodenum, testis, adrenal gland, ovary, lung and spleen [11, 12]. Recently an ACBP isoform (endozepine-like peptide) was identified in mouse testis [13]. In mammals the highest concentration of ACBP is found in liver, where it is evenly distributed in all hepatocytes [12]. In other tissues ACBP is reported to be high in specialized cells like steroid producing cells of the adrenal cortex and testis, and in epithelial cells specialised in secretion and in water and electrolyte transport, which all are characterized by a high energy metabolism. In *Drosophila melanogaster* ACBP has been found primarily expressed in tissues, which are associated with high energy production or fat metabolism [14]. ACBP has been purified, cloned and sequenced from a large number of different species and tissues and shows a high degree of similarity among the different species (Fig. 1). The broad range of distribution throughout the animal and plant kingdom and its high degree of similarity among the different tissues and species suggest that ACBP is a housekeeping protein. This was further supported by Mandrup *et al.* [15] who demonstrated that the genomic gene of ACBP has all the characteristics of a housekeeping gene.

ACBP is folded into a four α-helix bundle protein. The binding site is located in a hydrophobic grove on the surface of ACBP. The acyl chain is buried in the binding pocket and is completely protected from the aqueous solvent by the acylCoA head group which forms a lid on the binding pocket by interacting with specific residues on the rim of the binding cavity. The acylCoA head group is able to bind to ACBP with low affinity ($K_D = 2\ \mu M$ [16]) and plays an important role in determining binding specificity for acylCoA esters only [17].

ACBP binds medium and long chain acylCoA esters with very high affinity, with a preference for C_{14}-C_{22} acylCoA esters [18–20]. ACBP does not bind fatty acids, acyl carnitines, cholesterol and a number of nucleotides [20]. The binding affinities decrease with increasing ionic strength of the buffer used [18]. Estimates of the binding affinity by

Address for offprints: J. Knudsen, Biokemisk Institut, Odense Universitet, Campusvej 55, Odense M, Denmark

A. SQAE**F**DKA**A**EE**V**KH**L**KTK**P**ADEEM**LF**I**Y**SH[30]

B. SQVEFEMA**C**ASLKQLKGPVSDQEKLLVYSF[31]

A. **YKQA**TV**G**DINTERPGMLDFKGKA**KW**DA**W**NE[60]

B. YKQATQGD**C**NIPVPPATDVRAKAKYEAWMV[61]

A. LK**G**TSK**E**D**A**MKA**Y**IDK**V**EE**L**KKKYGI[86]

B. NKGMSKMDAMRIYIAKVEELKKKEP**C**[86]

Fig 1. Sequence comparison of bovine ACBP (A) and a rat testes specific ACBP isoform called Endozepine Like Peptide (ELP) (B). Bold letters in (A) indicate amino acids fully conserved among all of the following species: Man, pig, bovine, dog, rat, mouse, tortoise, duck, chicken, frog, larvae, fruit fly, yeast, arabidopsis and oilseed rape [40]. Bold letters in (B) show cysteines in ELP.

isotermal titration microcalorimetry of bovine ACBP, yeast ACBP and rat ACBP for palmitoyl-CoA in 25 mM ammonium acetate (pH 6.0), 150 mM NaCl or in 20 mM MOPS, 10 mM potassium phosphate (pH 7.2), 100 mM KCl, 20 mM NaCl, 3.5 mM $MgCl_2$, 3 mM ATP yielded K_D values of 2.02, 2.32 and 8.87 nM respectively (Knudsen, unpublished). However, it should be noted that the determined binding affinities are on the upper limit of what can be measured by direct titration calorimetry and may therefore be underestimated.

A number of *in vitro* and *in vivo* experimental results strongly indicate that ACBP is able to act as an intracellular acylCoA transporter and pool former. *In vitro,* ACBP has a strong attenuating effect on the inhibition of acetyl-CoA carboxylase and on the mitochondrial adenine nucleotide translocase by long chain acylCoA [21], it readily protects acylCoA against hydrolysis by microsomal hydrolases and stimulates the mitochondrial long chain acylCoA synthetase [21]. ACBP was found to desorb acylCoA esters immobilized in multilamellar liposomes on a nitrocellulose membrane and was able to transport and donate acylCoA to mitochondrial β-oxidation and to microsomal glycerolipid synthesis. Over-expression of either bovine or yeast ACBP in *Saccharomyces cerevisiae* led to an increased intracellular acylCoA level indicating that ACBP is able to act as an acylCoA pool former *in vivo* [22, 23].

Compelling evidence that ACBP participates in acylCoA transport *in vivo* has been obtained from yeast. Disruption of the ACBP-gene in *Saccharomyces cerevisiae* results in a dramatic perturbation of the acylCoA level and composition [24]. The level of total acylCoA and stearoyl-CoA was increased 2.5- and 7.0-fold respectively. Despite this, the Δ-9 desaturase mRNA level in the ACBP knockout strain was increased 3-fold. No change in the synthesis of mono–unsaturated fatty acids or the overall fatty acid composition in the knockout strain could be observed. These results strongly suggest that the increased stearoyl-CoA pool in the ACBP knockout strain was not available for the Δ-9 desaturase and that the ACBP knockout strain has a defect in intracellular acylCoA transport.

Intracellular acylCoA concentrations

The total cellular concentration of long chain acylCoA esters has been reported to be in the range of 5–160 μM, depending on the tissue and its metabolic state (Table 1). The levels of acylCoA esters are found to vary significantly in different metabolic conditions such as fasting [25, 29, 33], diabetes [33], fat/glucose feeding [33] and ingestion of hypolipidemic drugs [29, 34-36] e.g. in 48 h fasted rats the total level of acylCoA esters has been found to increase at least 2-4-fold [25, 29].

The compartmentation of long chain acylCoA esters is an important unsolved problem and the actual cytosolic concentration of free long chain acylCoA esters is not known for any tissue. Only few attempts to estimate the intracellular distribution of long chain acylCoA have been reported [21, 37, 38]. It has been suggested that 20–40% of the total acylCoA pool is cytosolic [39]. Deeney *et al.*, [32] estimated the cytosolic long chain acylCoA level in a clonal β-cell line to constitute approximately 78% of the total long chain acylCoA level, giving a cytosolic concentration of 90 μM. Sensitive methods for readily estimating the relative distributions of acylCoA esters in different cellular compartments are not currently available.

Regulation of the free concentration of acylCoA in cell cytosol

The most important factors in controlling the free concentration of cytosolic LCA is outlined in Fig. 2. The concentration of ACBP and long chain acylCoA in fed rat liver has been determined to be 40–50 nmole/g and 40–60 nmole/g tissue respectively [21, 29] indicating that the acylCoA/ACBP ratio might be close to one. At molar ratios of acylCoA/ACBP below one the calculated concentration of free LCA in the cytosol is < 10 nM. As the LCA concentration reaches the ACBP concentration (50 μM) the concentration of the unbound acylCoA will increase dramatically and approach the total acylCoA concentration. However, this situation will not appear because at this point SCP-2 and FABP will take over the LCA buffering function. SCP-2 will not play a significant role in buffering the LCA concentration because of its low concentration, 0.01–0.1% of the total protein [40]. When LCA exceeds the sum of ACBP plus SCP-2 concen-

Table 1. Levels of long-chain acyl-CoA esters in different tissues and cell types.

Tissue/cell type/organelle	Concentration	Measured compound	Reference
Rat liver	18.8 μM (fed state)	CoA by HPLC	Bortz & Lyner [25]
	53.9 μM (fed state)	LC-acyl-CoA by HPLC	Tardi *et al.* [26]
	64.4 μM (fed state)	LC-acyl-CoA by HPLC	Rosendal & Knudsen
[27]			
	55.0 μM cytosolic	LC-acyl-CoA by HPLC	Rasmussen *et al.* [21]
	23.6 μM (fed state)	LC-acyl-CoA by HPLC	Corkey & Deeney
[28]			
	54.6 μM (48 h-fasted)	LC-acyl-CoA by HPLC	
	60.0 μM (fed state)	CoA by HPLC	Sterchele *et al.* [29]
	164 μM (48 h-fasted)	CoA by HPLC	
Rat kidney	29.8 μM (fed state)	LC-acyl-CoA by HPLC	Rosendal & Knudsen
[27]			
Rat heart	27.4 μM (fed state)	LC-acyl-CoA by HPLC	
Rat skeletal muscle	12.3 μM (fed state)	LC-acyl-CoA by HPLC	
Rat brain	33.1 μM (fed state)	LC-acyl-CoA by HPLC	
Neutrophils	5–10 μM	Enzymatic meas. of CoA	Majumdar *et al.* [30]
Pancreatic β-cells	25 μM (nutrient stim.)	Enzyinatic meas. of CoA	Prentki *et al.* [31]
	100 μM (unstim.)	Enzyinatic meas. of CoA	
	115 μM	Enzymatic meas. of CoA	Deeney *et al.* [32]
	90 μm cytosolic	Enzymatic meas. of CoA	

tration FABP will be expected to take over the buffering function. L-FABP binds long chain acylCoA with a K_D of approximately 1.0 μM [41]. Assuming that the maximal obtainable acylCoA con–centration in liver cytosol is 150 μM and that the cytosolic FABP concentration is 0.3mM (0.6 mM binding sites) [42], the calculated free concentration of acylCoA will never be expected to exceed 200 nM.

Long chain acylCoA ester can also be expected to bind to a number of other proteins in the cell including the high affinity binding site on acylCoA synthetase and acylCoA-utilizing enzymes and other specific acylCoA binding proteins. Furthermore, by screening a bovine brain cDNA library using a degenerated ACBP-oligonucleotide probe, Webb *et al.* [43] isolated a related clone encoding a 533 amino acid protein containing an ACBP-like domain, termed bovine brain factor (bBF). Based on the amino acid sequence, it was suggested that bBF could be associated to either cell surface- or mitochondrial membrane [44]. It is tempting to speculate that membrane-associated bBF binds fatty acylCoA with high affmity, and thus by competing with the cytosolic ACBP is able to create a local pool of membrane bound acylCoA esters. LCAs also readily partition into membranes. The

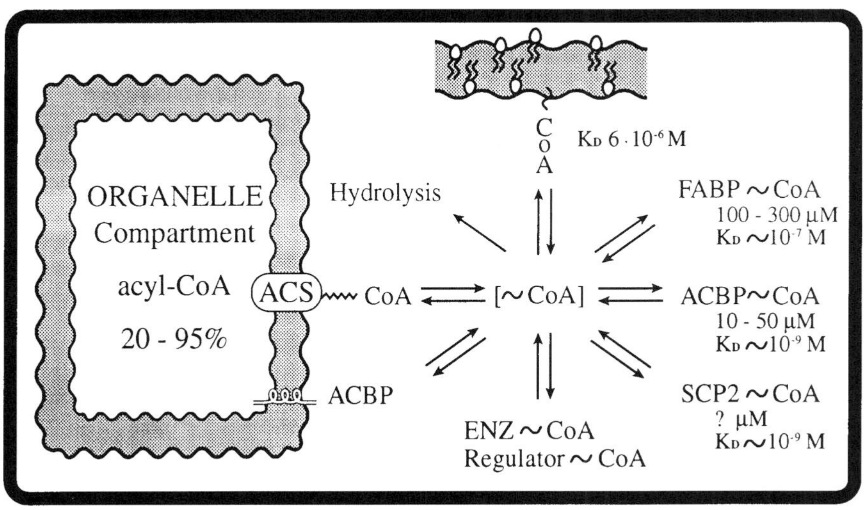

Table 2. Factors affecting the cytosol free acyl-CoA consentration.

partitioning constant for palmitoyl-CoA into phospholipid vesicles can be calculated from the original data of Peitzsch and McLaughlin [45] to 1.5×10^5 M^{-1} and to 5×10^5 M^{-1} by Requereo *et al.* [46]. If it is assumed that these values also can be applied to *in vivo* conditions the calculated free concentration of acylCoA in a liver cell in the absence of binding proteins would be about 1 µM if the LCA is allowed to partition freely into membranes.

An additional control factor for assuring low concentrations of LCAs under *in vivo* conditions is the high activity of acylCoA hydrolases, found in most subcellular compartments [47–49]. These enzymes include short-, medium-, and long chain acylCoA hydrolases, but the knowledge of the physiological role of these enzymes is sparse, although termination of fatty acid synthesis has been ascribed to a medium chain acylCoA hydrolase in the mammary gland [50, 51]. The acylCoA hydrolases usually display K_m values ranging from 0.1–6 µM for long chain acylCoA esters [47, 52, 53]. Whether the intracellular acylCoA concentration is regulated by acylCoA hydrolase activity is not known. Nevertheless, it is very likely that acylCoA hydrolases could act as 'scavengers', if the free long chain acylCoA pool increases to micromolar concentrations. Finally large fluctuations in the cytosolic concentration of free LCA esters will be expected to be prevented by feed back inhibition of the acylCoA synthetase (K_i = 4 µM [53]).

Cellular functions regulated by long chain acylCoA esters

In vitro experiments suggest that a large number of cellular functions may be regulated by long chain acylCoA esters. These functions include enzymes in carbohydrate and lipid metabolism, translocases, ion channels and pumps, protein kinases, nuclear transcription factors , proteases and protein transport (Table 2).

A major problem involved in making this type of experiments and evaluating their physiological significance is the concentration of LCA used and the physical chemical properties of these compounds. LCA is an amphipathic molecule and makes micelles at low concentrations. The micellar concentration ranges from 5–200 µM depending on chain length, number of double bonds in the acyl chain and salt concentration [13]. The actual free concentration is therefore unknown if the critical micellar concentration has not been determined under the experimental conditions used. A further critical problem is the fact that LCA readily partitions into membranes (see above). In a large number of the reported experiments with membrane-bound enzymes and ion channels acylCoA esters have been added directly to the membrane suspension without the addition of an acylCoA buffering protein. The added acylCoA esters will under these conditions accumulate to a very high concentration in/on the membrane. The concentration of LCA, which the enzyme or ion channel is exposed to, might therefore be much higher than the added concentration.

In vivo regulation of physiological functions by long chain acylCoA

Taking all of the above considerations into account we conclude that the intracellular free acylCoA concentration will be in the range of 0.1–200 nM under normal physiological conditions. If the cytosolic acylCoA/ SCP2 +ACBP ratio stays below 1 the free concentration will be in the 2–10 nM range. The fact that fatty acid synthesis occurs, although the K_i for inhibition of acetyl-CoA carboxylase is 5.5 nM, strongly indicates that the free concentration of liver cytosolic long chain acylCoA is below 5.5 nM during these conditions.

Immunohistochemical studies in yeast show that ACBP is not found in mitochondria and peroxisomes (Kal *et al.* personal communication). It can therefore not be excluded that large local changes in the free concentration of long chain acylCoA esters can occur, e.g. in the mitochondria where the total level of acylCoA can increase to extremely high levels (1 mM) [37], where processes like β-oxidation, the adenine nucleotide translocase, the citrate transporter and the pyruvate dehydrogenase could be inhibited. However, mitochondria also contains acylCoA hydrolases and this situation is therefore unlikely to occur [81].

If the total free acylCoA concentration under normal physiological conditions is well below 200 nM and most likely below 10 nM, the role of acylCoA as a physiological regulator of the cellular processes shown in Table 2 will be expected to be limited to the regulation of acetyl-CoA carboxylase, the AMP-activated kinase-kinase and gene expression in *E. coli*, unless the acylCoA/ACBP complex can donate acylCoA directly to the regulatory protein in question or that the complex can act as a regulator or enzyme substrate itself.

The ACBP acylCoA complex as enzyme substrate and regulator

The fact that the acylCoA/ACBP complex might indeed act as enzyme substrate is indicated by the observation that the acylCoA/ACBP complex at molar ratios below one can donate acylCoA for β-oxidation [55]. Changing the molar ratios of acylCoA/ACBP from 0.05–1.0 did not affect the rate of β-oxidation of palmitoylCoA by rat liver mitochondria [55]. ACBP also stimulated incorporation of arachidonic acid

Table 2. Effects of Acyl-CoA in cellular regulation and signal transduction

Acyl-CoA regulation of.	Effect	References
Lipid metabolism		
Fatty acid synthesis:		
Acetyl-CoA carboxylase	Inhibitory, K_i = 5.5 nM	Ogiwara *et al.* [54], Rasmussen *et al.* [55]
AMP-activated kinase-kinase	Stimulatory, nM-range	Carling *et al.* [56]
Mitochondrial Acyl-CoA synthetase	Inhibitory, K_i = 4 μM	Rasmussen *et al.* [21], Pande [5]
Citrate transporter	Inhibitory, K_i = 32 μM	Halperin *et al.* [57]
TG/PL/CE synthesis:		
HMG-CoA reductase	Inhibitory, IC_{30} - 1.9 μM	Lehrer *et al.* [58]
β-oxidation:		
Camitine palmitoyl transferase (CPT1)	Inhibitory, K_i 20 μM	Murthy & P ande. [59]
Long-chain acyl-CoA dehydrogenase	Inhibitory, K_i 0.2 μM	Powell *et al.* [60]
CE/TG-hydrolysis		
Hormone sensitive lipase	Inhibitory, IC_{50} = 0.1 μM	Jepson & Yeaman [61]
Energy metabolism		
Adenine nucleotide translocase	Inhibitory, K_i < 1 μM	Woldegiorgis *et al.* [62] Pasmussen *et al.* [21]
Glucoldnase	Inhibitory, K_i = 0.5 μM	Tippett *et al.* [36, 63]
Glucose-6-phosphatase	Inhibitory, K_i = 50 μM	Fulceri *et al.* [64]
Puryvate dehydrogenase	Inhibitory, K_i = 30 μm	Moore *et al.* [65]
Sienal transduction		
Ion channel pumps:		
Ca^{2+}-release from sarc. reticulum	Stimulatory, EC_{50} = 6 μM	Fulceri *et al.* [66]
Ca^{2+}-release from sea urchin eggs	Stimulatory	Chim & Dousa [67]
GTP-dep. Ca^{2+}-release from ER	Inhibitory, K_i 0.5 μM	Pys-Sikora *et al.* [68]
Ca^{2+}-induced Ca^{2+} accum.	Inhibitory, K_i 3 μM	Rys-Sikora *et al.* [68]
Ca^{2+}-release from ER (IP_3-insens)	Stimulatory, EC_{50} = 50 μM	Fulceri *et al.* [69]
Ca^{2+}-reuptake by Ca^{2+}-ATPases	StimulatoryEC_{50} = 0.5 μM	Deeney *et al.* [32]
ΔΨ and Mg^{2+} in mitochondria	Inhibitory	Siliprandy *et al.* [70]
Plasmamembrane Na^+/K^+-ATPase	Stimulatory, EC_{50} = 3 μM	Kakar *et al.* [71]
pH-dep. anion-cond. Channel	Inhibitory, K_i = 2.4 μM	Halle-Smith [72]
ATP-sensitive K^+-channel	Stimulatory	Larsson *et al.* [73]
Protein kinase C subtypes:		
Ca^{2+}/PL/DAG-dep. PKC	Stimulatory, $EC_{50} \leq$ 15 μM	Bronfman *et al.* [74]
PKCn	Inhibitory, K_i < 10 μM	Majurndar *et al.* [30]
Gene expression:		
Nuclear thyroid hormone receptor	Inhibitory, K_i = 0.12 μM	Li *et al.* [75]
FadR *E. coli* transcription factor	Inhibitory, K_i = 5 nM	Diptusso *et al.* [76]
Protein sorting:		
Vesicular transport in Golgi app.	Stimulatory	Pfanner *et al.* [77,78]
GTP-dep. vesicle fusion in ER	Inhibitory	Comerford & Dawson [79]
Proteolysis		
Proline endopeptidase	Inhibitory, K_i = 9 μM	Yamakawa *et al.* [80]

from arachidonoylCoA into phospholipids by the acylCoA-lysophospholipid acyltransferase in red blood cells at low arachidonoylCoA concentrations [10]. This situation is not unique for ACBP bound ligand. It has recently been demonstrated that the retinal/cellular retinol binding protein (CRBP) complex rather than free retinal is the preferred substrate for lecithin-retinol acyltransferase [82] and for the microsomal retinol dehydrogenase [83]. The ability of the acylCoA/ACBP complex to donate acylCoA to utilising or acylCoA regulated systems is not universal. Acetyl-CoA carboxylase was completely protected against inhibition by acylCoA at all concentrations up to 5 μM at acylCoA/ACBP ratios below 0.8 [55]. It is therefore tempting to speculate that ACBP by binding LCA creates a pool of long chain acylCoA available for specific purposes only.

The regulatory functions of the ACBP/acylCoA complex were investigated in experiments with the ryanodin receptor Ca^{2+} release channel from rabbit muscle terminal cisternae. This channel has been shown to be activated by palmitoylCoA in the micromolar range (Table 2). In these experiments palmitoylCoA was added to the rabbit muscle terminal cisternae without the addition of an acylCoA buffering protein. The concentration used (6 μM) is far above the expected concentration of free LCA in the rabbit muscle cyto sol. Furthermore the local concentration on/in the membrane would be even higher due to partitioning of palmitoylCoA into the membrane. However palmitoylCoA appeared to act directly on the channel. Both palmitoylCoA and its non-hydrolysable ether analog were able to activate it and the activation could be blocked by the specific channel blocker Ruthenium red [26]. We therefore decided to reinvestigate the activation of the ryanodine receptor in the presence of physiological concentrations of ACBP. Addition of 6 μM palmitoylCoA in the presence of 6.6 μM bovine ACBP to the terminal cisternae did not affect Ca^{2+} release significantly but significantly reduced the reuptake rate of an added Ca^{2+} pulse. However preincubation of the terminal cisternae membranes with increasing concentrations of palmitoylCoA/ACBP complex concentrations strongly potentiated caffeine-induced Ca^{2+} release. This effect was proportional to the complex concentration and independent of the calculated free palmitoylCoA concentration (Fulceri, Knudsen and Benedetti in press). These results strongly indicate that the acylCoA/ACBP complex can either donate acylCoA directly to the ryanodine receptor or act as a regulator of the receptor itself.

The role of long chain acylCoA ester and ACBP in regulation of gene expression

Acetyl-CoA carboxylase (ACC) catalyses the initial and key step in biosynthesis of long chain fatty acids [84]. The level of this enzyme in *Saccharomyces cerevisiae* is repressed by long chain fatty acids in the growth medium [85]. However, a mutant strain of *Saccharomyces cerevisiae* defective in acylCoA synthetase exhibits little repression of acetyl-CoA carboxylase by fatty acids, indicating that the activation of exogenously supplied fatty acids is required for repression of acetyl-CoA carboxylase [85]. Similar observations have been made with regard to repression of the Δ-9 desaturase by Δ-9 unsaturated fatty acids in yeast [86].

In *E. coli*, fatty acid biosynthesis and degradation are co-ordinately regulated at a transcriptional level by the product of the *fadR* gene, FadR [76]. Using DNA-protein gel retention assays, DiRusso *et al.* [76] demonstrated that binding of purified FadR to DNA containing the fadB-promoter was prevented by LCAs, but not by short chain acylCoA esters and fatty acids. The K_i for palmitoyl-CoA and oleoyl-CoA was approximately 5 nM, and for myristoyl-CoA and decanoyl-CoA 250 nM and 2 μM, respectively. These data provide strong evidence that long chain acylCoA esters bind to FadR and thereby inhibit DNA-binding activity of FadR. A direct interaction between long chain acylCoA and FadR was shown using a fluorescence quenching assay, the K_D for FadR binding of oleoyl CoA was determined to 12.1 nM [87].

The above-mentioned repression of acetylCoA carboxylase and Δ9-desaturase strongly indicates that a similar transcription factor to FadR might also exist in yeast and that ACBP may play a role in regulation of gene expression by delivering LCAs to or interact directly with this transcription factor. Differential display analysis shows that several genes are expressed differently in the above -mentioned yeast ACBP knockout strain including the Δ9-desaturase gene (OLE1, unpublished results). Preliminary DNA gel shift assays show that proteins binding to the Δ9-fatty acid response element far in the OLE1 promoter strain compared to the wild type parent strain (unpublished results).

Reduction of ACBP expression in 3T3-L1 cells by expression of an ACBP antisense cDNA has been shown to block differentiation into adipocytes [88]. The cause of this defect at the molecular level is unknown. Addition of low concentrations of prostaglandin synthetase inhibitors at different stages during the differentiation process relieve the block of differentiation. The block on differentiation may therefore be related to the increased synthesis of prostaglandins by the antisense cell lines.

Conclusion

Taking the concentration and the binding affinity of the major LCA binding proteins into consideration, the free concentration of LCA in cell cytosol will be expected to be in the low nM range (1–20 nM) and not higher than 200 nM under the most extreme conditions. This will require that the role of acylCoA as a physiological regulator of a number of cellular functions has to be reevaluated. ACBP has both by *in vivo* and *in vitro* experiments been shown to fulfil the necessary requirements to act as an intercellular acylCoA transporter and pool former. The results also show that ACBP is able to interact with or donate acylCoA esters to enzymes (i.e. carnitin-palmitoylCoA transferase) and acylCoA-regulated proteins (i.e. the ryanodine receptor). Strong indications have been obtained for the fact that ACBP plays a role in the fatty acid regulated gene expression.

Acknowledgment

This work was supported by CISFEM Carlsberg Foundation and The Danish Natural Science Research Council. We also thank Erling Knudsen, Jens Roswalld Andersen and Connic Gram for technical assistance.

References

1. Færgemann NJ, Knudsen J: The role of long-chain fatty acids in regulation of metabolism and in cell signalling. Biochem J 323: 1–12, 1997
2. Knudsen J, Mandrup S, Rasmussen JT, Andreasen PH, Poulsen, F, Kristiansen K: The function of acyl-CoA-binding protein (ACBP/diazepam binding inhibitor (DBI). Mol Cell Biochem 123: 129–138, 1993
3. Frolow A, Cho T, Billheimer JT, Schroeder F: Sterol carrier protein-2 a new fatty acid coenzyme A-binding protein. J Biol Chem 271: 31878–31884, 1996
4. Glatz JFC, van der Vusse GJ: Cellular fatty acid-binding proteins: Their function and physiological significance. Prog Lipid Res 35: 243–282, 1996
5. Pande SV: Reversal by CoA of palmityl-CoA inhibition of long chain acyl-CoA synthetase activity. Biochim Biophys Acta 306: 15–20, 1973
6. Berge RK, Slinde E, Farstad M: Intracellular localization of long-chain acyl-coenzyme A hydrolase and acyl-L-carnitine hydrolase in brown adipose tissue from guinea pigs. Biochem J 182: 347–351, 1979
7. Mogensen IB, Schulenberg H, Hansen HO, Spener F, Knudsen J: A novel acyl-CoA-binding protein from bovine liver. Effect on fatty acid synthesis. Biochem J 214: 189–192, 1987
8. Kragelund BB, Hojrup P, Jensen MS, Chjerling CK, Juul E, Knudsen J, Poulsen FM: Fast and one-step folding of closely and distantly related homologous proteins of four-helix bundle family. J Mol Biol 256: 187–200, 1996
9. Mikkelsen J, Knudsen J: Acyl-CoA-binding protein from cow. Binding characteristics and cellular and tissue distribution. Biochem J 248: 709–714, 1987
10. Fyrst H, Knudsen J, Schott MA, Lubin BH, Kuypers FA: Detection of acyl-CoA-binding protein in human red blood cells and investigation of its role in membrane phospholipid renewal. Biochem J 306: 793–799, 1995
11. Alho HE, Costa P, Ferrero M, Fujimoto D, Cosenza Murphy, Guidotti A: Diazepam-binding inhibitor: A neuropeptide located in selected neuronal populations of rat brain. Science 229: 179–182, 1985
12. Bovolin P, Schlichting J, Miyata M, Ferrarese C, Guidotti A, Alho H: Distribution and characterization of diazepam binding inhibitor (DBI) in peripheral tissues of rat. Regul Pept 29: 267–281, 1990
13. Pusch W, Balvers M, Hunt N, Ivell R: A novel endozepine-like peptide (ELP) is exclusively expressed in male germ cells. Mol Cell Endocrinol 122: 69–80, 1996
14. Kolmer M, Roos C, Tirronen M, Myohanen S, Alho H: Tissue-specific expression of diazepam-binding inhibitor in *Drosophila melanogaster*: Cloning, structure, and localization of the gene. Mol Cell Biol 14: 6983–6995, 1994
15. Mandrup S, Hummel R, Ravn S, Jensen G, Andreasen PH, Gregersen N, Knudsen J, Kristiansen K: Acyl-CoA-binding protein/diazepam-binding inhibitor gene and pseudogenes. A typical housekeeping gene gamily. J Mol Biol 228: 1011-1022, 1992
16. Robinson CV, Chung EW, Kragelund BB, Knudsen J, Aplin RT, Poulsen FM, Dobson CM: Probing the nature of noncovalent interations by mass spectrometry. A study of protein-CoA ligand binding and assembly. J Am Chem Soc 118: 8646–8653, 1996
17. Kragelund BB, Andersen KV, Madsen JC, Knudsen J, Poulsen FM: Three-dimensional structure of the complex between acyl-coenzyme A binding protein and palmitoyl-coenzyme A. J Mol Biol 230: 1260–1277, 1993
18. Færgeman NJ, Sigurskjold BW, Kragelund BB, Andersen KV, Knudsen J: Thermodynamics of ligand binding to acyl-coenzyme A binding protein studied by titration calorimetry. Biochemistry 35: 14118–14126, 1996
19. Rasmussen JT, Borchers T, Knudsen J: Comparison of the binding affinities of acyl-CoA-binding protein and fatty-acid-binding protein for long-chain acyl-CoA esters. Biochem J 265: 849–855, 1990
20. Rosendal J, Ertbjerg P, Knudsen J: Characterization of ligand binding to acyl-CoA-binding protein. Biochem J 290: 321–326, 1993
21. Rasmussen JT, Rosendal J, Knudsen J: Interation of acyl-CoA binding protein (ACBP) on processes for which acyl-CoA is a substrate, product or inhibitor. Biochem J 292: 907-913, 1993
22. Knudsen J, Færgemann NJ, Skott H, Hummel R, Borsting C, Rose TM, Andersen JS, Hojrup P, Roepstorff P, Kristiansen K: Yeast acyl-CoA-binding protein: Acyl-CoA-binding affinity and effect on intracellular acylo-CoA pool size. Biochem J 302: 479–485, 1994
23. Mandrup S, Jepsen R, Skott H, Rosendal J, Hojrup P, Kristiansen K, Knudsen J: Effect of heterologous expression of acyl-CoA-binding protein on acyl-CoA level and composition in yeast. Biochem J 290: 369–374, 1993
24. Schjerling CK, Hummel R, Hansen JK, Borsting C, Mikkelsen JM, Kristiansen K, Knudsen J: Disruption of the gene encoding the acyl-CoA-binding protein (ACB1) perturbs acyl-CoA metabolism in *Saccharomyces cerevisiae*. J Biol Chem 271: 22514–22521, 1996
25. Bortz WM, Lynen F: The inhibition of acetyl CoA carboxylase by long chaing acylCoA derivatives. Biochemische Zeitschrift 337: 505–509, 1963
26. Tardi PG, Man RY, Choy PC: The effect of methyl-lidocain on the biosynthesis of phospholipds de novo in the isolated hamster heart. Biochem J 285: 161–166, 1992
27. Rosendal J, Knudsen J: A fast and versatile method for extraction and quantitation of long-chain acyl-CoA esters from tissue: Content of individual long-chain acyl-CoA esters in various tissues from fed rat. Anal Biochem 207: 63–67, 1992
28. Corkey BE, Deeney JT: Acyl CoA regulation of metabolism and signal transduction. Prog Clin Biol Res 321: 217–232, 1990
29. Sterchele PF, van den Heuvel JP, Davis W, Shrago E, Knudsen J, Peterson RE: Induction of hepatic acyl-CoA-binding protein and liver fatty acid-binding protein by perfluorodecanoic acid in rats. Lack of correlation with hepatic long-chain acyl-CoA levels. Biochem Pharmacol 48: 955–966, 1994
30. Majumdar S, Rossi MW, Fujiki T, Phillips WA, Disa S, Queen CF, Johnston RB Jr, Rosen OM, Corkey BE, Korchak HM: Protein kinase C isotypes and signaling in neutrophils. Differential substrate specificities of a translocatable calcium- and phospholipid-dependent beta-protein kinase C and a phospholipid-dependent protein kinase which is inhibited by long chain fatty acyl coenzyme A. J. Biol Chem 266: 9285–9294, 1991
31. Prentki M, Vischer S, Glennon MC, Regazzi R, Deeney JT, Corkey BE: Malonyl-CoA and long chain acyl-CoA esters as metabolic coupling factors in nutrient-induced insulin secretion. J Biol Chem 267: 5802–5810, 1992
32. Deeney JT, Tornheim K, Korchak HM, Prentki M, Corkey BE: Acyl-CoA esters modulate intracellular Ca^{2+} handling by permeabilized clonal pancreatic beta-cells. J Biol Chem 267: 19840–19845, 1992
33. Tubbs PK, Garland PB: Variations in tissue contents of coenzyme A thio esters and possible metabolic implications. Biochem J 93: 550–557, 1964
34. Berge RK, Aarsland A, Bakke OM, Farstad M: hepatic enzymes, CoASH and long-chain acyl-CoA in subcellular fractions as affected by drugs inducing peroxisomes and smooth endoplasmic reticulum. Int J Biochem 15: 191–204
35. Constantinides PP, Steim JM: Micellization of fatty acid acyl-CoA mixtures and its relevance to the fatty acyl selectivity of acyl-transferases. Arch Biochem Biophys 261: 430–436, 1988
36. Tippet PS, Neet KE: Specific inhibition of glucokinase by long chain acyl coenzymes A below the critical micelle concentration. J Biol Chem 257: 12839–12845, 1982

37. Idell Wenger JA, Grotohann LW, Neely JR: Coenzyme A and carnitine distribution in normal and ischemic hearts. J Biol Chem 253: 4310–4318, 1978
38. Moore KH, Dandurand DM, Kiechle FL: Fasting induced alterations in mitochondrial palmitoyl-CoA metabolism may inhibit adipocyte pyruvate dehydrogenase activity. Int J Biochem 24: 809–814, 1992
39. Oram JF, Wenger JI, Neely JR: Regulation of long chain fatty acid activation in heart muscle. J Biol Chem 250: 73–78, 1975
40. Gossett RE, Frolov AA, Roths JB, Behnke WD, Kier AB, Schroeder F: Acyl-CoA binding proteins: Multiplicity and function. Lipids 31: 895–918, 1996
41. Rolf B, Oudenampsen Kruger E, Borchers T, Færgemann NJ, Knudsen J, Lezius A, Spener F: Analysis of the ligand binding properties of recombinant bovine liver-type fatty acid binding protein. Biochim Biophys Acta 1259: 245–253, 1995
42. Glatz JFC, van der Vusse GJ, Veerkamp JH: Fatty acid-binding proteins and their physiological significance. News Physiol Sci 35: 41–43, 1988
43. Webb NR, Rose TM, Malik N, Marquardt H, Shoyab M, Todaro GJ, Lee DC: Bovine and human cDNA sequences encoding a putative benzodiazepine receptor ligand. DNA 6: 71–79, 1987
44. Todaro GJ, Rose TM, Shoyab M: Human DBI (endozepine): Relationship to a homologous membrane associated protein (MA-DBI). Neuropharmacology 30: 1373–1380, 1991
45. Peitzsch RM, McLaughlin S: Binding of acylated peptides and fatty acids to phospholipid vesicles: Pertinence to myristoylated proteins. Biochemistry 32: 10436–10443, 1993
46. Requero MA, Gonzalez M, Goni FM, Alonso A, Fidelio G: Differential penetration of fatty acyl-coenzyme A and fatty acylcarnitines into phospholipid monolayers. FEBS Lett 357: 75–78, 1995
47. Berge RK, Farstad M: Purification and characterization of long-chain acyl-CoA hydrolase from rat liver mitochondria. Eur J Biochim 96: 393–401, 1979
48. Berge RK, Hosoy LH, Aarsland A, Bakke OM, Farstad M: Enzymatic changes in rat liver associated with low and high doses of a peroxisome proliferator. Toxicol Appl Pharmacol 73: 35–41, 1984
49. Waku K: Origins and fates of fatty acyl-CoA esters. Biochim Biophys Acta 1124: 101–111, 1992
50. Knudsen J, Clark S, Dils R: Acyl-CoA hydrolase(s) in rabbit mammary gland which control the chain length of fatty acids synthesised. Biochem Biophys Res Commun 65: 921–926, 1975
51. Knudsen J, Clark S, Dils R: Purficiation and some properties of a medium-chain acyl-thioester hydrolase from lactating-rabbit mammary gland which terminates chain elongation in fatty acid synthesis. Biochem J 160: 683–691, 1976
52. Broustas CG, Hajra AK: Purification, properties and specificity of rat brain cytosolic fatty acyl coenzyme A hydrolase. J Neurochem 64: 2345–2353, 1995
53. Yamada J, Furihata T, Tamura H, Watanabe T, Suga T: Long-chain acyl-CoA hydrolase from rat brain cytosol: Purification, characterization and immunohistochemical localization. Arch Biochem Biophys 326: 106–114, 1996
54. Ogiwara H, Tanabe T, Nikawa J, Numa S: Inhibition of rat-liver acetyl-coenzyme-A caboxylase by palmitoyl-coenzyme A. Formation of equimola enzyme-inhibitor complex. Eur J biochem 89: 33–41, 1978
55. Rasmussen JT, Færgemann NJ, Kristiansen K, Knudsen J: Acyl-CoA-binding protein (ACBP) can mediate intermembrane acyl-CoA transport and donate acyl-CoA for beta-oxidation and glycerolipid synthesis. Biochem J 299: 165–170, 1994
56. Carling D, Zammit VA, Hardie DG: A common bicyclic protein kinase cascade inactivates the regulatory enzymes of fatty acid and cholesterol biosynthesis. FEBS Lett 223: 217–222, 1987
57. Halperin ML, Pande SV: Fatty acyl group transport into mitochondria: Carnatine palmitoyl transferase EC 2.3.1.23 and the carnitine-acylcarnatine translocase. Methods Enzymol 56: 368–378, 1979
58. Lehrer G, Panini SR, Rogers DH, Rudney H: Modulation of rat liver 3-hydroxy-3-methylglutaryl coenzyme A reductase by lipid inhibitors, substrates and cytosolic factors. J Biol Chem 256: 5612–5619, 1981
59. Murthy MS, Pande SV: Some differences in the properties of carnitine palmitoyltransferase activities of the mitochondrial outer and inner membranes. Biochem J 248: 727–733, 1987
60. Powell PJ, Lau SM, Killian D, Thorpe C: Interaction of acyl coenzyme A substrates and analgues with pig kidney medium-chain acyl-CoA dehydrogenase. Biochemistry 26: 3704–3710, 1987
61. Jepson VA, Yeaman SJ: Inhibition of hormone-sensitive lipase by intermediary lipid metabolites. FEBS Lett 310: 197–200, 1992
62. Woldegoirgis G, Yousufzai SY, Shrago E: Studies on the interaction of palmutoyl coenzyme A with the adenine nucleotide translocase. J Biol Chem 257: 14783–14787, 1982
63. Tippett PS, Neet KE: An allosteric model for the inhibition of glucokinase by long chain acyl coenzyme A. J Biol Chem 257: 12846–12852, 1982
64. Fulceri R, Gamberucci A, Scott HM, Giunti R, Burchell A, Benedetti A: Fatty acyl-CoA esters inhibit glucose-6-phosphatase in rat liver microsomes. Biochem J 307: 391–397, 1995
65. Moore KH, Tsatso P, Staudacher DM, Kiechle FL: Counter modulation of adipocyte motochondrial processes by insulun and S-oxalyl-glutathione. Int J Biochem Cell Biol 28: 183–191, 1996
66. Fulceri R, Nori A, Gamberucci A, Volpe P, Giunti R, Benedetti A: Fatty acyl-CoA esters induce calcium release from terminal cisternae of skeletal muscle. Cell Calcium 15: 109–116, 1994
67. Chini EN, Dousa TP: Palmitoyl-CoA potentiates the Ca^{2+} release elicited by cyclic ADP-ribose. Am J Physiol 270: C530–C537, 1996
68. Rys Sikora Kem Ghosh TK, Gill DL: Modification of GTP-activated calcium translocation by fatty acyl-CoA esters. Evidence for a GTP-induced prefusion event. J Biol Chem 269: 31607–31613, 1994
69. Fulceri R, Gamberucci A, Bellomo G, Giunti R, Benedetti A: CoA and fatty acid acyl-CoA derivatives mobilize calcium from a liver reticular pool. Biochem J 295: 663–669, 1993
70. Siliprandi D, Biban C, Testa S, Toninello A, Siliprandi N: Effects of palitoyl CoA and palmitoyl carnitine on the mebrane potential and Mg^{2+} content of rat heart mitochondrial. Mol Cell Biochem 116: 117–123, 1992
71. Kaker SS, Huang WH, Askari A: Contro of cardiac sodium pump by long-chain acyl coenzymes A. J Biol Chem 262: 42–45, 1987
72. Hall Smith SC, Murray AG, Selwyn MJ: Palmyitoyl-CoA inhibits the mitochondrial inner membrane anion-conducting chanel. FEBD Lett 236: 155–158, 1988
73. Larsson O, Deeney JT, Branstrom R, Berggren PO, Corkey BE: Activation of ATP-sensitive K^+ channel by long chain acyl-CoA. A role in modulation of pancreatic beta-cell glucose sensitivity. J Biol Chem 271: 10623–10626, 1996
74. Bronfman M, Morales MN, Orellana A: Diacylglycerol activation of protein kinase C is modulated by long-chain acyl-CoA. Biochem Biophys Res Commun 152: 987–992, 1988
75. Li Q, Yamamoto N, Morisawa S, Inoue A: Fatty acyl-CoA binding activity of the nuclear thyroid hormone receptor. J Cell Biochem 51: 458–464, 1993
76. DiRusso CC, Heimert TL, Metzger AK: Characterization of FadR, a global transciptional regulator of fatty acid metabolism in *Escherichia coli*. Interaction with the FadB promoter is prevented by long chain fatty acy coenzyme A [published erratum appears in J Biol Chem 1992 Nov 5; 267(31): 22693] J Biol Chem 267: 8685–8691, 1992
77. Pfanner N, Glick BS, Arden SR, Rothman JE: Fatty acylation promotes fusion of transport vesicles with Golgi cisternae. J Cell Biol 110: 995–961, 1990

78. Pfanner N, Orci L, Glick BS, Amherdt M, Arden SR, Malhotra V, Rothman JE: Fatty acyl-coenzyme A is required for budding of transport vesicles from Golgi cisternae. Cell 59: 95–102, 1989
79. Comerford JG, Dawson AP: Effects of CoA and acyl-CoAs on GTP-dependent Ca^{2+} release and vesicle fusion in rat liver microsomal vesicles. Biochem J 289: 561–567, 1993
80. Yamakawa N, Shimeno H, Soeda S, Nagamutsu A: Inhibition of proline endopeptidase activity by acyl-coenzyme A esters. Biochim Biophys Acta 1037: 302–306, 1990
81. Svensson LT, Kilpelainen SH, Hiltunen JK, Alexson SE: Characterization and isolation of enzymes that hydrolyze short-chain acyl-CoA in rat-liver mitochondria. Eur J Biochem 239: 526–531, 1996
82. Ong DE: Cellular transport and metabolism of vitamin A: Roles of the cellular retinoid-binding proteins. Nutr Rev 52: S24–S31, 1994
83. Boerman MH, Napoli JL: Cellular retinol-binding protein-supported retinoic acid synthesis. Relative rolse of microsomes and cytosol. J Biol Chem 271: 5610–5616, 1996
84. Numa S, Yamashita S: Regulation of lipogenosis in animal tissues. Curr Top Cell Regul 8: 197–246, 1974
85. Kamiryo T, Parthasarathy S, Numa S: Evidence that acyl coenzyme A synthetase activity is required for repression of yeast acetyl coenzyme A carboxylase by exogenous fatty acids. Proc Natl Acad Sci USA 73: 386–390, 1976
86. Choi JY, Stukey J, Hwang, SY, Martin CE: Regulatory elements that control transcription activation and unsaturated fatty acid-mediated repression of the *Saccharomyces cerevisia* OLE1 gene. J Biol Chem 271: 3581–3589, 1996
87. Raman N, DiRusso CC: Analysis of acyl coenzyme A binding to the transcription factor FadR and identification of amino acid residues in the carboxyl terminus required for ligand binding. J Biol Chem 270: 1092–1097, 1995
88. Baldursson T, Gram C, Knudsen J, Krisiansen K, Mandrup S: Adipocyte differentiation is dependent onthe induction of the acyl-CoA binding protein. NATO ASI Series, Vol H92: 365374, 1995

Molecular and Cellular Biochemistry **192:** 105–108, 1999.

Structure and function of cytoplasmic retinoid binding proteins

Ellen Li

Departments of Medicine, Biochemistry and Molecular Biophysics, Washington University, School of Medicine, St. Louis, MO, USA

Abstract

We examined the ligand protein interactions of two highly homologous cellular retinol binding proteins, CRBP and CRBP-II, and two highly homologous cellular retinoic acid binding proteins, CRABP-I and CRABP-II. While the crystal structures of all four have been determined, nuclear magnetic resonance studies provide a means for observing dynamic aspects of ligand protein interactions of these proteins in solution. The cellular functions of these proteins are less well understood. We have modeled retinoid flux between cytoplasmic retinoid proteins and model membranes and with nuclear receptors. Based on our *in vitro* studies, we propose that certain retinoids may indirectly influence retinoid signaling by displacing endogenous retinoids from the cytoplasmic proteins to the nuclear receptors. (Mol Cell Biochem **192**: 105–108, 1999)

Key words: retinal binding proteins, cytoplasmic retinoid proteins

Abbreviations: CRBP – cellular retinol binding protein; CRABP – cellular retinoic acid binding protein; iLBP – intracellular lipid binding protein; RAR – retinoic acid receptor; RXR – retinoid-X-receptor; NMR – nuclear magnetic resonance; NOESY – nuclear Overhauser effect spectroscopy

Introduction

We have examined the ligand protein interactions of four members of the intracellular lipid binding protein family that bind retinoids. These include two cytoplasmic retinol binding proteins, cellular retinol binding protein and cellular retinol binding protein II, which bind all-trans-retinol and all-trans retinaldehyde but not retinoic acid, and two cytoplasmic retinoic acid binding proteins, cellular retinoic acid binding protein-I and cellular retinoic acid binding protein-II, which bind all-trans-retinoic acid but not retinol [1, 2]. Retinol, the alcohol form of vitamin A, is an essential dietary nutrient necessary for vision, growth, development, reproduction and differentiation of epithelial tissues, and is the form that is absorbed by the intestinal mucosa and delivered to peripheral tissues. Within the cell, retinol can be esterified with long chain fatty acids for storage within the cell or oxidized to the aldehyde, which is required for the visual process, and subsequently to retinoic acid. Retinoic acid, or the acid form of vitamin A exerts many of its biological effects by interaction with heterocomplexes of nuclear retinoic acid receptors (RARs) and retinoid X receptors (RXRs) [3]. The cellular functions of the cytoplasmic retinoid proteins remain to be precisely defined. Since these proteins bind their respective ligands with high affinity, they are likely to influence retinoid signalling pathways by modulating intracellular retinoid metabolism and by influencing the ligand occupancy of the nuclear receptors. Analysis of retinoid metabolism in the presence or absence of these proteins provides support for the hypothesis that these proteins are involved in modulating intracellular retinoid metabolism. In contrast, there is significantly more known about the three dimensional structures of these proteins. The crystal structures of all four cytoplasmic retinoid binding proteins has been determined [4–7]. These proteins share the β-barrel motif observed in the crystalline structures of other members

Address for offprints: E. Li, Departments of Medicine, Biochemistry and Molecular Biophysics, Washington University School of Medicine, Campus Box 8124, 660 South Euclid Avenue, St. Louis, MO 63110, USA

of the iLBP family. This motif consists of a 10-stranded, anti parallel β-barrel with a helix-turn-helix between the first and second β-strands, with the bound ligand sandwiched inside the barrel with the polar moiety situated innermost. Since the polar moiety of the ligand bound to the cytoplasmic retinoid binding proteins appears to be relatively inaccessible, the structural basis for how these proteins present the bound ligand as substrate for modifying enzymes presents an interesting problem. Furthermore, the bound conformation of retinoic acid in holo-CRABP-I could not be directly determined from the X-ray data [6].

To further understand how these proteins potentially interact with membrane bound enzymes and with the nuclear receptors, the ability of these proteins to compete for ligand relative to model membranes and nuclear receptors was examined. We have employed nuclear magnetic resonance techniques to observe the dynamic aspects of ligand protein interactions of these proteins in solution.

Cellular retinoic acid binding proteins

CRABP-I and CRABP-II are highly homologous, sharing 72% amino acid identity in humans (reviewed in refs [1, 2]). They both bind all-*trans*-retinoic acid with high affinity. CRABP-I binds all-*trans*-retinoic acid with somewhat higher affinity than does CRABP-II. The apparent dissociation constant K_d' of mouse CRABP-I was measured by fluorometric titration as less than 0.4 nM and the K_d' of mouse CRABP-II is 2 nM [8]. Both CRABPs bind the synthetic retinoic acid analog acetretin, which is used in the treatment of psoriasis, with high affinity. The two CRABPs exhibit distinct tissue-specific and developmental patterns of expression. In adult animals, CRABP-I is found in many tissues, whereas CRABP-II appears to be relatively restricted to the skin, suggesting that these two proteins subserve different physiological functions. While the CRABP's do not directly mediate retinoic acid signaling, they potentially influence retinoid signaling pathways by competing for the same ligand. The relative ability of the CRABPs to compete for retinoic acid vs. the nuclear RARs is an important question. To address this question, complexes of recombinant RAR-RXR ligand binding domains retaining full ligand binding capacity were purified using a glutathione S-transferase (GST) fusion protein containing the ligand binding domain of human RXRα to copurify the ligand-binding domain of human RARγ by affinity chromatography over glutathione-agarose [9]. The complexes were subsequently eluted by thrombin cleavage of the matrix bound GST fusion protein. Size-exclusion chromatography of purified recombinant proteins were used to directly compare the relative binding affinities of all-*trans*-retinoic acid for the CRABPs and RAR-RXR ligand binding domain complexes. When [^{3}H]-all-*trans*-retinoic acid was added to a mixture comprised of equal binding equivalents of CRABP-II and the nuclear receptor complexes, 8% of the total radioactivity was recovered in the CRABP-II fraction where as 78% of the total radioactivity was recovered with the nuclear receptors. When the CRABP-II fraction was increased 10-fold, 69% of the radioactivity was associated with CRABP-II and the amount associated with the nuclear receptors was reduced to 17%. When [^{3}H] all-*trans*-retinoic acid was added to equal binding equivalents of CRABP-I, 49% of the total radioactivity was recovered in the CRABP-I fraction and 40% was recovered with the nuclear receptors. These results suggest demonstrate that all-*trans*-retinoic acid binds to the nuclear receptor heterocomplexes and CRABP-I with comparable affinity, but binds to CRABP-II with approximately 10-fold lower affinity. These results suggest that CRABP-I will more effectively sequester all-*trans*-retinoic acid from the nuclear receptors than will an equivalent amount of CRABP-II.

That CRABP-I could potentially perturb retinoid signaling is supported by the observation that stably transfected F9 teratocarcinoma cells overexpressing CRABP-I exhibited a reduced response to exogenously added retinoic acid [10]. However transfected Cos-1 cells overexpressing CRABP-I did not exhibit an altered response to retinoic acid [11]. The *in vivo* function of CRABP-I and CRABP-II has been investigated by the creation of null mutants in CRABP-I and CRABP-II [12–15]. While the CRABP-I null mutant mice had no detectable phenotype, both the CRABP-II null mutant mice and the CRABP-I and CRABP-II double knockout mice exhibited an extra, postaxial digit in the forelimb. The penetrance of this effect is 10–50% in the CRABP-II knockout mice, and is 83% in the double knockout mice. The double knockout mice also exhibit significantly reduced viability, with 9% mortality by six weeks of age. It is interesting that loss of the CRABP-II protein, which binds less tightly to retinoic acid than does CRABP-I exhibits the more severe phenotype.

Acitretin, a synthetic aromatic retinoid analog is used in the treatment of psoriasis. It however binds poorly to the nuclear receptors although it binds to both CRABPs with high affinity. The mechanism by which acitretin exerts its biological effect may therefore differ significantly from the retinoids that directly transactivate the RARs. A mechanism by which acitretin exerts its biological effects could potentially involve an indirect signaling pathway by displacing endogenous retinoic acid complexed with the CRABPs. The effect of acitretin on the partitioning of [^{3}H] all-*trans*-retinoic acid was therefore examined. There was a large shift in the distribution of radioactivity between CRABP-I and the nuclear receptor complexes upon the addition of a stoichiometric amount of acitretin, increasing the amount associated with the nuclear receptors to 82% and reducing the amount associated with CRABP-I to 14%. A similar shift was observed from CRABP-II to the nuclear receptors. Mutant mice lacking CRABP-I and/or CRABP-II [12–15] probably represent the most definitive experimental system for examining whether the CRABP's play a significant role in acitretin signaling and toxicity. While these mice have been shown not to have increased susceptibility to the teratogenic effects of excess retinoic acid, they may exhibit reduced susceptibility to the teratogenic effects of acitretin.

Differences in the interaction of ^{13}C labeled all-*trans*-retinoic acid with CRABP-I and CRABP-II were detected using isotope-directed NMR methods [16], which could potentially assist in the rational design of selective ligands. Specifically, there were differences noted in the 6-s bond torsion angle between the ring moiety and the polyene chain and in protein-ligand close contacts at the ring moiety. These differences were observed in ^{13}C-filtered ^{1}H NOESY spectra of CRABP-I and CRABP-II which reveal close through-space interactions between protons directly attached to the enriched carbon of the bound retinoic acid, and other protons on the ligands and protons of neighboring amino-acid side-chains within the binding pocket. The measured cross-peak volumes in the NOESY spectra for CRABP-II bound ^{13}C-labeled retinoic acid were well predicted by a single, static conformation having a 6-s torsion angle of –60°C skewed from a *cis*- conformation, which is in good agreement with the skewed *cis* conformation found in the crystal structure of CRABP-I [6]. In contrast no single static bound conformation was able to match the pattern of cross-peaks observed in the NOESY spectrum of CRABP-I. The pattern suggests that the 6-s-bond torsion angle exhibits time-dependent fluctuations, similar to that observed for retinoids in organic solution at room temperature [17]. There were also differences in the pattern of intermolecular NOESY cross-peaks between ^{13}C-labeled protons in the ring portion of retinoic acid and protein protons in the NOESY spectra of CRABP-I and CRABP-II which are consistent with amino acid substitutions between the two proteins located in the second helix, which has been implicated as the putative portal region for entry and exit of the ligand into the binding cavity.

Cellular retinol binding proteins

CRBP and CRBP-II are highly homologous, sharing 56% amino acid identity [1, 2]. CRBP is found in a wide variety of rat tissues, notably liver, kidney and testes but is not present in the intestinal epithelium. The localization of CRBP-II is restricted to the villus-associated columnar absorptive cells of the proximal intestinal epithelium where it represents approximately 1% of the cytosolic protein and in the perinatal liver. The question arises as to why the intestinal epithelium expresses a distinct retinol binding protein from that found in other tissues. Our approach to examining this question has been to embark on a comparative study of the ligand binding interactions of these two proteins.

Although both CRBP and CRBP-II appear to bind retinol with comparable dissociation constants in the range of 10^{-8} M by fluorometric titration in direct competition experiments, subsequent ^{19}F nuclear magnetic resonance studies of CRBP and CRBP-II isotopically labeled with 6-fluorotryptophan demonstrated that retinol complexed to CRBP-II is readily transferred to CRBP, whereas retinol complexed to CRBP is not transferred to CRBP-II. Retinal, in contrast is readily transferred in both directions [18]. Furthermore ^{19}F-NMR studies of retinol transfer between CRBP and CRBP-II and phospholipid vesicles, using either fluorine-labeled ligand or protein, demonstrated that there was significantly more transfer of retinol from CRBP-II to lipid vesicles than from CRBP [19]. That CRBP-II retinol dissociates more readily than does CRBP retinol may facilitate rapid transport of retinol across the intestinal cytosol. Since bound retinol is oriented within the binding pockets of both CRBP and CRBP-II with the alcohol moiety situated innermost, this alcohol moiety of bound retinol is not immediately accessible for modification. Differences in how readily retinol is released from the binding pocket to other proteins and lipid bilayers may potentially relate to differences observed in how these two proteins modulate retinol esterification and oxidation [1, 2].

There is increasing evidence that ligand binding is a dynamic process. ^{19}F NMR studies revealed that fluororetinol and fluororetinaldehyde, when bound to CRBP-II, and fluororetinaldehyde, when bound to CRBP, exchanges between magnetically distinct binding sites [18]. The distinct binding sites may correlate with the multiple conformations of bound retinol observed in crystalline holo-CRBP-II [5]. The dissociation constants for CRBP-retinaldehyde, CRBP-II-retinaldehyde, and CRBP-II-retinol are estimated to be in the range of 10^{-7} to 10^{-1} M . In contrast, for CRBP-retinol, which has a dissociation constant of 10^{-10} M, exchange between magnetic subsites was not detected. Thus the conformational dynamics of retinol binding differ between CRBP-retinol and CRBP-II-retinol, and may correlate with the avidity of binding.

Similarly, the conformation of the protein may also undergo time-dependent fluctuations about an average structure. We have embarked on a more detailed NMR analysis of the solution structures of apo and holo CRBP-II, using heteronuclear multidimensional NMR methods to study uniformly enriched ^{13}C, ^{15}N enriched recombinant rat CRBP-II. Preliminary results indicate that the chemical shift-derived secondary structures of holo- and apo-CRBP-

II are predominantly beta-strand. Of note, we were unable to detect the second helix from the chemical shift-derived secondary structure of either holo or apo CRBP-II.

Nuclear magnetic resonance studies of the ligand binding interactions of the cytoplasmic retinoid binding proteins is a dynamic process. There are a number of differences in the ligand binding interactions of CRBP and CRBP-II which could potentially relate to how these two proteins modulate intracellular trafficking.. The question remains to be answered as to why the intestinal epithelium contains a retinol binding protein that is distinct from the retinol binding protein found in other tissues. Experiments overexpressing these proteins in intestinal epithelial cells indicate that both these proteins facilitate uptake and esterification of retinol [20, 21]. Induction of overexpression of CRBP in the intestinal epithelium and the liver in transgenic mice under the control of the metallothionine promoter produced no detectable alteration in retinyl ester storage [22]. However loss-of function experiments parallel to those performed for the CRABP's remain to be reported. Efforts are currently underway in our laboratory to target disruption of the CRBP-II gene to determine its role in the intestinal absorption and esterification of vitamin A.

Acknowledgements

This work was supported by grants DK 40172 and DK 49684 from the National Institutes of Health. E.L. is a Burroughs Wellcome Scholar in Toxicology.

References

1. Li E, Norris AW: Structure/function of cytoplasmic vitamin A-binding proteins. Ann Rev Nutr 16: 205–234, 1996
2. Ong DE, Newcomer ME, Chytil F: Cellular retinoid-binding proteins. In: M.B. Sporn, A.B. Roberts, D.S. Goodman (eds). The Retinoids: Biology, Chemistry, and Medicine, Raven Press, New York, 1994, pp 283–317
3. Mangelsdorf DJ, Evans RM: The RXR heterodimers and orphan receptors (review). Cell 83: 841–850, 1995
4. Cowan SW, Newcomer ME, Jones TA: Crystallographic studies on a family of cellular lipophilic transport proteins. J Mol Biol 230: 1225–1246, 1993
5. Winter NS, Bratt JM, Banaszak LJ: The crystal structure of holo- and apo-cellular retinol binding protein II. J Mol Biol 230: 1247–1259, 1993
6. Kleywegt GJ, Bergfors T, Senn H, Le Motte P, Gsell B, Studo K, Jones TA: Crystal structure of cellular retinoic acid binding proteins I and II in complex with all-*trans*-retinoic acid and a synthetic retinoid. Structure 2: 1241–1258, 1994
7. Thompson JR, Bratt JM, Banaszak LJ: Crystal structures of cellular retinoic acid binding protein I show increased access to the binding cavity due to formation of an intermolecular β-sheet. J Mol Biol 252: 433–436, 1995
8. Norris AW, Cheng L, Giguère V, Rosenberger M, Li, E: Measurement of sub-nanomolar retinoic acid binding affinities for cellular retinoic acid binding proteins by fluorometric titration. Biochem Biophys Acta 1209: 10–18, 1994
9. Tian K-P, Norris AW, Lin C-LS, Li E: The isolation and characterization of purified heterocomplexes of recombinant retinoic acid receptor and retinoid X receptor ligand binding domains. Biochemistry 36: 5669–5676, 1997
10. Boylan JF, Gudas LJ: Overexpression of the cellular retinoic acid binding protein I (CRABP-I) results in a reduction in differentiation-specific gene expression in F9 teratocarcinoma stem cells. J Cell Biol 112: 965–979, 1991
11. Venepally P, Reddy LG, Sani BP: Analysis of the effects of CRABP-I expression on the RA-induced transcription mediated by retinoid receptors. Biochemistry 35: 9974–9982, 1996
12. De Bruijn DRH, Oerlemans F, Hendriks W, Baats E, Ploemacher R, Wieringa B, Guerts van Kessel: Normal development, growth and reproduction in cellular retinoic acid binding protein-I (CRABP-I) null mutant mice. Differentiation 58: 141–148, 1994
13. Gorry P, Luffin T, Dierich A, Rochette-Egly C, Décimo D, Dollè P, Mark M, Durand B, Chambon P: The cellular retinoic acid binding protein I is dispensable. Proc Natl Acad Sci USA 91: 9032–9036, 1994
14. Fawcett D, Pasceri P, Fraser R, Colbert M, Rossant J, Gigurère V: Postaxial polydactyly in forelimbs of CRABP-II mutant mice. Development 121: 671–679, 1995
15. Lampron C, Rochette-Egly C, Gorry P, Dollé P, Mark M, Lufkin T, LeMeur M, Chambon P: Mice deficient in cellular retinoic acid binding protein II (CRABP-II) or in both CRABP-I and CRABP-II are essentially normal. Development 121: 539–548, 1995
16. Norris AW, Rong D, d'Avignon DA, Rosenberger M, Tasaki K, Li E: Nuclear magnetic resonance studies demonstrate differences in the interaction of retinoic acid with two highly homologous cellular retinoic acid binding proteins. Biochemistry 34: 15564–15573, 1995
17. Honig B, Hudson B, Sykes BD, Karplus M: Ring orientation in β-ionone and retinals. Proc Natl Acad Sci USA 68: 1289–1293, 1971
18. Li E, Qian S, Yang NC, d'Avignon A, Levin MS, Gordon JI: Fluorine nuclear magnetic resonance analysis of the ligand binding properties of two homologous rat cellular retinol binding proteins expressed in *Escherichia coli*. J Biol Chem 265: 11549–11554, 1991
19. Rong D, Lin C-LS, d'Avignon DA, Lovey AJ, Rosenberger M, Li E: ^{19}F-NMR studies of retinol transfer between cellular retinol binding proteins and phospholipid vesicles. FEBS Lett 402: 116–120
20. Levin MS: Cellular retinol binding proteins are determinants of retinol uptake and metabolism in stably transfected Caco-2 cells. J Biol Chem 268: 8267–8276, 1993
21. Lissoos TW, Davis AE, Levin MS: Vitamin A trafficking in Caco-2 cells stably transfected with cellular retinol binding proteins. Am J Physiol 268: G224–G231, 1995
22. Troen G, Eskild W, Fromm SH, Reppe S, Nilsson A, Norum KR, Blomhoff R: Retinyl ester storage is normal in transgenic mice with enhanced expression of cellular retinol binding protein I. J Nutr 123: 2709–2719, 1996

Molecular and Cellular Biochemistry **192:** 109–121, 1999.

A comparative study of the backbone dynamics of two closely related lipid binding proteins: Bovine heart fatty acid binding protein and porcine ileal lipid binding protein

Christian Lücke,[1] David Fushman,[2] Christian Ludwig,[1] James A. Hamilton,[3] James C. Sacchettini[4] and Heinz Rüterjans[1]

[1]*Institut für Biophysikalische Chemie, J.W. Goethe-Universität, Frankfurt, Germany;* [2]*The Rockefeller University, New York, NY;* [3]*Department of Biophysics, Boston University School of Medicine, Boston, MA;* [4]*Department of Biochemistry and Biophysics, Texas A&M University, College Station, TX, USA*

Abstract

The backbone dynamics of bovine heart fatty acid binding protein (H-FABP) and porcine ileal lipid binding protein (ILBP) were studied by ^{15}N NMR relaxation (T_1 and T_2) and steady state heteronuclear $^{15}N\{^1H\}$ NOE measurements. The microdynamic parameters characterizing the backbone mobility were determined using the 'model-free' approach. For H-FABP, the non-terminal backbone amide groups display a rather compact protein structure of low flexibility. In contrast, for ILBP an increased number of backbone amide groups display unusually high internal mobility. Furthermore, the data indicate a higher degree of conformational exchange processes in the μsec-msec time range for ILBP compared to H-FABP. These results suggest significant differences in the conformational stability for these two structurally highly homologous members of the fatty acid binding protein family. (Mol Cell Biochem **192**: 109–121, 1999)

Key words: lipid binding proteins, ^{15}N relaxation, protein backbone dynamics, model-free approach

Abbreviations: H-FABP – heart fatty acid binding protein; HSQC – heteronuclear single-quantum coherence; HTQC – heteronuclear triple-quantum coherence; I-FABP – intestinal fatty acid binding protein; ILBP – ileal lipid binding protein; LBP – lipid binding protein; NOE – nuclear Overhauser effect; NOESY – nuclear Overhauser enhancement and exchange spectroscopy; TOCSY – total correlation spectroscopy

Introduction

Protein dynamics is an important aspect of the relationship between structure and function found for most biochemical processes. Intracellular lipid binding proteins (LBPs) comprise a large number of relatively small (15 kDa) ligand-binding proteins with diverging amino acid sequences, yet very homologous tertiary structures (for more recent, comprehensive reviews see [1, 2]). Many of these structures have been solved by x-ray crystallography [3–15], and more recently several solution structures have been determined by NMR spectroscopy [16–19]. While some of the differences in binding specificity could be explained in part from the crystal structure data by side-chain replacements inside the binding cavity [20–21], the mechanisms of ligand entry and exit still remain poorly understood.

The solution structure of ILBP [17], a bile acid-binding member of the LBP family [22], displays the same tertiary fold that characterizes the entire protein family. A 10-stranded antiparallel β-sheet forms the clam-like structure that is

Address for offprints: H. Rüterjans, Institut für Biophysikalische Chemie, J.W. Goethe-Universität, Biozentrum, N230, 1. OG., Marie-Curie-Straße 9, 60439 Frankfurt a.M., Germany

covered on one side by two short α-helices. However, compared to other fatty-acid binding proteins (FABPs) that have been studied in solution, a significant difference in the hydrogen exchange rates of the backbone amide protons of ILBP indicated differences in protein stability and dynamics [17]. These results suggested that, in addition to variations in the amino acid sequence, also dynamical aspects of the LBP structures could affect their respective binding properties. A systematic comparative study of the variations in protein dynamics among LBPs is thus required to obtain a better insight into the relationship between their function and dynamics.

Molecular dynamics influence nuclear spin relaxation properties and can therefore be investigated by NMR spectroscopy. In recent years, isotope labeling made these techniques available for studies of large biomolecules such as proteins (e.g. [23, 24]) and fragments of nucleic acids [25, 26]. In this study we have focused on determining and comparing the backbone dynamics of bovine H-FABP and porcine ILBP using ^{15}N relaxation measurements, since the H-D exchange data [17] had suggested differences in structural stability for these two proteins.

Materials and methods

Both proteins were expressed in M9 minimal medium with $^{15}NH_4Cl$ as sole nitrogen source and purified according to procedures reported previously [16, 17]. The NMR samples consisted of nondelipidated protein (2–3 mM) in phosphate buffer (10% D_2O) at pH 5.8 and 5.0 for H-FABP and ILBP, respectively.

All NMR experiments were performed on Bruker DMX-800, DMX-600 and DMX-500 spectrometers at 310 K. The spectra were recorded in the phase-sensitive mode using time-proportional phase incrementation. Unless stated otherwise, the ^{1}H carrier was set to the H_2O resonance frequency and the ^{15}N carrier to the center of the amide region. Decoupling of the ^{15}N nuclei during acquisition was achieved by a GARP sequence [27]. Chemical shifts were referenced to external 3-trimethylsilyl[2,2,3,3-$^{2}H_4$]-propionate for protons and to external 2.9 M $^{15}NH_4Cl$ in 1 M HCl (24.93 ppm relative to $^{15}NH_3$ at 310 K) for nitrogen.

For the ^{15}N assignment of ILBP, heteronuclear 2D $^{1}H/^{15}N$ HSQC, HTQC, HSQC-TOCSY and HSQC-NOESY experiments were collected at 800, 600 and 500 MHz. The TOCSY spin-lock times ranged from 60–80 msec, the NOESY mixing time was set to 200 msec. The water resonance was suppressed during the relaxation delay either by low-power presaturation or by an off-resonance DANTE sequence [28], depending on the actual ^{1}H carrier frequency. The HSQC and HTQC experiments at 600 and 500 MHz were performed with the ^{1}H carrier in the center of the NH and NH_2 regions, respectively. At 800 MHz, the spectral widths were 11161 Hz (^{15}H) and 4865 Hz (^{15}N). At 600 MHz, the spectral widths were 8389 Hz (^{1}H) and 4257 Hz (^{15}N), except for the HSQC (4195 Hz/4257 Hz) and HTQC (2097 Hz/1216 Hz) spectra. At 500 MHz, the spectral widths were 7062 Hz (^{1}H) and 3041 Hz (^{15}N), except for the HSQC (3754 Hz/3041 Hz) and HTQC (1751 Hz/1115 Hz) spectra. All experiments were acquired with 512 t_1 increments of 2048 data points in t_2, except for the HTQC spectra (512 t_1 increments with 1024 data points at 600 MHz; 256 t_1 increments with 512 data points at 500 MHz).

The T_1- and T_2-measurements were performed according to known pulse schemes [29] with a modified water suppression using the water flip-back technique. For the relaxation rates, a series of 8 spectra with relaxation times between 20–800 msec for T_1 and 15–185 msec for T_2 were collected at 600 and 500 MHz. The steady-state heteronuclear $^{15}N\{^{1}H\}$ NOE measurements were performed at 600 MHz only: two 2D spectra, one with and the other without presaturation of the amide hydrogens, were recorded applying the water flip-back method of Grzesiek and Bax [30]. In order to suppress time- or temperature-dependent effects, the spectra at 500 MHz were acquired as pseudo-3D experiments with incremented relaxation delays. All experiments were recorded with 256 t_1 increments of 2048 data points in t_2. The spectral widths were 7003 Hz (^{1}H) and 1773 Hz (^{15}N) at 500 MHz and 8389 Hz (^{1}H) and 2129 Hz (^{15}N) at 600 MHz.

Data acquisition and processing was achieved by using the Bruker software packages XWINNMR and AURELIA. The analysis of the relaxation data was performed with the computer program RELAXFIT [31]. Subsequently, microdynamic parameters of the backbone NH vectors (the generalized order parameter S and the correlation time of local motion τ_{loc}) and the conformational exchange contribution to the transverse relaxation rate (R_{ex}) were determined with the computer program DYNAMICS [31], which performs a screening of various models for the spectral density function based on the 'model-free' approach of Lipari and Szabo [32–34]. The order parameter represents (on a scale from 1 to 0) an amplitude of the nanosecond-subnanosecond backbone mobility: S = 1 corresponds to a completely restricted reorientation of an amide NH vector, while S = 0 is expected when the motion is unrestricted. Nonzero values of R_{ex} are expected in those cases where processes of conformational and/or chemical exchange are present on a μsec-msec time scale.

The following general model-free form of the correlation function of local motion was used (see [31] for a detailed description of all models for the spectral density function):

$$C(t) = S^2 + (S_{fast}^2 - S^2)\exp(-t/\tau_{loc}) + (1 - S_{fast}^2)\exp(-t/\tau_{fast}),$$

where S_{fast} and τ_{fast} refer to the order parameter and the correlation time of a 'fast' component of local motion in those cases where the 'extended' approach [34] was selected ($\tau_{fast} << \tau_{loc} << \tau_c$). Note that $S_{fast} = 1$ corresponds to the original two-parameter model-free treatment [32, 33].

Results and discussion

The ^{15}N and ^{1}H resonance assignments for bovine H-FABP have been reported previously [16]. Prior to determining the ^{15}N relaxation rates for porcine ILBP, the ^{15}N resonance assignments were obtained with the help of heteronuclear 2D ^{1}H/^{15}N HSQC (Fig. 1), HTQC, HMQC-TOCSY and HSQC-NOESY experiments at 500, 600 and 800 MHz. The amide ^{15}N and ^{1}H chemical shift values for non-delipidated porcine ILBP are listed separately in Table 1.

From the relaxation experiments, the ^{15}N T_1- and T_2-relaxation times and the steady-state ^{15}N{^{1}H} heteronuclear nuclear Overhauser effect were determined for both bovine H-FABP and porcine ILBP. A total of 119 and 115 sufficiently well resolved ^{1}H/^{15}N-correlation cross peaks were observed in H-FABP and ILBP, respectively.

The three-dimensional structures of the members of the LBP family have all been described as β-clams' or 'flattened β-barrels' with a water-filled cavity inside. This implies that their overall molecular shape is not spherical, but rather

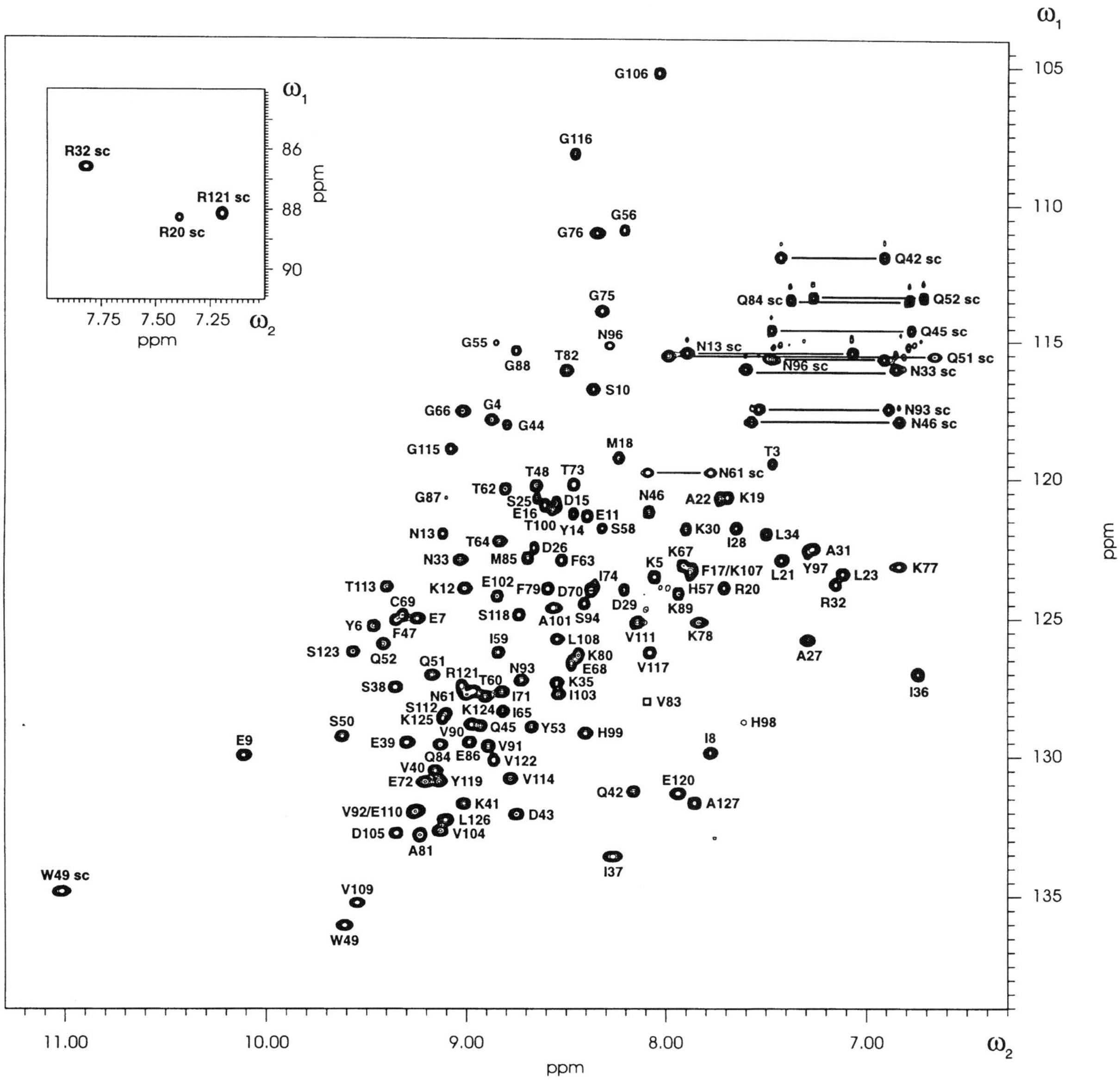

Fig. 1. ^{1}H/^{15}N HSQC spectrum of porcine ILBP at 600 MHz. Both backbone and side-chain (sc) assignments are shown. The proton resonances of each side-chain NH_2 group are connected by a straight line. The insert in the upper left corner shows the arginine NεH correlations.

Table 1. ^{15}N and ^{1}H assignments for the amide groups of ILBP. Side-chain amides are marked with 'sc'. Non-assigned resonances are marked 'n.a.'

Residue	N ppm	HN ppm	Residue	N ppm	HN ppm	Residue	N ppm	HN ppm
A1	n.a.	n.a.	G44	118.0	8.79	M85	122.8	8.68
F2	n.a.	8.18	Q45	128.9	8.92	E86	129.4	8.98
T3	119.5	7.47	Q45 sc	114.5	7.46/6.77	G87	120.7	9.09
G4	117.8	8.87	N46	121.2	8.08	G88	115.4	8.74
K5	123.5	8.05	N46 sc	117.9	7.57/6.83	K89	124.1	7.93
Y6	125.3	9.46	F47	125.0	9.33	V90	128.8	8.98
E7	125.0	9.23	T48	120.2	8.64	V91	129.6	8.87
I8	129.8	7.76	W49	136.0	9.62	V92	132.0	9.26
E9	129.9	10.12	W49 sc	134.8	11.03	N93	127.2	8.72
S10	116.7	8.36	S50	129.2	9.63	N93 sc	117.5	7.53/6.87
E11	121.3	8.39	Q51	127.0	9.15	S94	124.4	8.41
K12	123.9	9.02	Q51 sc	115.6	8.01/6.64	P95	n.a.	–
N13	122.0	9.11	Q52	125.9	9.40	N96	115.1	8.26
N13 sc	115.4	7.89/7.06	Q52 sc	113.4	7.26/6.71	N96 sc	115.7	7.47/6.90
Y14	121.2	8.46	Y53	128.9	8.70	Y97	122.6	7.29
D15	120.8	8.54	P54	n.a.	–	H98	128.8	7.63
E16	121.0	8.60	G55	115.1	8.85	H99	129.2	8.41
F17	123.4	7.88	G56	110.9	8.20	T100	121.1	8.55
M18	119.3	8.23	H57	123.3	7.88	A101	124.6	8.57
K19	120.7	7.69	S58	121.8	8.26	E102	124.2	8.85
R20	123.9	7.69	I59	126.2	8.83	I103	127.7	8.52
R20 sc	88.3	7.39	T60	127.6	8.95	V104	132.7	9.12
L21	122.9	7.40	N61	127.7	8.99	D 105	132.7	9.34
A22	120.7	7.74	N61 sc	119.8	8.13/7.80	G 106	105.2	8.02
L23	123.4	7.11	T62	120.3	8.80	K107	123.2	7.86
P24	n.a.	–	F63	122.9	8.51	L108	125.7	8.53
S25	120.7	8.64	T64	122.3	8.82	V109	135.2	9.55
D26	122.5	8.65	I65	128.3	8.82	EI 10	132.0	9.24
A27	125.7	7.29	G66	117.5	9.02	V111	125.1	8.16
I28	121.8	7.64	K67	123.1	7.90	S112	128.5	9.09
D29	124.0	8.20	E68	126.6	8.48	T113	123.8	9.43
K30	121.8	7.89	C69	124.9	9.31	V114	130.7	8.79
A31	122.5	7.27	D70	124.0	8.35	G115	118.9	9.08
R32	123.8	7.14	I71	127.7	8.82	G116	108.2	8.45
R32 sc	86.6	7.82	E72	130.9	9.22	V117	126.2	8.08
N33	122.9	9.03	T73	120.2	8.46	S118	124.9	8.73
N33 sc	116.0	7.59/6.85	I74	123.6	8.33	Y119	130.9	9.14
L34	122.0	7.50	G75	113.8	8.31	E120	131.3	7.93
K35	127.3	8.53	G76	111.0	8.33	R121	127.4	9.02
I36	127.0	6.74	K77	123.1	6.84	R121 sc	88.1	7.20
I37	133.5	8.26	K78	125.1	7.85	V122	130.1	8.85
S38	127.5	9.35	F79	123.9	8.60	S123	126.2	9.57
E39	129.4	9.29	K80	126.3	8.43	K124	127.8	8.89
V40	130.5	9.16	A81	132.8	9.22	K125	128.6	9.11
K41	131.7	9.02	T82	116.1	8.48	L126	132.2	9.11
Q42	131.2	8.16	V83	128.2	8.10	A127	131.6	7.85
Q42 sc	111.9	7.42/6.90	Q84	129.6	9.12			
D43	132.0	8.74	Q84 sc	113.4	7.37/6.78			

resembles an oblate ellipsoid. The ratio of the principal components of the rotational diffusion tensor obtained from hydrodynamic calculations using the bead method [35] is 1.00:0.92:0.88 for H-FABP and 1.00:0.91:0.81 for ILBP. This suggests a small degree of anisotropy of the overall rotation in both cases. Although such an anisotropy can, in principle, be detected by NMR relaxation measurements [36], the precision of the present data is probably not sufficient to distinguish between the isotropic and anisotropic models of the overall motion. Therefore, the overall rotation was assumed to be isotropic for both proteins throughout the calculations. An overall rotational correlation time (τ_c) of 5.8 nsec was derived for both proteins from the relaxation data at both 500 and 600 MHz by using the optimization procedure described in [31] and taking into account only those residues which belong to well-defined elements of the secondary structure.

The relaxation data (T_1, T_2, and NOE) are listed in Table 2. A comparison of the T_1 values between H-FABP (excluding the terminal residues Met0, Val1, Gln131 and Ala132) and ILBP (excluding the 'extreme' residues Ile74, Val83 and the terminal residue Ala127) shows similar average T_1 values in H-FABP (534 ± 29 msec) and in ILBP (544 ± 49 msec). However, in ILBP a greater number of non-terminal residues display T_1 values > 600 msec (Ile8, Ile36, Glu68, Ile74, Val83, Tyr97 and Val111), while in H-FABP this is

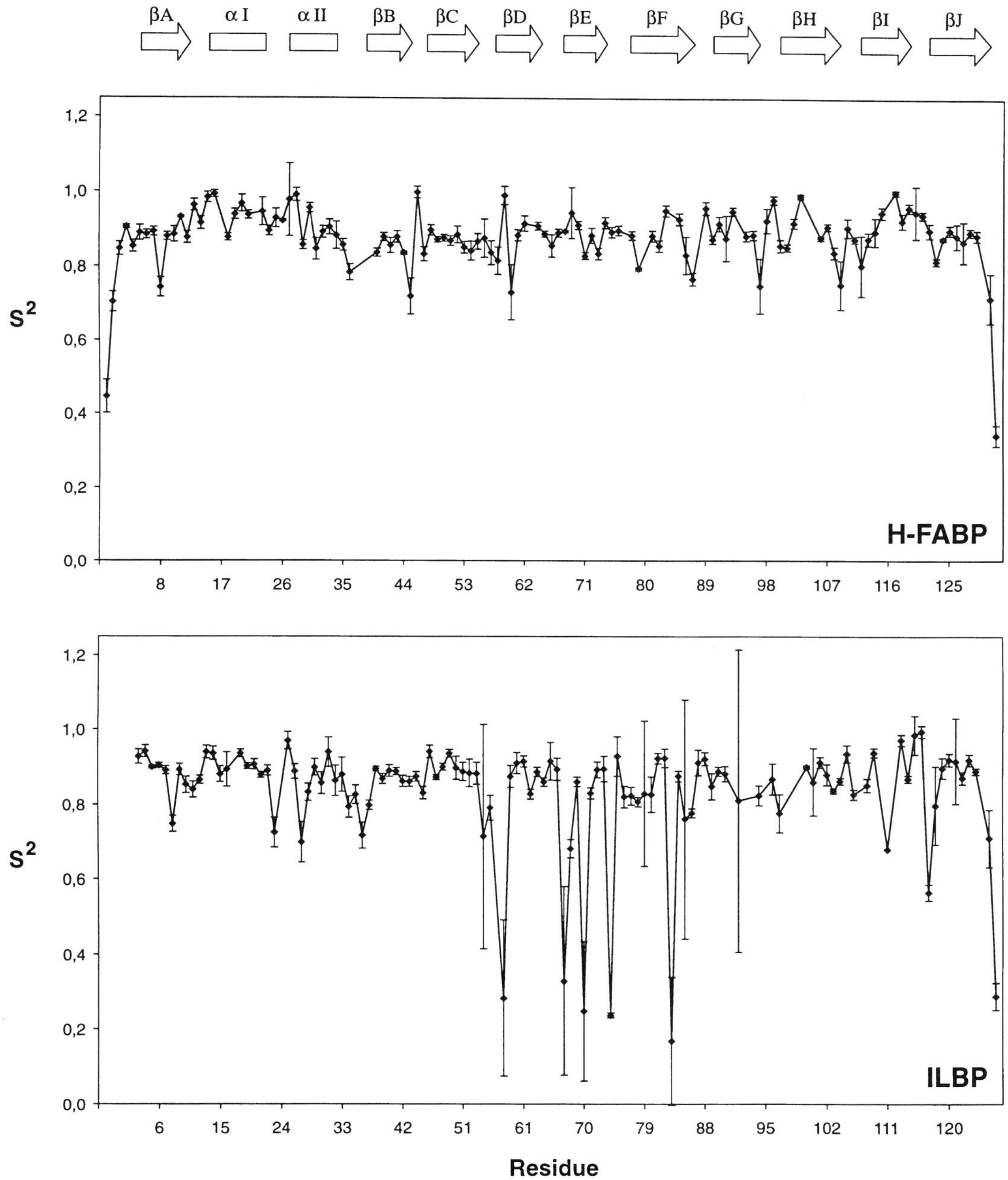

Fig. 2. Backbone order parameter (S^2) vs. amino acid sequence of H-FABP (upper panel) and ILBP (lower panel). The error bars represent standard deviations. The secondary structure elements (β-strands A through J as well as α-helices I and II) are representative for the H-FABP sequence. The residues in ILBP have been aligned with the corresponding residues in H-FABP, as indicated in Table 2. The y-axis scaling is the same for both graphs. For ILBP, a larger number of non-terminal residues show rather low S^2 values, indicating increased internal mobility. H-FABP, on the other hand, displays a rather uniform stability throughout the backbone structure.

Table 2. [15]N relaxation data for bovine H-FABP and porcine ILBP. The numbers in parentheses indicate standard deviations. Sequentially corresponding residues (AA) have been aligned. Residues for which no value could be calculated are marked 'n.v.'

	FABP				ILBP		
AA	T_1 msec	T_2 msec	NOE –	AA	T_1 msec	T_2 msec	NOE –
M0	919.2 (273.1)	220.3 (24.1)	0.24 (0.14)				
V1	588.4 (44.5)	157.6 (7.7)	0.51 (0.12)				
D2	567.8 (33.6)	125.6 (5.7)	0.72 (0.02)				
A3	537.0 (30.7)	123.6 (3.9)	0.82 (0.02)	A1			
F4	547.8 (10.7)	138.4 (2.8)	0.80 (0.02)	F2			
V5	528.4 (12.7)	130.2 (7.2)	0.85 (0.02)	T3	513.0 (48.5)	121.3 (7.1)	0.76 (0.02)
G6	528.9 (8.6)	131.4 (3.8)	0.84 (0.03)	G4	509.0 (18.9)	117.5 (5.7)	0.78 (0.03)
T7	551.1 (17.7)	130.3 (4.1)	0.80 (0.02)	K5	539.1 (8.6)	122.9 (2.1)	0.81 (0.02)
W8	546.5 (16.0)	141.2 (4.4)	0.76 (0.03)	Y6	527.2 (12.1)	124.1 (4.1)	0.78 (0.03)
K9	542.0 (7.4)	133.1 (3.6)	0.82 (0.03)	E7	524.2 (16.6)	128.8 (4.4)	0.78 (0.03)
L10	524.4 (12.6)	138.0 (4.5)	0.82 (0.02)	I8	613.5 (18.6)	142.2 (6.7)	0.54 (0.04)
V11	518.4 (13.6)	119.9 (2.4)	0.80 (0.03)	E9	503.9 (18.7)	129.9 (3.2)	0.80 (0.03)
D12	558.5 (20.2)	123.4 (4.1)	0.73 (0.02)	S10	555.0 (14.7)	123.2 (1.7)	0.79 (0.03)
S13	533.3 (54.6)	114.3 (5.9)	0.79 (0.02)	E11	566.2 (28.8)	117.2 (2.1)	0.83 (0.02)
K14	537.3 (26.4)	124.5 (6.6)	0.83 (0.03)	K12	560.4 (13.3)	126.9 (4.3)	0.79 (0.02)
N15	489.6 (66.6)	105.8 (3.2)	0.79 (0.03)	N13	525.3 (18.3)	115.5 (4.1)	0.78 (0.02)
F16	477.9 (31.7)	115.6 (4.9)	0.83 (0.03)	Y14	495.1 (29.1)	112.3 (6.1)	0.81 (0.02)
D17	456.5 (44.7)	108.9 (4.0)	0.86 (0.03)	D15	526.9 (13.0)	111.4 (1.8)	0.82 (0.02)
D18	577.8 (19.3)	129.4 (2.9)	0.81 (0.02)	E16	527.5 (11.0)	115.5 (2.0)	0.76 (0.02)
Y19	510.3 (15.0)	120.4 (5.2)	0.82 (0.03)	F17			
M20	521.0 (38.2)	119.4 (4.2)	0.85 (0.07)	M18	508.9 (9.2)	115.1 (4.4)	0.87 (0.02)
K21	492.9 (10.7)	120.5 (2.9)	0.80 (0.03)	K19	519.6 (9.1)	117.2 (5.7)	0.85 (0.03)
S22				R20	513.3 (7.6)	121.5 (2.5)	0.75 (0.03)
L23	441.4 (70.1)	92.9 (2.5)	0.77 (0.05)	L21	540.3 (9.9)	122.9 (4.3)	0.80 (0.03)
G24	525.5 (16.0)	133.8 (10.3)	0.76 (0.04)	A22	529.6 (9.5)	92.1 (1.7)	0.72 (0.03)
V25	491.4 (11.6)	126.9 (4.3)	0.82 (0.06)	L23	544.4 (7.1)	147.4 (2.5)	0.69 (0.03)
G26	519.9 (27.2)	120.7 (2.7)	0.78 (0.05)	P24			
F27	480.1 (109.2)	102.4 (5.8)	0.83 (0.05)	S25	511.5 (51.6)	113.1 (5.2)	0.72 (0.05)
A28	464.8 (70.7)	993 (4.2)	0.85 (0.03)	D26	491.5 (38.3)	117.7 (6.2)	0.69 (0.02)
T29	555.7 (35.6)	131.8 (1.7)	0.81 (0.03)	A27	535.8 (15.8)	148.8 (2.2)	0.71 (0.03)
R30	533.1 (23.8)	120.5 (6.6)	0.86 (0.04)	I28	559.9 (12.7)	135.6 (3.3)	0.73 (0.02)
Q31	536.6 (18.8)	136.9 (2.2)	0.84 (0.02)	D29	523.8 (14.5)	120.4 (1.9)	0.68 (0.01)
V32	543.4 (17.7)	129.4 (5.1)	0.78 (0.06)	K30	525.1 (12.3)	112.0 (5.7)	0.69 (0.02)
G33	520.7 (14.4)	129.1 (4.8)	0.98 (0.08)	A31	545.6 (14.6)	112.8 (2.5)	0.76 (0.03)
N34	549.2 (20.2)	119.2 (4.6)	0.80 (0.07)	R32	536.0 (22.3)	109.3 (5.1)	0.73 (0.02)
M35	564.6 (17.3)	134.9 (6.6)	0.84 (0.03)	N33	539.9 (23.8)	108.2 (2.2)	0.76 (0.02)
T36	582.2 (29.7)	151.4 (6.1)	0.77 (0.06)	L34	574.1 (15.2)	127.6 (4.7)	0.70 (0.02)
K37				K35	555.3 (17.1)	113.5 (2.7)	0.72 (0.02)
P38				I36	633.2 (16.5)	120.0 (4.8)	0.63 (0.02)
T39				I37	587.7 (22.7)	136.0 (7.8)	0.65 (0.03)
T40	555.0 (27.2)	135.8 (2.1)	0.77 (0.05)	S38	525.2 (22.4)	122.8 (2.8)	0.80 (0.03)
I41	549.5 (18.6)	127.8 (2.8)	0.81 (0.04)	E39	528.8 (12.6)	131.9 (4.1)	0.77 (0.02)
I42	537.6 (13.2)	140.2 (3.6)	0.81 (0.03)	V40	525.7 (8.5)	133.5 (7.0)	0.81 (0.03)
E43	546.3 (6.0)	134.2 (5.1)	0.77 (0.02)	K41	534.8 (14.5)	124.3 (4.3)	0.79 (0.03)
V44	579.4 (9.6)	134.3 (4.3)	0.79 (0.02)	Q42	540.6 (28.4)	127.2 (5.1)	0.77 (0.06)
N45	578.6 (14.0)	148.7 (3.2)	0.72 (0.02)	D43	542.4 (11.1)	131.4 (3.3)	0.78 (0.02)
G46	502.3 (99.1)	109.1 (2.9)	0.82 (0.03)	G44	514.4 (69.0)	112.7 (5.5)	0.77 (0.02)
D47	520.0 (59.1)	135.7 (6.0)	0.78 (0.02)	Q45	546.7 (35.0)	131.9 (4.7)	0.76 (0.02)
T48	546.5 (33.3)	127.0 (4.7)	0.76 (0.02)	N46	515.2 (27.4)	117.5 (4.3)	0.79 (0.05)
V49	548.8 (7.8)	134.2 (4.3)	0.82 (0.03)	F47	549.1 (15.4)	125.9 (6.1)	0.79 (0.01)
I50	547.6 (5.3)	133.4 (3.0)	0.79 (0.02)	T48	548.9 (12.2)	122.6 (2.6)	0.81 (0.03)
I51	548.1 (15.6)	134.0 (2.5)	0.82 (0.02)	W49	516.7 (14.1)	115.1 (3.3)	0.81 (0.02)
K52	548.0 (17.4)	130.3 (3.9)	0.88 (0.05)	S50	532.1 (14.2)	87.4 (2.5)	0.79 (0.05)
T53	552.5 (33.7)	137.9 (3.9)	0.78 (0.03)	Q51	541.7 (19.3)	111.2 (4.9)	0.80 (0.02)
Q54	538.7 (24.8)	140.4 (8.3)	0.75 (0.07)	Q52	536.3 (16.4)	79.7 (1.8)	0.79 (0.06)
S55	500.5 (33.8)	136.9 (5.9)	0.85 (0.07)	Y53	530.1 (15.5)	112.0 (3.4)	0.75 (0.03)
				P54			

Table 2. Continued

FABP				ILBP			
AA	T_1 msec	T_2 msec	NOE –	AA	T_1 msec	T_2 msec	NOE –
T56	506.0 (149.6)	108.6 (14.5)	0.69 (0.14)	G55	523.9 (110.1)	116.5 (6.7)	0.68 (0.02)
F57	504.1 (43.3)	132.6 (6.1)	0.72 (0.04)	G56	513.7 (62.6)	135.2 (8.7)	0.66 (0.03)
K58	566.0 (60.2)	141.6 (7.5)	0.70 (0.06)	H57			
N59	503.3 (35.6)	111.8 (5.6)	0.73 (0.04)	S58	536.0 (67.3)	124.6 (5.4)	0.65 (0.02)
T60	513.2 (14.4)	139.4 (6.0)	0.77 (0.07)	I59	540.2 (15.9)	119.3 (2.8)	0.77 (0.05)
E61	554.2 (16.9)	121.0 (2.9)	0.78 (0.02)	T60	540.4 (32.7)	106.8 (2.1)	0.81 (0.02)
I62	556.9 (26.8)	134.3 (5.0)	0.79 (0.04)	N61	538.5 (19.9)	124.1 (3.4)	0.79 (0.02)
S63				T62	583.7 (17.1)	126.3 (2.7)	0.82 (0.02)
F64	531.7 (8.8)	122.8 (3.7)	0.85 (0.02)	F63	539.7 (16.1)	116.8 (4.5)	0.84 (0.03)
K65	547.2 (14.0)	125.5 (2.4)	0.80 (0.03)	T64	567.3 (18.1)	129.8 (3.4)	0.75 (0.03)
L66	569.4 (46.4)	127.8 (6.3)	0.69 (0.10)	I65	512.8 (11.1)	112.8 (4.0)	0.75 (0.03)
G67	550.1 (10.3)	135.0 (7.6)	0.83 (0.03)	G66	529.9 (13.7)	115.2 (4.1)	0.78 (0.02)
V68	534.4 (28.3)	125.2 (3.7)	0.79 (0.02)	K67	533.7 (15.6)	116.9 (3.8)	0.67 (0.02)
E69	565.3 (10.5)	125.1 (3.4)	0.88 (0.02)	E68	621.1 (21.9)	75.3 (7.2)	0.46 (0.05)
F70	513.3 (17.2)	126.4 (4.2)	0.81 (0.02)	C69	525.6 (13.1)	129.4 (0.9)	0.74 (0.02)
D71	574.4 (11.9)	135.5 (1.3)	0.79 (0.02)	D70	578.4 (11.6)	120.7 (2.9)	0.70 (0.02)
E72	528.0 (12.9)	136.7 (5.1)	0.76 (0.03)	I71	566.2 (6.3)	138.2 (3.5)	0.76 (0.03)
773	564.9 (23.2)	137.1 (2.4)	0.85 (0.03)	E72	532.7 (12.2)	114.7 (3.2)	0.80 (0.05)
U4	544.9 (13.8)	129.1 (3.3)	0.83 (0.03)	T73	545.0 (18.2)	50.8 (2.7)	0.80 (0.02)
A75	521.2 (25.5)	125.8 (2.9)	0.73 (0.04)	I74	869.9 (96.2)	384.3 (30.8)	–1.39 (0.03)
D76	516.9 (20.9)	129.2 (7.4)	0.77 (0.05)	G75	508.3 (20.6)	98.1 (4.7)	0.74 (0.03)
D77				G76	545.1 (15.7)	142.9 (4.8)	0.69 (0.04)
R78	530.2 (14.9)	127.9 (2.5)	0.82 (0.03)	K77	570.6 (30.3)	130.9 (4.5)	0.69 (0.02)
K79	603.6 (10.2)	142.0 (2.6)	0.80 (0.03)	K78	581.3 (17.9)	135.7 (2.4)	0.69 (0.09)
V80				F79	545.7 (8.8)	125.7 (1.9)	0.73 (0.03)
K81	525.3 (11.9)	132.1 (2.8)	0.83 (0.03)	K80	584.0 (11.2)	82.7 (4.5)	0.69 (0.06)
S82	528.7 (15.3)	131.6 (2.8)	0.77 (0.01)	A81	519.2 (15.2)	52.1 (4.4)	0.78 (0.07)
I83	535.7 (27.7)	120.1 (3.6)	0.80 (0.02)	T82	552.7 (16.8)	119.2 (1.5)	0.73 (0.03)
V84				V83	972.1 (148.5)	238.5 (34.3)	0.55 (0.02)
T85	532.5 (19.3)	119.0 (1.9)	0.80 (0.03)	Q84	543.7 (12.6)	128.0 (4.3)	0.79 (0.07)
L86	501.2 (44.9)	128.5 (7.8)	0.57 (0.08)	M85	569.4 (8.4)	129.0 (2.2)	0.79 (0.02)
D87	596.9 (13.3)	152.0 (5.7)	0.67 (0.02)	E86	592.4 (12.8)	145.5 (6.8)	0.67 (0.02)
G88				G87	509.9 (113.7)	100.6 (6.4)	0.77 (0.03)
G89	493.1 (82.9)	110.9 (3.8)	0.71 (0.03)	G88	499.9 (62.2)	116.6 (3.0)	0.76 (0.04)
K90	540.3 (23.2)	130.8 (3.5)	0.79 (0.02)	K89	531.0 (14.5)	130.3 (4.0)	0.75 (0.04)
L91	542.8 (14.5)	127.2 (3.9)	0.88 (0.03)	V90	539.3 (12.6)	113.4 (4.8)	0.78 (0.09)
V92	551.0 (18.5)	97.2 (2.0)	0.84 (0.03)	V91	539.6 (8.1)	96.1 (4.3)	0.77 (0.03)
H93	521.3 (25.8)	119.4 (3.0)	0.81 (0.04)	V92			
V94				N93	538.6 (9.3)	120.0 (2.0)	0.79 (0.04)
Q95	516.2 (18.9)	132.3 (4.7)	0.81 (0.03)				
K96	561.5 (13.8)	130.6 (5.2)	0.77 (0.03)				
W97	540.8 (13.0)	137.2 (3.0)	0.77 (0.02)	S94	559.8 (22.4)	122.8 (1.9)	0.71 (0.02)
N98	481.7 (87.5)	104.2 (5.1)	0.74 (0.03)	P95			
G99	513.3 (70.5)	111.0 (4.7)	0.80 (0.03)	N96	549.1 (77.5)	89.9 (6.3)	0.69 (0.03)
Q100	522.2 (19.5)	130.5 (3.9)	0.73 (0.02)	Y97	604.2 (41.7)	79.6 (8.8)	0.67 (0.04)
E101	550.6 (10.5)	131.5 (2.8)	0.73 (0.03)				
T102	526.9 (12.8)	125.4 (6.4)	0.81 (0.03)				
S103	503.0 (37.1)	112.7 (3.5)	0.83 (0.04)	H98	n.v.	n.v.	n.v.
L104				H99	536.2 (25.9)	102.4 (4.1)	0.82 (0.20)
V105				T100	598.8 (43.8)	99.5 (10.01)	0.94 (0.05)
R106	537.1 (20.4)	128.8 (6.2)	0.81 (0.03)	A101	432.0 (20.3)	118.8 (3.8)	0.80 (0.06)
E107	541.6 (15.9)	122.6 (3.0)	0.79 (0.03)	E102	535.3 (16.5)	121.4 (3.8)	0.79 (0.02)
M108	566.8 (7.9)	141.9 (3.2)	0.79 (0.02)	I103	556.0 (14.8)	134.1 (1.4)	0.74 (0.12)
V109	559.3 (7.3)	143.1 (4.3)	0.77 (0.03)	V104	545.5 (12.6)	131.0 (5.5)	0.79 (0.05)
D110	493.4 (49.5)	120.0 (7.5)	0.77 (0.02)	D105	505.7 (47.8)	113.7 (4.4)	0.77 (0.02)
G111	539.2 (12.7)	123.4 (3.8)	0.81 (0.02)	G106	553.3 (15.9)	131.5 (5.7)	0.75 (0.02)
K112	527.9 (9.6)	135.6 (3.8)	0.78 (0.02)	K107			
L113	537.8 (10.5)	136.4 (3.9)	0.81 (0.03)	L108	527.8 (16.2)	132.1 (2.4)	0.73 (0.02)

Table 2. Continued

AA	FABP T_1 msec	T_2 msec	NOE –	AA	ILBP T_1 msec	T_2 msec	NOE –
I114	535.3 (13.2)	128.0 (5.9)	0.87 (0.02)	V109	536.0 (17.6)	116.0 (2.5)	0.78 (0.03)
L115	511.8 (18.1)	121.1 (4.6)	0.82 (0.03)	E110			
T116				V111	640.3 (35.9)	160.0 (9.9)	0.38 (0.03)
L117	497.9 (30.4)	111.0 (2.1)	0.81 (0.03)	S112			
T118	548.9 (37.6)	124.0 (4.3)	0.82 (0.03)	T113	431.4 (29.4)	113.8 (4.4)	0.80 (0.03)
H119	526.7 (35.5)	115.2 (3.3)	0.79 (0.03)	V114	514.7 (15.3)	129.4 (1.1)	0.82 (0.03)
G120	497.2 (93.1)	93.1 (4.5)	0.76 (0.08)	G115	505.1 (37.0)	103.2 (3.3)	0.78 (0.02)
T121	527.1 (96.7)	113.6 (5.1)	0.77 (0.04)	G116	491.6 (48.1)	107.9 (4.5)	0.78 (0.02)
A122	509.0 (28.8)	125.7 (3.4)	0.87 (0.04)	V117	531.1 (13.5)	142.7 (3.2)	0.73 (0.03)
V123	589.2 (8.1)	140.1 (2.1)	0.79 (0.02)	S118	537.6 (12.1)	134.4 (3.0)	0.81 (0.03)
C124	546.3 (14.4)	127.1 (4.5)	0.81 (0.03)	Y119	558.7 (17.9)	114.7 (2.7)	0.77 (0.04)
T125	564.0 (28.0)	123.6 (3.0)	0.81 (0.03)	E120	525.2 (11.8)	112.1 (2.5)	0.81 (0.02)
R126	552.3 (22.0)	94.6 (3.8)	0.84 (0.03)	R121	529.7 (21.6)	87.0 (1.9)	0.81 (0.05)
T127	577.2 (28.1)	113.9 (3.0)	0.82 (0.03)	V122	543.8 (19.6)	77.4 (2.1)	0.83 (0.02)
Y128	532.3 (15.9)	126.8 (6.6)	0.80 (0.03)	S123	539.8 (29.2)	121.5 (3.1)	0.81 (0.04)
E129	555.7 (9.0)	132.8 (2.0)	0.78 (0.03)	K124	524.6 (13.3)	126.6 (1.9)	0.78 (0.02)
K130				K125			
Q131	551.2 (7.8)	152.0 (3.7)	0.71 (0.02)	L126	225.6 (7.1)	140.4 (4.2)	0.75 (0.03)
A132	663.6 (7.1)	254.9 (9.6)	0.42 (0.01)	A127	548.8 (6.8)	227.5 (11.5)	0.56 (0.01)

the case only for Lys79. In H-FABP on the other hand, 13 non-terminal residues show T_1 values < 500 msec versus only 4 residues in ILBP. Thus, there appears to be a tendency to higher internal backbone mobility (hence higher T_1) in individual residues throughout the ILBP sequence. A similar effect can be observed for the average T_2 values of H-FABP (126 ± 12 msec) and ILBP (119 ± 21 msec). Again, even though the average T_2 values indicate no significant difference, in ILBP the non-terminal residues show a clearly greater spread of T_2 values than in H-FABP. Finally, the same trends were observed for the NOE values (0.80 ± 0.05 and 0.75 ± 0.07 for H-FABP and ILBP, respectively).

The microdynamic parameters (S^2, τ_{loc}, R_{ex}, and S_{fast}^2) derived from the relaxation data are shown in Table 3. The relaxation data collected at both 500 and 600 MHz were analyzed simultaneously to provide a reliable estimate of the (magnetic-field dependent) R_{ex} contribution. None of the models for the spectral density function provided a reasonably good fit for the relaxation data of Asp17 in H-FABP. For several amide groups in ILBP (residues Lys5, Ile8, Ile28, Ile37, Lys80, Thr82, His99, Val111, and Glu120), only the relaxation data at 600 MHz were used for the analysis. The NH group of His98 in ILBP was excluded from analysis because a rapid decrease in the signal intensity with longer T_2 relaxation delays (probably due to conformational exchange) yielded insufficient data.

In bovine H-FABP all non-terminal backbone amide groups show S^2 values > 0.7, with an average value of 0.89 ± 0.06 (0.89 ± 0.05, if only those residues belonging to the secondary structure elements are considered). This data suggests a rather compact protein structure of low flexibility. In contrast, even though most residues in ILBP also have S^2 > 0.7, with an average value of 0.84 ± 0.13 (0.85 ± 0.14 for the elements of secondary structure), the larger spread in the order parameter values and several residues with S^2 values well below 0.7 (a cone semiangle of > 28° assuming the wobbling-in-a-cone model for the NH-vector motion) indicate higher internal mobilities in certain regions of the ILBP structure.

A direct comparison of the order parameters of H-FABP and ILBP with respect to the secondary structure elements (Fig. 2) shows that residues of high internal mobility are distributed throughout the entire ILBP sequence, including helix II (Ala27) and β-strands A (Ile8), B (Ile36), D (Ser58), E (Asp70), F (Val83) and I (V111). Remarkably, this unusually large difference in the backbone mobilities of H-FABP and ILBP in the psec-nsec time range is consistent with the significant differences in the backbone flexibility postulated previously [17] on the basis of the extremely high divergence in the hydrogen exchange rates of ILBP and H-FABP.

A striking difference in the dynamics of the backbone structures of H-FABP and ILBP is also indicated by the conformational exchange parameter R_{ex} (Fig. 3). While for H-FABP only a few amide groups display R_{ex} values > 0, for ILBP an increased number of residues throughout the amino acid sequence show much larger exchange contributions. Thus, the data indicates a much higher degree of conformational exchange motion in the μsec-msec time range for ILBP

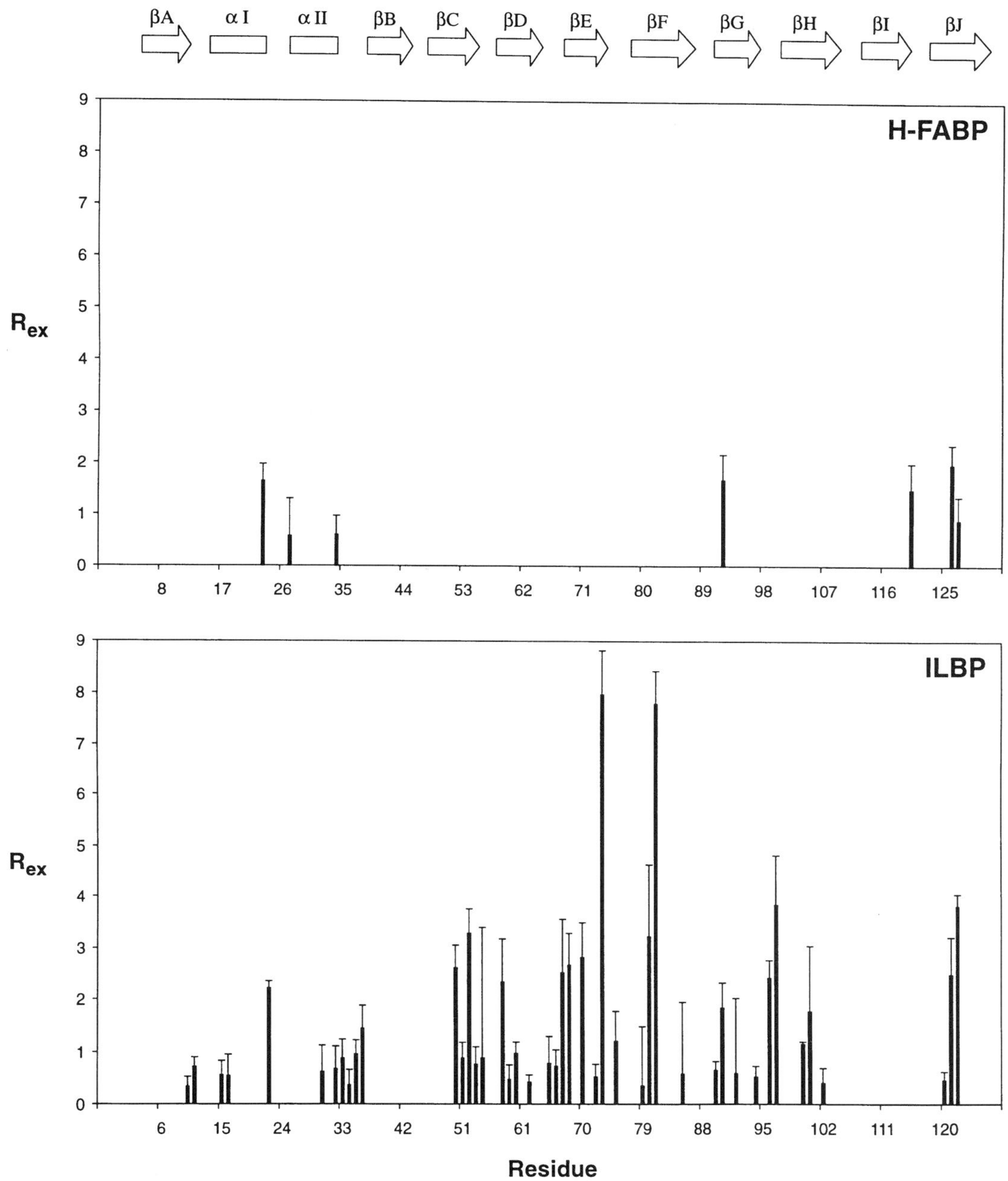

Fig. 3. Conformational exchange parameter (R_{ex}) vs. amino acid sequence of H-FABP (upper panel) and ILBP (lower panel). Standard deviations are shown as positive error bars only. The H-FABP secondary structure representation and ILBP residue alignment are the same as in Fig. 2. The y-axis scaling is the same for both graphs. The larger number and size of exchange contributions (R_{ex}) in ILBP compared to H-FABP indicate a greater extent of conformational exchange-type motions in the ILBP structure.

compared to H-FABP. This could be related to the increased hydrogen exchange rates observed for the backbone amide protons in the β-sheet structure of ILBP [17].

It should be pointed out that one of the few residues in H-FABP with a significant R_{ex} value is Arg126, which is directly involved in the ligand binding via electrostatic interactions between its own guanidinium group and the carboxylate group of the fatty acid. The corresponding residue in the ILBP sequence, Arg121, also shows a significant exchange contribution. The same appears to be true for holo I-FABP [18]. Thus, a direct correlation between ligand binding and conformational exchange motion in the backbone amide group of this arginine residue (which is highly conserved in FABPs) appears plausible.

Table 3. Microdynamic parameters for bovine H-FABP and porcine ILBP. The numbers in parentheses indicate standard deviations. Sequentially corresponding residues (AA) have been aligned. Only nonzero values of τ_{loc} and R_{ex} are shown in the table. The R_{ex} values presented here correspond to a spectrometer frequency of 500 MHz; the conformation exchange contribution at 600 MHz is greater by a factor of 1.44. In those cases where the extended model [30] was selected, a nonzero value of τ_{fast} was obtained only for G115 in ILBP. Standard deviations were left blank (–) when they exceeded the value of the microdynamic parameter. Residues for which no value could be calculated are marked 'n.v.'

		FABP					ILBP		
AA	S^2 –	τ_{loc} nsec	R_{ex} s^{-1}	S_{fast}^2 –	AA	S^2 –	τ_{loc} nsec	R_{ex} s^{-1}	S_{fast}^2 –
M0	0.45 (0.05)	0.05 (0.02)		1					
V1	0.70 (0.03)	0.08 (0.03)		1					
D2	0.85 (0.02)	0.04 (0.01)		1					
A3	0.91 (0.01)			1	A1				
F4	0.85 (0.02)			1	F2				
V5	0.89 (0.02)			1	T3	0.93 (0.02)			1
G6	0.89 (0.01)			1	G4	0.94 (0.02)			1
T7	0.89 (0.01)			1	K5	0.90 (0.00)			1
W8	0.74 (0.03)	2.78 (1.16)		0.86 (0.01)	Y6	0.90 (0.01)			1
K9	0.88 (0.01)			1	E7	0.89 (0.01)			1
L10	0.89 (0.02)			1	I8	0.75 (0.02)	0.07 (0.01)		1
V11	0.93 (0.00)			1	E9	0.89 (0.02)			1
D12	0.88 (0.02)	0.05 (0.01)		1	S10	0.85 (0.02)		0.34 (0.18)	1
S13	0.96 (0.02)			1	E11	0.84 (0.02)		0.72 (0.18)	1
K14	0.92 (0.02)			1	K12	0.87 (0.01)			1
N15	0.98 (0.01)			1	N13	0.94 (0.02)			1
F16	0.99 (0.01)			1	Y14	0.94 (0.02)			1
D17	n.v.	n.v.	n.v.	n.v.	D15	0.88 (0.02)		0.56 (0.28)	1
D18	0.88 (0.01)			1	E16	0.89 (0.05)	0.03 (0.04)	0.54 (0.41)	1
Y19	0.94 (0.01)			1	F17				
M20	0.97 (0.02)			1	M18	0.94 (0.01)			1
K21	0.94 (0.01)			1	K19	0.90 (0.01)			1
S22					R20	0.91 (0.01)			1
L23	0.95 (0.04)		1.64 (0.33)	1	L21	0.88 (0.01)			1
G24	0.90 (0.01)			1	A22	0.89 (0.01)	0.05 (0.02)	2.21 (0.14)	1
V25	0.93 (0.02)			1	L23	0.72 (0.04)	1.39 (1.00)		0.86 (0.04)
G26	0.92 (0.00)			1	P24				
F27	0.98 (0.10)		0.57(–)	1	S25	0.97 (0.02)			1
A28	0.99 (0.02)			1	D26	0.89 (0.02)	0.08 (0.03)		1
T29	0.86 (0.01)			1	A27	0.70 (0.05)	1.94 (1.23)		0.85 (0.04)
R30	0.96 (0.01)			1	I28	0.83 (0.02)	0.03 (0.01)		1
Q31	0.85 (0.03)			1	D29	0.90 (0.02)	0.11 (0.04)		1
V32	0.89 (0.02)			1	K30	0.86 (0.03)	0.06 (0.04)	0.61 (0.52)	1
G33	0.91 (0.02)			1	A31	0.94 (0.04)			1
N34	0.88 (0.04)		0.59 (0.37)	1	R32	0.86 (0.04)	0.04 (0.02)	0.67 (0.44)	1
M35	0.86 (0.02)			1	N33	0.88 (0.05)		0.88 (0.36)	1
T36	0.78 (0.02)			1	L34	0.79 (0.03)	0.04 (0.02)	0.37 (0.28)	1
K37					K35	0.83 (0.03)	0.03 (0.01)	0.96 (0.27)	1
P38					I36	0.72 (0.03)	0.04 (0.01)	1.46 (0.42)	1
T39					I37	0.80 (0.01)	0.05 (0.01)		1
T40	0.84 (0.01)			1	S38	0.90 (0.01)			1
I41	0.88 (0.01)			1	E39	0.87 (0.01)			1
I42	0.86 (0.02)			1	V40	0.89 (0.02)			1
E43	0.88 (0.02)			1	K41	0.89 (0.01)			1
V44	0.84 (0.00)			1	Q42	0.86 (0.02)			1
N45	0.72 (0.05)	1.59 (0.74)		0.83 (0.02)	D43	0.86 (0.01)			1
G46	1.00 (0.02)			1	G44	0.88 (0.01)			1
D47	0.83 (0.02)			1	Q45	0.83 (0.02)			1
T48	0.90 (0.01)			1	N46	0.94 (0.02)			1
V49	0.87 (0.01)			1	F47	0.87 (0.01)			1
I50	0.88 (0.01)			1	T48	0.90 (0.01)			1
I51	0.87 (0.01)			1	W49	0.94 (0.01)			1
K52	0.89 (0.02)			1	S50	0.90 (0.03)	0.13 (–)	2.61 (0.45)	0.91 (0.12)
T53	0.85 (0.02)			1	Q51	0.89 (0.02)		0.89 (0.29)	1

Table 3. Continued

	FABP					ILBP			
AA	S^2	τ_{loc}	R_{ex}	S_{fast}^2	AA	S^2	τ_{loc}	R_{ex}	S_{fast}^2
	–	nsec	s^{-1}	–		–	nsec	s^{-1}	
Q54	0.84 (0.03)			1	Q52	0.88 (0.04)		3.29 (0.47)	1
S55	0.87 (0.02)			1	Y53	0.88 (0.03)		0.76 (0.35)	1
					P54				
T56	0.88 (0.05)			1	G55	0.71 (0.30)	0.03 (0.04)	0.89 (–)	1
F57	0.84 (0.03)			1	G56	0.79 (0.03)	0.05 (0.02)		1
K58	0.82 (0.04)			1	H57				
N59	0.99 (0.02)			1	S58	0.28 (0.21)	3.19 (0.83)	2.34 (0.85)	0.78 (0.07)
T60	0.73 (0.07)	3.60 (–)		0.88 (0.04)	I59	0.88 (0.03)		0.47 (0.28)	1
E61	0.88 (0.02)			1	T60	0.91 (0.03)		0.98 (0.21)	1
I62	0.92 (0.02)			1	N61	0.92 (0.02)			1
S63					T62	0.83 (0.01)		0.42 (0.13)	1
F64	0.91 (0.01)			1	F63	0.89 (0.01)			1
K65	0.89 (0.01)			1	T64	0.86 (0.01)			1
L66	0.86 (0.03)			1	I65	0.92 (0.05)	0.05 (0.09)	0.78 (0.51)	1
G67	0.89 (0.01)			1	G66	0.89 (0.03)		0.73 (0.32)	1
V68	0.89 (0.00)			1	K67	0.33 (0.25)	3.22 (1.17)	2.53 (1.03)	0.77 (0.07)
E69	0.94 (0.07)	1.41 (1.14)		0.86 (0.03)	E68	0.68 (0.02)	0.07 (0.01)	2.67 (0.62)	1
F70	0.91 (0.01)			1	C69	0.86 (0.01)	0.04 (0.02)		1
D71	0.83 (0.01)			1	D70	0.25 (0.19)	4.04 (1.35)	2.82 (0.67)	0.68 (0.05)
E72	0.88 (0.02)			1	I71	0.83 (0.01)	0.02 (0.02)		1
T73	0.83 (0.02)			1	E72	0.89 (0.02)		0.53 (0.24)	1
T74	0.92 (0.01)			1	T73	0.90 (0.03)		7.97 (0.84)	1
A75	0.89 (0.01)			1	T74	0.24 (0.01)	0.17 (0.02)		1
D76	0.90 (0.01)			1	G75	0.93 (0.05)		1.21 (0.56)	1
D77					G76	0.82 (0.03)	0.05 (0.04)		1
R78	0.88 (0.01)			1	K77	0.82 (0.02)	0.05 (0.01)		1
K79	0.79 (0.00)			1	K78	0.81 (0.01)			1
V80					F79	0.83 (0.19)	0.84 (–)	0.35 (–)	0.88 (0.06)
K81	0.88 (0.01)			1					
K80	0.83 (0.05)		3.23 (1.40)	1					
S82	0.86 (0.02)	0.02 (0.01)		1	A81	0.92 (0.01)		7.80 (0.63)	1
I83	0.95 (0.01)			1	T82	0.92 (0.02)			1
V84					V83	0.17 (0.17)	2.10 (1.24)		0.43 (0.15)
T85	0.93 (0.02)			1	Q84	0.88 (0.01)			1
L86	0.83 (0.05)	0.13 (0.10)		1	M85	0.76 (0.32)	2.93(–)	0.58 (1.37)	0.81 (0.10)
D87	0.77 (0.02)	0.04 (0.01)		1	E86	0.78 (0.01)	0.04 (0.01)		1
G88					G87	0.91 (0.03)			
1									
G89	0.96 (0.02)	0.34(–)		1	G88	0.92 (0.02)			1
K90	0.87 (0.01)			1	K89	0.85 (0.04)			1
L91	0.92 (0.02)			1	V90	0.89 (0.01)		0.65 (0.17)	1
V92	0.88 (0.06)		1.67 (0.49)	1	V91	0.88 (0.02)		1.84 (0.49)	1
H93	0.95 (0.01)			1	V92				
V94					N93	0.81 (0.40)	3.11 (–)	0.59 (–)	0.87 (0.13)
Q95	0.88 (0.01)			1					
K96	0.89 (0.01)			1					
W97	0.75 (0.07)	4.01 (–)		0.87 (0.03)	S94	0.82 (0.03)	0.04 (0.01)	0.51 (0.21)	1
N98	0.92 (0.03)			1	P95				
G99	0.98 (0.01)			1	N96	0.87 (0.04)	0.07 (0.03)	2.42 (0.35)	1
Q100	0.86 (0.02)	0.04 (0.02)		1	Y97	0.78 (0.05)	0.04 (0.05)	3.85 (0.97)	1
E101	0.85 (0.01)	0.04 (0.01)		1					
T102	0.92 (0.01)			1					
S103	0.99 (0.01)			1	H98	n.v.	n.v.	n.v	n.v.
L104					H99	0.90 (0.00)		1.16 (0.04)	1
V105					T100	0.86 (0.09)		1.77 (1.28)	1
R106	0.88 (0.01)			1	A101	0.91 (0.01)			1
E107	0.91 (0.01)			1	E102	0.88 (0.03)		0.40 (0.29)	1

Table 3. Continued

AA	S^2	FABP τ_{loc}	R_{ex}	S^2_{fast}	AA	S^2	ILBP τ_{loc}	R_{ex}	S^2_{fast}
	–	nsec	s^{-1}	–		–	nsec	s^{-1}	
M108	0.84 (0.02)			1	I103	0.84 (0.01)			1
V109	0.75 (0.07)	3.02 (–)		0.83 (0.03)	V104	0.86 (0.01)			1
D110	0.91 (0.02)			1	D105	0.94 (0.02)			1
G111	0.87 (0.01)			1	G106	0.83 (0.01)	0.02 (0.02)		1
K112	0.80 (0.08)	3.03 (–)		0.88 (0.04)	K107				
L113	0.88 (0.02)			1	L108	0.85 (0.02)	0.04 (0.02)		1
I114	0.89 (0.04)			1	V109	0.94 (0.01)			1
L115	0.95 (0.02)			1	E110				
T116					V111	0.68 (0.00)	0.09 (0.00)		1
L117	1.00 (0.01)			1	S112				
T118	0.92 (0.02)			1	T113	0.97 (0.02)			1
H119	0.96 (0.01)			1	V114	0.87 (0.01)			1
G120	0.95 (0.07)		1.48 (0.51)	1	G115	0.99 (0.05)	0.21(–)		1.00 (0.07)
T121	0.94 (0.01)			1	G116	0.99 (0.02)			1
A122	0.90 (0.02)			1	V117	0.56 (0.02)	2.83 (0.66)		0.79 (0.01)
V123	0.82 (0.01)			1	S118	0.80 (0.11)	4.57 (–)		0.86 (0.03)
C 124	0.88 (0.00)			1	Y119	0.90 (0.03)			1
T125	0.90 (0.01)			1	E120	0.92 (0.02)		0.46 (0.16)	1
R126	0.88 (0.03)		1.97 (0.39)	1	R121	0.92 (0.11)	0.03 (–)	2.49 (0.74)	0.87 (0.67)
T127	0.87 (0.06)		0.89 (0.45)	1	V122	0.87 (0.02)		3.81 (0.24)	1
Y128	0.89 (0.01)			1	S123	0.92 (0.02)			1
E129	0.88 (0.01)			1	K124	0.89 (0.01)			1
K130					K125				
Q131	0.72 (0.07)	1.73 (1.13)		0.85 (0.04)	L126	0.71 (0.08)	2.50 (1.95)		0.86 (0.04)
A132	0.34 (0.03)	1.38 (0.10)		0.71 (0.01)	A127	0.29 (0.04)	2.24 (0.15)		0.77 (0.01)

In conclusion the present data demonstrate significantly different backbone dynamics for bovine H-FABP and porcine ILBP, despite their highly homologous tertiary fold. On the one hand, H-FABP displays a rather compact backbone structure of low flexibility. On the other hand, in ILBP individual residues exhibit high backbone mobility and a large number of residues show conformational exchange. Thus, a greater flexibility in certain parts of the backbone structure of ILBP might account for its preferential binding of bile acid ligands, which compared to fatty acids are more bulky and would require more space to enter and exit the protein. Furthermore, the largest effects with respect to both S^2 and R_{ex} are observed in β-strands D, E, F, G and H, near the proposed entry portal for the bile acid ligand [17]. This again indicates that conformational flexibility may be a critical factor for ligand binding in ILBP.

Interestingly, the H-FABP data presented here does not exhibit low order parameters or significant conformational exchange in the region of the fatty acid portal, as reported for rat I-FABP in a similar ^{15}N relaxation study by Hodsdon and Cistola [19]. However, such a significant difference in the backbone mobilities of H-FABP and I-FABP around the portal region would certainly emphasize the importance of structural dynamics in the functional aspects of the lipid binding proteins. Whether there actually is an inherent difference between H-FABP and I-FABP with regard to the dynamics in the fatty acid entry portal region, or whether these inconsistencies are just due to differences in the experimental procedures or data analysis, needs to be examined more closely in the future.

Acknowledgements

The authors would like to thank Dr. Azita Mesgarzadeh (Frankfurt) for preparing the ^{15}N-labeled bovine H-FABP sample, Sangita Patel (New York) for helping to purify ^{15}N-labeled porcine ILBP and Dr. Fengli Zhang (Boston) for expert technical assistance at the NMR spectrometer.

References

1. Veerkamp JH, Maatman RGHJ: Cytoplasmic fatty acid-binding proteins: their structure and genes. Prog Lipid Res 34: 17–52, 1995
2. Banaszak L, Winter N, Xu Z, Bernlohr DA, Cowan S, Jones TA: Lipid-binding proteins: A family of fatty acid and retinoid transport proteins. Adv Prot Chem 45: 89–151, 1994
3. Sacchettini JC, Gordon JI, Banaszak LJ: The structure of crystalline *Escherichia coli*-derived rat intestinal fatty acid-binding protein at 2.5-Å resolution. J Biol Chem 263: 5815–5819, 1988

4. Sacchettini JC, Gordon JI, Banaszak LJ: Refined apoprotein structure of rat intestinal fatty acid binding protein produced in *Escherichia coli*. Proc Natl Acad Sci USA 86: 7736–7740, 1989
5. Scapin G, Gordon JI, Sacchettini JC: Refinement of the structure of recombinant rat intestinal fatty acid-binding apoprotein at 1.2-Å resolution. J Biol Chem 267: 4253–4269, 1992
6. Scapin G, Spadon P, Mammi M, Zanotti G, Monaco HL: Crystal structure of chicken liver basic fatty acid-binding protein at 2.7 Å resolution. Mol Cell Biochem 98: 95–99, 1990
7. Thompson J, Winter N, Terwey D, Bratt J, Banaszak LJ: The crystal structure of the liver fatty acid-binding protein. A complex with two bound oleates. J Biol Chem 272: 7140–7150, 1997
8. Müller-Fahrnow A, Egner U, Jones TA, Rüdel H, Spener F, Saenger W: Three-dimensional structure of fatty-acid-binding protein from bovine heart. Eur J Biochem 199: 271–276, 1991
9. Zanotti G, Scapin G, Spadon P, Veerkamp JH, Sacchettini JC: Three-dimensional structure of recombinant human muscle fatty acid-binding protein. J Biol Chem 267: 18541–18550, 1992
10. Young ACM, Scapin G, Kromminga A, Patel SB, Veerkamp JH, Sacchettini JC: Structural studies on human muscle fatty acid binding protein at 1.4 Å resolution: Binding interactions with three C18 fatty acids. Structure 2: 523–534, 1994
11. Xu Z, Bernlohr DA, Banaszak LJ: Crystal structure of recombinant murine adipocyte lipid binding protein. Biochemistry 31: 3484–3492, 1992
12. Jones TA, Bergfors T, Sedzik J, Unge T: The three-dimensional structure of P2 myelin protein. EMBO J 7: 1597–1604, 1988
13. Cowan SW, Newcomer ME, Jones TA: Crystallographic studies on a family of cellular lipophilic transport proteins: The refinement of P2 myelin protein and the structure determination and refinement of cellular retinol-binding protein in complex with all-trans retinol. J Mol Biol 230: 1225–1246, 1993
14. Benning MM, Smith AF, Wells MA, Holden HM: Crystallization, structure determination and least-squares refinement to 1.75 Å resolution of the fatty acid binding protein isolated from Manducta sexta L. J Mol Biol 228: 208–219, 1992
15. Winter NS, Bratt JM, Banaszak LJ: Crystal structures of holo and apo-cellular retinol-binding protein II. J Biol Chem 230: 1247–1259, 1993
16. Lassen D, Lücke C, Kveder M, Mesgarzadeh A, Schmidt JM, Specht B, Lezius A, Spener F, Rüterjans H: Three-dimensional structure of bovine heart fatty-acid-binding protein with bound palmitic acid, determined by multidimensional NMR spectroscopy. Eur J Biochem 230: 266–280, 1995
17. Lücke C, Zhang F, Rüterjans H, Hamilton JA, Sacchettini JC: Flexibility is a likely determinant of binding in the case of ileal lipid binding protein. Structure 4: 785–800, 1996
18. Zhang F, Ücke C, Baier LJ, Sacchettini JC, Hamilton JA: Solution structure of human intestinal fatty acid binding protein: Implications for ligand entry and exit. J Biomol NMR 9: 213–228, 1997
19. Hodsdon ME, Cistola DR: Ligand binding alters the backbone mobility of intestinal fatty acid binding protein as monitored by ^{15}N relaxation and ^{1}H exchange. Biochemistry 36: 2278–2290, 1997
20. Jakoby MG IV, Miller KR, Toner JJ, Bauman A, Cheng L, Li E, Cistola DR: Ligand-protein electrostatic interactions govern the specificity of retinol- and fatty acid-binding proteins. Biochemistry 32: 872–878, 1993
21. Eads J, Sacchettini JC, Kromminga A, Gordon R: *Escherichia coli*-derived rat intestinal fatty acid binding protein with bound myristate at 1.5 Å resolution and I-FABP$^{Arg106\to Gln}$ with bound oleate at 1.74 Å resolution. J Biol Chem 268: 26375–26385, 1993
22. Sacchettini JC, Hauft SM, Van Camp SL, Cistola DP, Gordon JI: Developmental and structural studies of an intracellular lipid binding protein expressed in the ileal epithelium. J Biol Chem 265: 19199–19207, 1990
23. Kay LE, Torchia DA, Bax A: Backbone dynamics of proteins as studied by nitrogen-15 inverse detected heteronuclear NMR spectroscopy: Application to staphylococcal nuclease. Biochemistry 28: 8972–8979, 1989
24. Fushman D, Weisemann R, Thüring H, Rüterjans H: Backbone dynamics of ribonuclease T_1 and its complex with 2′GMP studied by two-dimensional heteronuclear NMR spectroscopy. J Biomol NMR 4: 61–78, 1994
25. Gaudin F, Paquet F, Chanteloup L, Beau JM, Nguen TT, Lancelot G: Selectively ^{13}C-enriched DNA: Dynamics of the C1′-H1′ vector in d(CGAAATTTVG)3. J Biomol NMR 5: 49–58, 1995
26. King GC, Harper JW, Xi Z: Isotope labeling for ^{13}C relaxation measurements on RNA. Meth Enzymol 261: 436–50, 1995
27. Shaka AJ, Barker PB, Freeman R: Computer-optimized decoupling scheme for wideband applications and low-level operation. J Magn Res 64: 547–552, 1985
28. Morris GA, Freeman R: Selective excitation in fourier transform nuclear magnetic resonance. J Magn Res 29: 433–462, 1978
29. Kay LE, Nicholson LK, Delaglio F, Bax A, Torchia DA: Pulse sequences for removal of the effects of cross correlation between dipolar and chemical-shift anisotropy relaxation mechanisms on the measurement of heteronuclear T_1 and T_2 values in proteins. J Magn Res 97: 359–375, 1992
30. Grzesiek S, Bax A: The importance of not saturating H_2O in protein NMR. Application to sensitivity enhancement and NOE measurements. J Am Chem Soc 115: 12593–12594, 1993
31. Fushman D, Cahill S, Cowburn D: The main-chain dynamics of the dynamin pleckstrin homology (PH) domain in solution: Analysis of ^{15}N relaxation with monomer/dimer equilibration. J Mol Biol 266: 173–194, 1997
32. Lipari G, Szabo A: Model-free approach to the interpretation of nuclear magnetic resonance relaxation in macromolecules. 1. Theory and range of validity. J Am Chem Soc 104: 4546–4559, 1982
33. Lipari G, Szabo A: Model-free approach to the interpretation of nuclear magnetic resonance relaxation in macromolecules. 2. Analysis of experimental results. J Am Chem Soc 104: 4559–4570, 1982
34. Clore GM, Szabo A, Bax A, Kay LE, Driscoll PC, Wingfield PT, Gronenborn AM: Deviations from the simple two-parameter model-free approach to the interpretation of ^{15}N nuclear magnetic relaxation of proteins. J Am Chem Soc 112: 4989–4991, 1990
35. Garcia de la Torre J, Bloomfield V: Hydrodynamic properties of complex rigid, biological macromolecules: Theory and applications. Quart Rev Biophys 14: 81–139, 1981
36. Tjandra N, Feler SE, Pastor RW, Bax A: Rotational diffusion anisotropy of human ubiquitin from ^{15}N NMR relaxation. J Am Chem Soc 117: 12562–12566, 1995

Molecular and Cellular Biochemistry **192**: 123–128, 1999.

Probable interaction between S100A7 and E-FABP in the cytosol of human keratinocytes from psoriatic scales

Gerry Hagens,[1] Karen Roulin,[1] Raymonde Hotz,[1] Jean-Hilaire Saurat,[1] Ulf Hellman[2] and Georges Siegenthaler[1]

[1]*Department of Dermatology and DHURDV, University Hospital, Geneva, Switzerland;* [2]*Ludwig Institute for Cancer Research, Uppsala, Sweden*

Abstract

The overexpression of E-FABP and S100A7 in lesional psoriatic skin suggests a possible link with this hyperproliferative skin disease. In order to investigate a role for the proteins in this disease, the purifications for both proteins were re-analyzed. Moreover, a specific antiserum directed against purified human S100A7 was generated. By SDS-PAGE immunoblotting we show that E-FABP and S100A7 are expressed in cultured human differentiating keratinocytes and confirm their overexpression in psoriatic scales. Gel filtration and non-denaturing PAGE revealed that S100A7 co-purified with E-FABP, indicating an association between the two proteins. Ion-exchange chromatography resulted in the dissociation of the complex. Finally, immunoprecipitations using antiserum against E-FABP revealed that S100A7 co-immunoprecipitated with E-FABP from protein extracts of psoriatic scales. These data indicate that E-FABP and S100A7 might form a complex in the cytosol of human keratinocytes. (Mol Cell Biochem **192**: 123–128, 1999)

Key words: psoriasis, calcium-binding protein, protein-protein interaction

Introduction

In normal skin and in culture, keratinocyte differentiation is recognized by the formation of different epidermal cell layers. From the basal proliferating cell layer, differentiating keratinocytes migrate to the surface of epidermis, where terminally differentiated cells form a cornified cell envelope. These horny layers contain many lipids organized as a lipid barrier (for a review see [1]). These lipids are released into the intercellular domains through lamellar body secretion by the last living keratinocyte layer (stratum granulosum) [2]. Skin represents a very active lipid-synthesizing tissue in mammals. However, marked changes in lipid composition in successive epidermal cell layers have been described, which might indicate a relation between fatty acid (FA) transport/metabolism and keratinocyte differentiation [3].

Recently, two FA carriers have been identified in human skin. The epidermal FABP (E-FABP) and the FA-p34 complex, which is composed of the two Ca^{2+}-binding proteins (CaBP) S100A8 and S100A9. The former has a high affinity for stearic acid a FA important for membrane formation and energy delivery [4–6], whereas FA-p34 binds specifically unsaturated FA in a Ca^{2+}-dependent way [7]. In contrast to FA-p34, which is highly expressed in abnormally differentiated keratinocytes and polymorphonuclear leukocytes, E-FABP is detected in the stratum granulosum of normal human skin, throughout psoriatic skin except the basal layer [6], in human endothelial cells of the microvasculature and in secretory cells of the lung [8].

In psoriatic scales, characterized by hyperproliferative and abnormally differentiated keratinocytes, an altered transport/metabolism of FA has been described [9–11]. These findings are in line with the up-regulated expression of E-FABP [4–6, 12] and FA-p34 [7], whereby both FA carriers are expressed through all differentiating cell layers of the diseased skin.

Address for offprints: G. Siegenthaler, Department of Dermatology, University Hospital, CH-1211 Geneva 14, Switzerland

S100A7, another CaBP, previously characterized in psoriatic skin is expressed in differentiating cell layers of normal skin and is highly up-regulated in psoriatic scales [13, 14]. S100A7 belongs to the S100 CaBP family, characterized by EF-hand structural motifs able to bind Ca^{2+} [15, 16]. S100A7 is composed of 100 amino acids and its measured molecular mass is 11,368 Da [14]. In psoriatic scales a cytosolic and a membrane-attached form of S100A7 co-exist and both can be distinguished by their slightly different electrophoretic mobility in SDS-PAGE. However, since their molecular mass is almost identical it has been suggested that conformational changes, occuring at the membrane, might be responsible for the observed differences [14]. Members of the S100 protein family are believed to mediate a variety of functions in eukaryotic cells, including cell cycle progression and differentiation [16]. Nevertheless, the precise functions could not yet been attributed to the individual members of this large protein family.

The up-regulation of E-FABP and S100A7 in psoriatic skin might implicate a link with this skin disease affecting ~2% of the population. In this report, we analyzed more in detail the purification steps of these two proteins and show indications suggesting that E-FABP and S100A7 might form a complex.

Materials and methods

Cell cultures and tissue isolations

Normal human keratinocytes from foreskin were cultured in Dulbecco's modified Eagle's medium (Life Technologies Inc., Switzerland) containing 1.3 mM Ca^{2+} and 10% fetal calf serum [17]. Psoriatic scales were obtained by gentle scraping of lesional skin from volunteer psoriatic patients.

Protein extraction

Proteins were extracted as described previously [7] using 50 mM Tris-HCl, 25 mM NaCl, 2.5 mM EDTA, 1 mM DTT, pH 7.5 as homogenization buffer. Homogenates were subsequently centrifuged at 100,000 g for 60 min at 4°C. The supernatant, containing cytoplasmic proteins was concentrated using centrifugal microconcentrators (Amicon) with a molecular mass cutt-off of 3000 Da and stored at –20°C or directly used for further experiments.

The pellet constituted of cellular debris is called the membrane fraction. The pellet was washed 3 times in 1 × PBS (phosphate buffered saline) and treated for 1 h at 4°C as described [7] with a high salt KCl buffer (10 mM Tris/HCl, 0.8 M KCl, 10 mM monothioglycerol, 10% glycerol, 1 mM phenylmethylsulfonyl fluoride, 10 units/ml aprotinin, 10 µg/ml leupeptin, pH 8.0) and centrifuged at 20,000 g. This new supernatant, containing membrane-attached proteins, was dialyzed against 20 mM Tris/HCl buffer at pH 8.0 and concentrated as above.

Simultaneous purification of E-FABP and cytosolic S100A7

Purification of E-FABP and cytosolic S100A7 was performed as described [5, 14]. Briefly, 143 mg of cytoplasmic proteins incubated with [9,10-^{3}H(N)]oleic acid (1.2 µM; specific activity, 9.2 Ci/mmol; DuPont NEN) were separated by gel filtration on a Sephadex G100 column (Pharmacia LKB, Uppsala, Sweden), equilibrated with 5 mM sodium phosphate buffer containing 0.2 M NaCl, 2 mM DTT, 1 mM PMSF and 0.02% sodium azide at pH 7.5. The elution profile of proteins was monitored by a U.V.-absorbance detector set at 280 nm. The radioactive elution was assessed in a liquid-scintilation counter by counting the radioactivity in 100 µl aliquots of each eluted fraction diluted in 4 ml of Ultima-Gold (Packard). Eluted fractions containing E-FABP, as identified by SDS-PAGE immunoblotting, were pooled and further separated by a semipreparative non-denaturing PAGE (7.5%) [5]. The gel was Coomassie Blue-stained. The protein bands at $R_f = 0.34$ corresponding to E-FABP were excised and proteins were extracted from the gel slice with 100 mM Tris-HCl, 0. 5 M KCl and 0.05% Triton X-100 at pH 7.5. This solution was centrifuged at 5000 g and the supernatant dialysed against 20 mM Tris-HCl, 1 mM DTT, pH 8.0 using a Spectra/Por dialysis membrane (Spectrum, Houston TX, USA), with a molecular mass cut-off of 3500 Da. 1.68 mg of this concentrated protein solution was injected into an ion-exchange Mono Q HR5/5 column (Pharmacia LKB, Uppsala, Sweden), which has been equilibrated with 20 mM Tris-HCl at pH 8.0. The column was connected to a HPLC (Varian) and the proteins were eluted with a linear gradient of equilibration buffer containing up to 0.5 M NaCl. Elution of proteins was monitored by U.V.-absorbance at 280 nm. S100A7 was not retained by the gel matrix, whereas E-FABP was obtained after 9 min of elution time. Final purification was achieved by filtrating the concentrated protein peaks through a Superose 12 column (Pharmacia LKB, Uppsala, Sweden), equilibrated with PBS containing 0.2 M NaCl. Eluted fractions were concentrated as described above and stored at –20°C until use.

Generation of polyclonal anti-S100A7 antiserum

Purified S100A7 (70 µg) was mixed with Freund's complete adjuvant and injected subcutaneously at multiple points on the back of a rabbit. S100A7 (70 μg) was mixed with Freund's incomplete adjuvant and injected twice into the rabbit in 3-week intervals after the initial injection. The antiserum was collected 3 weeks after the second booster injection. The resulting immunserum was stored at –80°C until use. The specificity of the antiserum was checked by SDS-PAGE (15%) immunoblotting.

Immunoblotting and immunoprecipitations

Proteins were separated by SDS-PAGE (15%) and subsequently transfered onto a nitrocellulose membrane (Electran, BDH Laboratory Supplies, United Kingdom). The membrane was blocked with PBS containing 5% skimmed dry milk. After blocking, the membranes were incubated in PBS containing 0.5% skimmed dry milk, 0.2% Tween 20 and antiserum directed against E-FABP [5] or S100A7 (each diluted at 1:1000). Immunoreactive bands were visualized using a peroxidase-labeled goat anti-rabbit IgG Fab′ fragment (Cappel) and 3,3′-diaminobenzidine dihydrochloride (Sigma) and H_2O_2 as substrates. As described earlier, the E-FABP antiserum was highly specific [5].

For immunoprecipitations, 1 mg cytoplasmic protein extracts were precleared for 2 h at 4°C using the corresponding preimmunserum and a 50 % Protein A-Sepharose (Pharmacia) slurry (in PBS). After a short centrifugation, precleared protein extracts were incubated for 2 h at 4°C with either anti-E-FABP, anti-S100A7 or an irrelevant rabbit IgG antiserum. Antibody-protein complexes were precipitated for 1 h at 4°C with Protein A-Sepharose as above. Precipitates were washed 10 times with homogenization buffer containing 200 mM NaCl and 0.1% Triton X-100 and subsequently resuspended in Laemmli's sample buffer [18] and boiled for 5 min before separation by SDS-PAGE.

Results

Co-purification of E-FABP and S100A7

Cytosolic proteins from psoriatic scales incubated with [^{3}H]oleic acid were fractionated on Sephadex G100. The radioactive elution profile (Fig. 1) shows a large radioactive peak eluting at 15 kDa. This peak is not symetric and contains a shoulder of higher molecular mass. The fractions of this peak were analyzed by immunoblotting using anti-E-FABP and anti-S100A7 antisera. The presence of E-FABP was confirmed in all radioactive fractions constituting the peak at 15 kDa (closed box). S100A7 (open box) appears earlier and overlaps E-FABP elution. The fractions containing E-FABP were pooled and concentrated before subjecting to

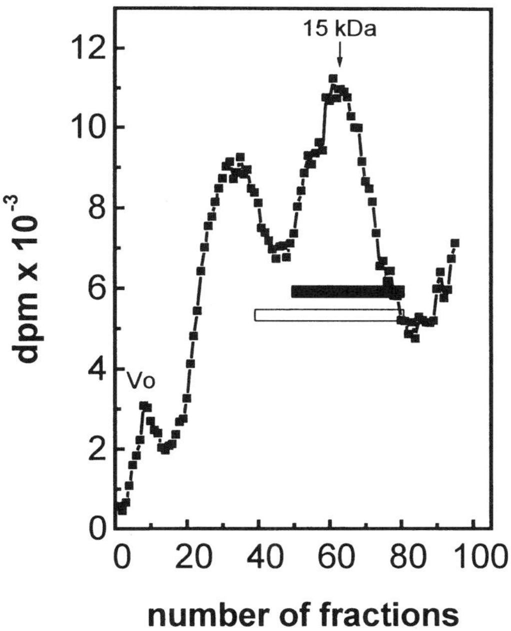

Fig. 1. Radioactive elution profile of protein extract from psoriatic scales. Protein sample was incubated with [^{3}H]oleic acid as a tracer as described in the methods section and fractionated on a Sephadex G-100 column. Fractions containing E-FABP and/or S100A7 as identified by SDS-PAGE immunoblotting are indicated by a closed or an open box, respectively.

PAGE under non-denaturing conditions. A strong Coomassie Blue-stained band was observed at $R_f = 0.34$ that corresponds to E-FABP mobility (Fig. 2, lane 1) [5]. This band was excised and proteins extracted and re-analyzed. Using non-denaturing PAGE only one Coomassie Blue-stained band was detected (Fig. 2, lanes 2 to 4). In contrast, SDS-PAGE revealed minor high molecular weight contaminants and two prominent bands composed of E-FABP and the recently identified S100A7 (Fig. 2, lane 7) [14]. The gel extraction step shows that E-FABP and S100A7 were enriched compared to proteins samples before and after gel filtration (lane 5 and 6). Compared to the initial sample (lane 5) the intensity of the E-FABP band was increased, whereas S100A7 intensity was decreased.

Gel-extracted proteins were subjected to ion-exchange chromatography on a Mono Q column to analyze their possible interaction. S100A7 eluted in the first minute, whereas E-FABP eluted at about 9 min (Fig. 3A). Both peaks have a similar height reflecting the intensities of the Coomassie Blue-stained bands in the SDS-PAGE analysis. The fractions containing S100A7 and E-FABP, respectively were pooled, concentrated and homogeneity was monitored by SDS-PAGE. Both pools were considered homogeneous on the basis of the presence of only one Coomassie-stained band (Fig 3B).

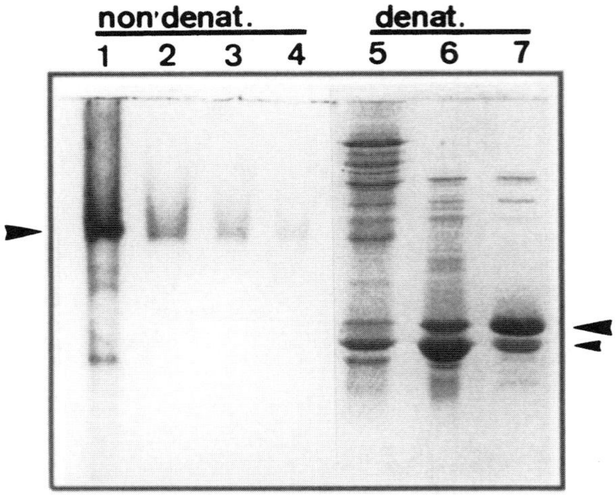

Fig. 2. PAGE and SDS-PAGE analysis of E-FABP and S100A7 after the different chromatography steps. Aliquots after the different purification steps were separated under non-denaturing conditions (non denat.) by PAGE (lanes 1–4) or under denaturing (denat.) conditions by SDS-PAGE (lanes 5–7). Lanes 1 and 6: 17 µg protein extract after Sephadex G-100 gel filtration; lanes 2 and 7: 10 µg of the gel-extracted E-FABP band migrating at an R_f of 0.34. Lanes 3 and 4: aliquots of 5 and 2 µg, respectively, of the gel-extracted E-FABP band. Lane 5: 20 µg of cytosolic protein extract from psoriatic scales. Thick arrow heads indicate the position of the E-FABP bands in PAGE and SDS-PAGE. Thin arrow head indicates the position of the S100A7 band.

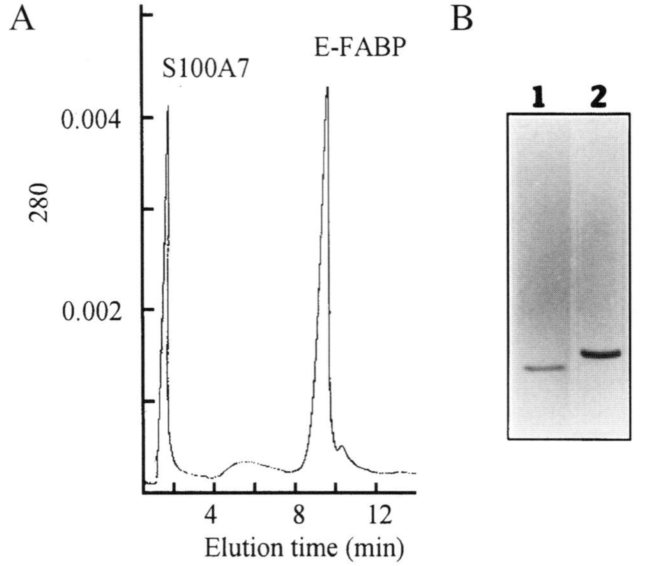

Fig. 3. (A) Elution profile of proteins obtained after semipreparative PAGE subjected to ion-exchange chromatography. Proteins extracted from the semipreparative PAGE were purified by H.P.L.C. on a Mono Q column as described in the Materials and methods section. S100A7 was not retained and eluted with the first fractions. E-FABP eluted at about 9 min. The peaks corresponding to the two proteins are indicated. (B) Analysis of S100A7 and E-FABP purity. Mono Q-eluted fractions containing S100A7 or E-FABP, respectively, were pooled and an aliquot separated by SDS-PAGE (5 µg of each). Proteins were stained with Coomassie Blue. S100A7 (lane 1) and E-FABP (lane 2) were considered pure on the basis of only one Coomassie Blue-stained band.

Characterization of the anti-S100A7 antiserum and expression studies on S100A7

The antiserum against S100A7 generated in a rabbit was tested by SDS-PAGE immunoblotting using protein extracts from psoriatic scales. The antiserum recognized only proteins co-migrating with cytosolic and membrane-attached S100A7 from psoriatic scales [14] (Fig. 4A). In a next step we investigated the expression pattern of S100A7 in cultured human keratinocytes. Similar to E-FABP, S100A7 expression was restricted to differentiating keratinocytes and to psoriatic scales (Fig. 4B, lanes 3–4, and 5–6, respectively), whereas non-differentiating keratinocytes did not express the proteins (Fig. 4B, lanes 1 and 2).

Furthermore, S100A7 was also weakly expressed in the membrane fraction of differentiating keratinocytes (Fig. 4B, lane 4). An increased S100A7 expression could be observed in psoriatic scales in parallel with the appearance of the membrane-attached form (Fig. 4B, lane 6), thus confirming observations described in earlier reports [13, 14].

Co-immunoprecipitation of S100A7 with E-FABP

Co-localization and co-migration in non-denaturing PAGE indicated a possible interaction between E-FABP and S100A7. Therefore, immunoprecipitations were performed using the specific antisera directed against the two proteins. Both antisera were able to immunoprecipitate the corresponding proteins from protein extracts of psoriatic scales under non-denaturing conditions, as shown by subsequent SDS-PAGE immunoblotting of the immunoprecipitates (Fig. 5, lanes 1 and 5). Not surprisingly, the E-FABP-immunoprecipitations contained also S100A7 (Fig. 5, lane 4). The negative control using an irrelevant antibody, revealed the specificity of this observation, since neither E-FABP nor S100A7 were detectable in the control experiments (Fig. 5, lanes 3 and 6). On the other hand, the anti-S100A7 antiserum was not able to co-immunoprecipitate E-FABP (Fig. 5, lanes 2 and 5).

Discussion

In this work, we present results assuming that E-FABP and S100A7 might form a complex in the cytosol of psoriatic scales. First indications for a possible association were obtained by purifying both molecules. Since the cytosolic protein E-FABP and S100A7 are over-expressed in psoriatic scales [5, 13], this tissue was used to investigate this hypo-

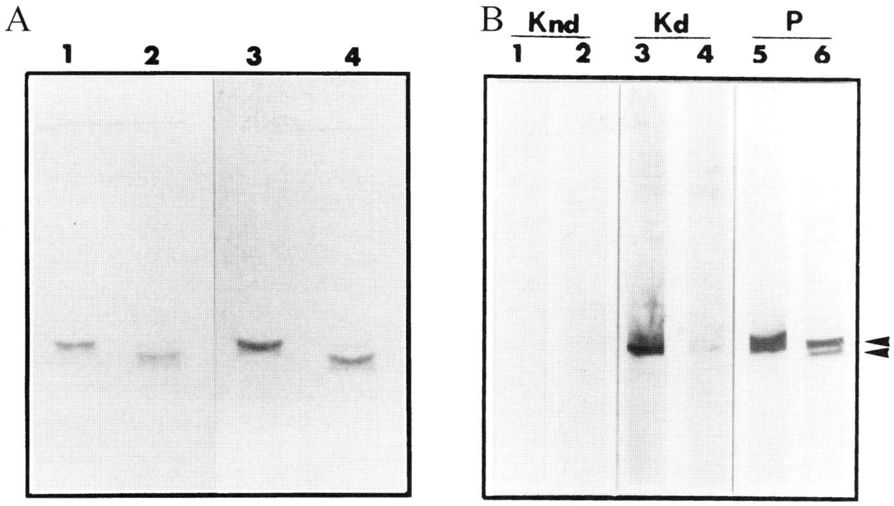

Fig. 4. (A) Specificity test of the anti-S100A7 antiserum. Cytosolic (lane 1), membrane protein extracts (lane 2), (20 μg each) and 1 μg each homogenous cytosolic (lane 3) or membrane (lane 4) S100A7 from psoriatic scales were separated by SDS-PAGE (15%) and subsequently immunoblotted onto a nitrocellulose membrane. Using protein extracts the antiserum recognized only proteins co-migrating with cytosolic or membrane S100A7 and was therefore considered specific. (B) Expression of S100A7 in cultured and psoriatic keratinocytes. Cytosolic (odd lane numbers) and membrane proteins (even lane numbers) (20 μg each) were separated by SDS-PAGE (15%) and immunoblotted onto a nitrocellulose membrane. Non-differentiating keratinocytes (Knd) do not express S100A7 (lanes 1 and 2). Differentiating keratinocytes (Kd) express in the cytosol and at the membrane S100A7 (lanes 3 and 4). In psoriatic scales (P) S100A7 is overexpressed in the cytosol and at the membrane (lanes 5 and 6). Arrow heads indicate the cytosolic and the faster migrating membrane-attached S100A7 form.

thesis. The observed assymetric radioactive peak obtained by gel filtration contained both proteins. The shoulder of higher molecular mass could be explained by an association of E-FABP and S100A7, whereas the highest point of the peak at 15 kDa corresponds to free E-FABP. Significant amounts of FA-p34 in the shoulder, which might contribute to the radioactive peak, can be excluded since this complex is found mainly in the membrane fraction of psoriatic scales [7]. The fact that E-FABP and S100A7 migrate as a single entity at an R_f of 0.34 in a non-denaturing PAGE is another argument re-inforcing the hypothesis of a complex formation. A complete separation of E-FABP and S100A7 was achieved using ion-exchange chromatography suggesting that both proteins might have distinct pI values. The observation that S100A7 was not retained by the MonoQ indicates that its pI value may be higher than the one of E-FABP measured previously at 5.6 [5]. The pI of S100A7 under denaturing conditions has already been measured at 6.77 [13]. Data concerning the native protein have still to be determined. Taking in account that the putative pI of S100A7 is at least one pH unit higher than the one of E-FABP suggests different electrophoretic migrations for the two proteins in PAGE.

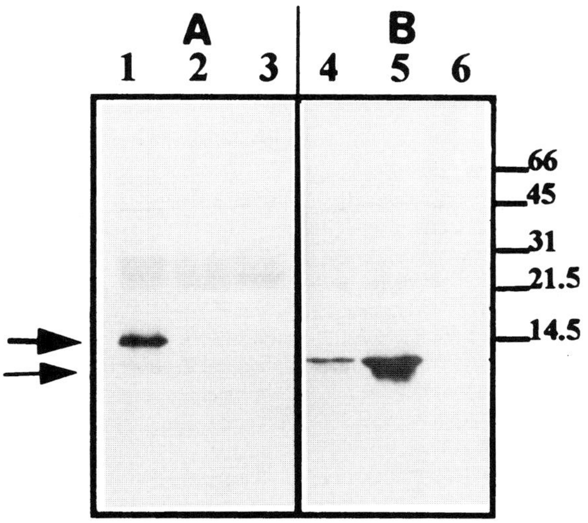

Fig. 5. Co-immunoprecipitation of S100A7 with E-FABP. Immunoprecipitations (IP) were performed as described in Materials and methods. The precipitates were divided and subsequently a SDS-PAGE immunoblotting was performed. Lanes 1 and 4: iP using anti-E-FABP antiserum; lanes 2 and 5: IP using anti-S100A7 antiserum; lanes 3 and 6: IP using irrelevant antiserum. Membrane A was revealed with anti-E-FABP antiserum, membrane B was revealed with anti-S100A7 antiserum. Thick and thin arrows indicate the positions of E-FABP and S100A7, respectively.

By SDS-PAGE immunoblotting we confirm that S100A7 parallels the already described E-FABP expression pattern [5]. S100A7 and E-FABP expressions were studied in an *in vitro* model of human cultured keratinocytes and in psoriatic scales. Only differentiating keratinocytes express S100A7 and E-FABP which is in contrast to non-differentiating keratinocytes. Psoriatic scales are the result of abnormally differentiated hyperproliferative keratinocytes, resembling cultured keratinocytes. In these psoriatic cells, S100A7 and

E-FABP are overexpressed. Using the specific antisera we could further show by immunoprecipitation and subsequent SDS-PAGE immunoblotting that S100A7 co-precipitates with E-FABP. This indicates that E-FABP and S100A7 can form a complex at least *in vitro*. The fact that Triton X-100 was used as a detergent to wash the precipitates, suggests an ionic interaction between the two proteins.

Two forms of S100A7 have previously been described [14]. No significant differences have been detected between these two forms suggesting that the forms might be conformationally different. The membrane-attached S100A7 could never been detected in cytosolic extracts of psoriatic scales by SDS-PAGE immunoblotting. Since E-FABP is a cytosolic protein, it is reasonable to speculate that the membrane form of S100A7, however, has lost the capability to associate with E-FABP.

Immunohistochemical and biochemical data from earlier reports using healthy and diseased skin suggest that E-FABP might play a role in FA-mediated mechanism(s) regulating differentiation of epidermal cells [5, 6, 12, 19]. A similar role has been proposed for S100A7 [20]. How a possible complex formation between the two proteins could contribute to FA-mediated processes in keratinocytes is currently being investigated.

Acknowledgments

This investigation was supported, in part, by the Swiss National Science Foundation (grant no. 32-47057.96).

References

1. Holbrook K: Ultrastructure of the epidermis. In: E.B. Lane, F.M. Watt (eds). The Keratinocyte Handbook. Cambridge University Press, Cambridge, 1994, pp 3–42
2. Elias PM, Menon GK: Structural and biochemical correlates of the epidermal permeability barrier. Adv Lipid Res 24: 1–26, 1991
3. Rawlings AV, Scott IR, Harding CR, Browser PA: Stratum corneum moisturization at the molecular level. J Invest Dermatol 103: 731–740, 1994
4. Siegenthaler G, Hotz R, Chatellard-Gruaz D, Jaconi S, Saurat J-H: Characterization and expression of a novel human fatty acid-binding protein: The epidermal type (E-FABP). Biochem Biophys Res Commun 190: 482–487, 1993
5. Siegenthaler G, Hotz R, Chatellard-Gruaz D, Didierjean L, Hellman U, Saurat J-H: Purification and characterization of the human epidermal fatty acid-binding protein: Localization during epidermal differentiation *in vivo* and *in vitro*. Biochem J 302: 363–371, 1994
6. Madsen P, Rasmussen HH, Leffers H, Honoré B, Celis LE: Molecular cloning and expression of a novel keratinocyte protein [psoriasis-associated fatty acid binding protein (PA-FABP)] that is highly up-regulated in psoriatic skin and that shares similarity to fatty acid-binding proteins. J Invest Dermatol 99: 299–305, 1992
7. Siegenthaler G, Roulin K, Chatellard-Gruaz D, Hotz R, Saurat L-H, Hellman U, Hagens G: A heterocomplex formed by the calcium-binding proteins MRP8 (S100A8) and MRP14 (S100A9) binds unsaturated fatty acids. J Biol Chem 272: 9371–9377, 1997
8. Masouyé I, Hagens G, van Kuppevelt TH, Madsen P, Saurat JH, Veerkamp JH, Pepper MS, Siegenthaler G: Endothelial cells of the human microvasculature express epidermal fatty acid-binding protein (E-FABP). Circ Res 81: 297–303, 1997
9. Kragballe K, Voorhees JJ: Eicosanoids in psoriasis-15-HETE on the stage. Dermatologica 174: 209–213, 1987
10. Fogh K, Herlin T, Kragballe K: Eicosanoids in acute and chronic psoriatic lesions: Leukotriene B4, but not 12-hydroxy-eicosatetraenoic acid, is present in biologically active amounts in acute guttate lesions. J Invest Dermatol 92: 837–841, 1989
11. Duell EA, Ellis CN, Voorhees LL: Determination of 5,12, and 15-lipoxygenase products in keratomed biopsies of normal and psoriatic skin. J Invest Dermatol 91: 446–450, 1988
12. Masouyé L, Saurat J-H, Siegenthaler G: Epidermal fatty acid-binding protein in psoriasis, basal and squamous cell carcinomas: An immunohistological study. Dermatology 192: 208–213, 1996
13. Madsen R, Rasmussen HH, Leffers H, Honoré B, Dejgaard K, Olsen E, Kiil J, Walbum E, Andersen AH, Basse B, Lauridsen JB, Ratz GP, Celis A, Vanderkerckhove L, Celis JE: Molecular cloning, occurrence, and expression of a novel partially secreted protein 'psoriasin' that is highly up-regulated in psoriatic skin. J Invest Dermatol 97: 701–712, 1991
14. Bürgisser D, Siegenthaler G, Kuster T, Hellman U, Hunziker R, Birchler N, Heizmann CW: Amino acid sequence analysis of human S100A7 (Psoriasin) by tandem mass spectrometry. Biochem Biophys Res Com 217: 257–263, 1995
15. Hoffmann HJ, Oisen E, Etzerodt M, Madsen R, Thogersen HC, Kruse T, Celis LE: Psoriasin binds calcium and is upregulated by calcium to levels that resemble those observed in normal skin. J Invest Dermatol 103: 370–375, 1994
16. Kligman D, Hilt DC: The S100 protein family. Trends Biochem Sci 13: 437–443, 1988
17. Rheinwald JH, Green H: Epidermal growth factor and the multiplication of cultured epidermal keratinocytes. Nature 265: 421–424, 1977
18. Laemmli UK: Cleavage of structural proteins during the assembly of the head of bacteriophage T4. Nature 277: 680–685, 1979
19. Hagens G, Roulin K, Hotz R, Saurat J-H, Siegenthaler G: The MRP8/14 complex and not the epidermal fatty acid-binding protein is responsible for the translocation of arachidonic acid in human keratinocytes. In: L.Y. Vanderhoek (ed). Frontiers in Bioactive Lipids. Plenum Press, New York, 1996, pp 61–66
20. Hardas BD, Zhao X, Zhang J, Longqing X, Stoll S, Elder M: Assignment of psoriasin to the chromosomal band 1q12: Coordinate overexpression of clustered genes in psoriasis. J Invest Dermatol 106: 753–758, 1996

Molecular and Cellular Biochemistry **192:** 129–135, 1999.

CD36 mediates long-chain fatty acid transport in human myocardium: Complete myocardial accumulation defect of radiolabeled long-chain fatty acid analog in subjects with CD36 deficiency

Shuichi Nozaki,[1] Takao Tanaka,[2] Shizuya Yamashita,[1] Koichi Sohmiya,[1] Tohru Yoshizumi,[3] Fumio Okamoto,[2] Yasushi Kitaura,[2] Chikao Kotake,[4] Hiroyuki Nishida,[4] Atsuyuki Nakata,[1] Tsutomu Nakagawa,[1] Kengo Matsumoto,[1] Kaoru Kameda-Takemura,[1] Seiji Tadokoro,[1] Yoshiyuki Kurata,[5] Yoshiaki Tomiyama,[1] Keishiro Kawamura[2] and Yuji Matsuzawa[1]

[1]*The Second Department of Internal Medicine, Osaka University Medical School;* [2]*The Third Division, Department of Internal Medicine, Osaka Medical College;* [3]*Minoo City Hospital;* [4]*Saiseikai Nakatsu Hospital;* [5]*Department of Blood Transfusion, Osaka University Hospital, Osaka, Japan*

Abstract

Long-chain fatty acids (LCFA) are the major energy substrate for heart and their oxidation is important for achieving maximal cardiac work. However, the mechanism of uptake of LCFA by myocardium has not been clarified. We previously reported that bovine myocardial LCFA transporter has a sequence homology to human CD36. Clinically, total defect of myocardial uptake of radiolabeled long-chain fatty acid analog [^{123}I-BMIPP: Iodine-123 15-(p-iodophenyl)-(R,S)-methylpentadecanoic acid] has been reported in some restricted cases, but the etiology has not been clarified. In the present study, we analyzed CD36 expression and CD36 gene in subjects who showed total lack of myocardial ^{123}I-BMIPP accumulation, and, vice versa, evaluated myocardial ^{123}I-BMIPP uptake in subjects with CD36 deficiency. Four unrelated subjects were evaluated; Two were found to have negative myocardial LCFA accumulation by ^{123}I-BMIPP scintigraphy, after which the expression of CD36 on their platelets and monocytes was analyzed. Remaining two subjects were identified as CD36 deficiency by screening, then ^{123}I-BMIPP scintigraphy was performed. Expression of CD36 on platelets and monocytes was measured by flow cytometric analysis. The molecular defects responsible for CD36 deficiency was detected by allele-specific restriction enzyme analysis. CD36 expression was totally deficient in all 4 subjects on both platelets and monocytes. Two subjects were homozygous for a ^{478}C→T mutation. One was heterozygous for the dinucleotide deletion of exon V and single nucleotide insertion of exon X, and remaining one was considered to be heterozygous for the dinucleotide deletion of exon V and an unknown gene abnormality. All cases demonstrated a completely negative accumulation of myocardial LCFA despite of normal myocardial perfusion, which was evaluated by thallium scintigraphy. In addition, all cases demonstrated apparently normal hepatic LCFA accumulation Thus, these findings suggested that CD36 acts as a major myocardial specific LCFA transporter in humans. (Mol Cell Biochem **192**: 129–135, 1999)

Key words: CD36 deficiency, myocardial long-chain fatty acid uptake, mutation of CD36 gene

Address for offprints: S. Nozaki, The Second Department of Internal Medicine, Osaka University Medical School, 2-2, Yamadaoka, Suita, Osaka 565 Japan.

Introduction

Long-chain fatty acids (LCFA) are the major energy substrate for heart and their oxidation is important for achieving maximal cardiac work. Although LCFA is thought to be took up via specific transporter, the myocardial LCFA uptake mechanism(s) has not yet been clarified. Abumrad *et al.* cloned fatty acid transporter (FAT) in rats which has high homology with CD36 in humans [1]. We also previously isolated a myocardial LCFA transporter from bovine heart [2], which showed a sequence homology to human CD36.

CD36 is a glycoprotein with a molecular weight of 88 kDa and is expressed on platelets, monocytes/macrophages, capillary endothelial cells and adipocytes [1]. It has been proposed that CD36 is a multifunctional molecule [3], including receptors for thrombospondin [4], collagen [5], oxidized LDL [6], or a membrane bound fatty acids transporter [1, 2].

A lack of CD36 expression on platelets was first identified in a thrombocytopenic patient with refractoriness to HLA-matched platelet transfusion [7]. The patient had anti-platelet antibody, Nak^a antibody. Tomiyama *et al.* revealed that Nak^a antigen was identical to CD36 [8]. Since then, we found several cases with CD36 deficiency and have clarified the molecular defects responsible for CD36 deficiency [9–12]. We and others also identified two types of CD36 deficiency [13, 14]. In type I CD36 deficiency, neither platelets nor monocytes express CD36; in type II CD36 deficiency, monocytes express CD36, but platelets do not. The frequency of type I CD36 deficiency is much rarer than that of type II CD36 deficiency [13]. However, proposed roles of CD36 have not been clarified, especially for LCFA uptake in human.

The clinical availability of radiolabeled LCFA analogues has made it possible to investigate myocardial LCFA metabolism in patients with heart disease. Among them, Iodine-123 15-(p-iodophenyl)-(R,S)-methylpentadecanoic acid (^{123}I-BMIPP) has the character to show normal myocardial extraction and no readily catabolization through the oxidative pathway [15]. In clinical studies, many researches have described the clinical usefulness of LCFA analog scintigraphy in ischemic heart disease [16, 17] and hypertrophic cardiomyopathy [18, 19]. Recently total lack of accumulation has been reported in very restricted cases, in whom the etiology has not yet been identified [20, 21]. We also experienced a case with CD36 deficiency [22] with hypertrophic cardiomyopathy (HCM) who showed absence of myocardial LCFA accumulation.

Based on these background, in this study we investigated the relationship between the abnormalities of CD36 molecule and myocardial LCFA uptake. CD36 expression levels and CD36 gene were analyzed in two individuals who showed no LCFA accumulation by myocardial ^{123}I-BMIPP scintigraphy. In addition, we also studied the myocardial LCFA accumulation using ^{123}I-BMIPP scintigraphy in two individuals who were identified as CD36 deficiency at the screening of normal volunteers and outpatient clinics. The present study shows that all these 4 individuals were type I CD36 deficiency with a complete absence of myocardial LCFA accumulation despite an apparently normal hepatic LCFA accumulation. These findings suggest that abnormalities of CD36 molecule might cause the total myocardial accumulation deficiency of ^{123}I-BMIPP and that CD36 might be a myocardial specific LCFA transporter.

Materials and methods

Subjects

Four unrelated subjects were enrolled in this study. In two subjects (Cases 1 and 2), who were found to have absence of myocardial LCFA accumulation by ^{123}I-BMIPP scintigraphy, the expression of CD36 on their platelets and monocytes was analyzed. The remaining two subjects were identified to be CD36 deficiency by screening. The expression of CD36 on platelets was screened in 629 normal volunteers and 285 outpatients by flow cytometric analysis. Two out of 4 type I CD36-deficient subjects found in the screening were investigated in this study. The subjects' clinical profiles are as follows:

Case 1: K.S. (63-year-old male) was diagnosed as having HCM at the age of 39 years, and had been followed up at the outpatient clinic of Osaka Medical College Hospital. At the age of 61 years, his myocardial scintigram demonstrated a negative myocardial LCFA accumulation in spite of an apparently normal myocardial perfusion by thallium scintigraphy.

Case 2: H.Y. (56-year-old male) first visited Saiseikai Nakatsu Hospital because of chest pain. Coronary angiography demonstrated 90% stenosis at the left anterior descending coronary artery and myocardial scintigraphy revealed the absence of myocardial accumulation of ^{123}I-BMIPP.

Case 3: A.T. (51-year-old female) visited Osaka Red Cross Blood Center at the age of 48 years for donating blood. Flow cytometric analysis on screening revealed that both her platelets and monocytes lacked CD36 expression on the cell surface. She suffered from hypertension, hyperlipidemia and obesity, hence she was referred to the outpatient clinic of Osaka University Hospital.

Case 4: S.A. (63-year-old female) was admitted for the examination of exertional chest pain. Flow cytometric analysis revealed that both her platelets and monocytes lacked CD36 expression on the cell surface. Thallium (T1)-scintigraphy at rest showed no defect in myocardium although cardiac catheterization revealed 90% stenosis in the left anterior descending coronary artery and 90% stenosis in the right coronary artery.

Flow cytometric detection of CD36 on platelets and monocytes

Expression of CD36 on platelets and monocytes was analyzed by flow cytometry as previously described [13]. In brief, a 50 μl suspension of platelets (2×10^5/μl) or mononuclear cells (2×10^4/μl) was incubated for 30 min with 5 μl of FITC-labeled anti-human CD36 monoclonal antibody (OKM5; Ortho Diagnostic Systems, Raritan, NJ, USA) or FITC-labeled control IgG. Cells were then diluted to 400 μl in TBS (Tris-buffered saline) and analyzed on a FACScan (Becton Dickinson, Sunnyvale, CA, USA).

Detection of mutations in the CD36 gene

Detection of the mutations leading to CD36 deficiency was performed as previously described [9–12]. In brief, the regions of the CD36 gene around the previously identified mutations were amplified by polymerase chain reaction (PCR) using the following primers: primers NAK386(+)-SAU and ID(–) for the region around the C→T mutation at nucleotide (nt) 478 located in exon IV of the CD36 gene; primers ID(+) and SSPI(–) for the region around the dinucleotide deletion in exon V; and primers XMN(+) and XMN(–) for the region around the single nucleotide insertion in exon X. The amplified fragments were directly digested with *Sau*96 I, *Ssp* I, and *Xmn* I for the detection of the ^{478} C→T mutation, the dinucleotide deletion, and the single nucleotide insertion, respectively. The resulting fragments were subjected to electrophoresis on nondenatured polyacrylamide gels and stained with ethidium bromide.

Iodine-^{123} BMIPP and thallium-201 (^{201}Tl) myocardial single photon emission computed tomography

Myocardial LCFA accumulation and myocardial perfusion were respectively evaluated using a radiolabeled LCFA analog (^{123}I-BMIPP) and ^{201}Tl scintigraphy. After overnight fasting, ^{123}I-BMIPP (Nihon Medi-Physics, Japan) (148MBq) was administered intravenously, then planar and singlephoton emission computed tomography (SPECT) imaging was carried out 20 min after the injection. Thallium-201 SPECT imaging was performed on another day.

Results

Flow cytometric analyses demonstrated that all cases showed totally negative expression of CD36 on both platelets and monocytes (Fig. 1). Thus, they were diagnosed as type I CD36 deficiency. Then, genetic analysis for these four type I CD36 deficiency was performed. Figure 2 depicts the

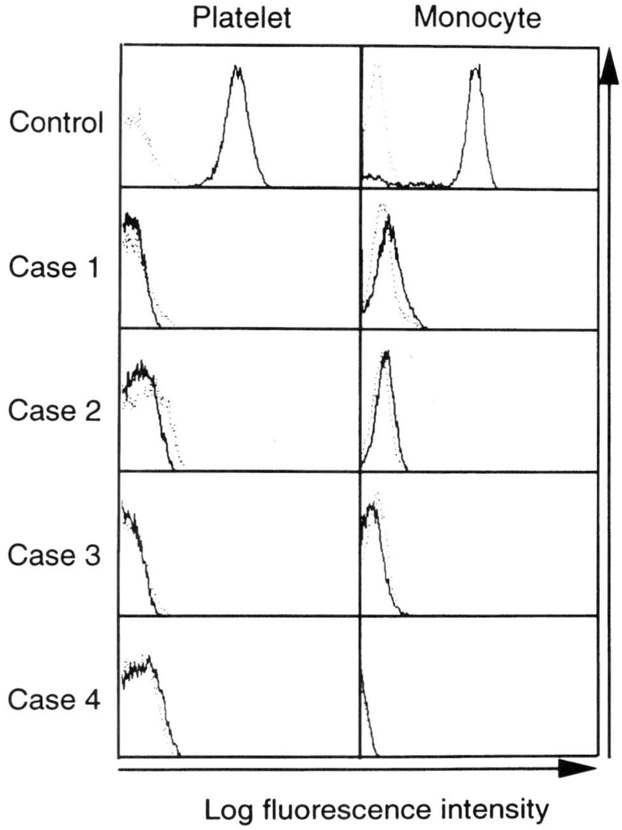

Fig. 1. Flow cytometric analysis for surface expression of CD36 on platelets and monocytes from Cases 1–4 and a normal control. Dotted line shows the fluorescence intensity expression level of control platelets or monocytes incubated with control FITC-conjugated IgG1. Solid line shows the fluorescence intensity of platelets or monocytes incubated with FITC-conjugated OKM5.

mutations in the CD36 gene leading to CD36 deficiency. Lanes 1, 2, 3, 4, and C in Fig. 2 contain the digested PCR fragments from Case 1, Case 2, Case 3, Case 4, and a normal control, respectively. The undigested 182 bp fragments in (A), the digested 107-bp fragments in (B), and the undigested 111-bp fragments in (C) represent the presence of the ^{478}C→T mutation, the dinucleotide deletion, and the single nucleotide insertion, respectively. These data showed that Case 1 and Case 2 were the homozygous for a ^{478}C→T mutation. Case 3 was heterozygous for the dinucleotide deletion of exon V and the single nucleotide insertion of exon X. Case 4 was heterozygous for the dinucleotide deletion of exon V and an unknown gene abnormality.

Echocardiography revealed normal cardiac wall motion in all cases and normal myocardium thickness in Cases 2, 3, and 4 (data not shown). Case 1 showed hypertrophic myocardium by echocardiogram. All subjects showed a total absence of ^{123}I-BMIPP accumulation in the myocardium in spite of apparently normal hepatic accumulation (Fig. 3). On the other

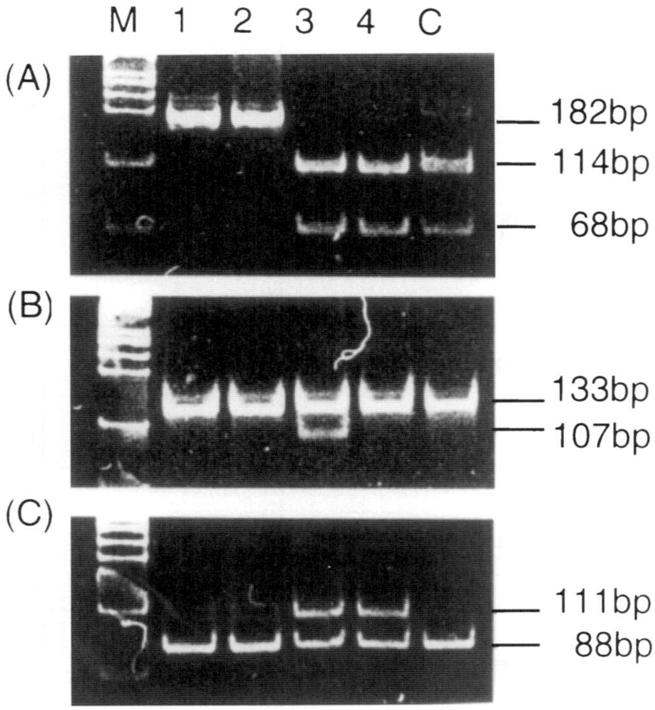

Fig. 2. Detection of the mutations in the CD36 gene. The regions of the CD36 gene around the previously identified mutations were amplified by PCR, followed by (A) *Sau96I* digestion for the detection of the C - T mutation in exon IV, (B) *Ssp I* digestion for the dinucleotide deletion in exon V, or (C) *Xmn I* digestion for the single nucleotide insertion in exon X. The resulting fragments were subjected to electrophoresis on an 8% polyacrylamide gel. Lane M contains fX 174 digested with *Hae III* as a size marker. Lanes 1, 2, 3, 4, and C contain the digested PCR fragments from Case 1, Case 2, Case 3, Case 4, and a normal control, respectively. The undigested 182-bp fragments in (A), the digested 107-bp fragments in (B), and the undigested 111-bp fragments in (C) represent the presence of the ^{478}C→T mutation, the dinucleotide deletion, and the single nucleotide insertion, respectively.

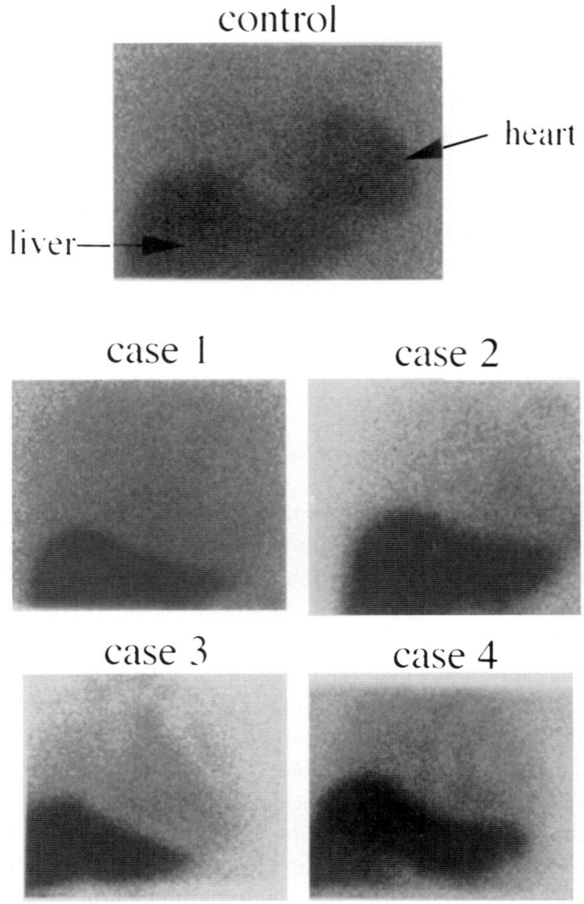

Fig. 3. Anterior planar imaging of myocardium by ^{123}I-BMIPP cardiac scintigraphy. ^{123}I-BMIPP (148 MBq) was injected intravenously under resting conditions after an overnight fast. Planar imaging revealed no myocardial ^{123}I-BMIPP accumulation in the subjects with CD36 deficiency in spite of apparently normal hepatic accumulation.

hand, thallium-201 single photon emission computed tomographic (SPECT) imaging showed an apparently normal myocardial perfusion at the resting stage in all cases. Figure 4 shows representative imaging of ^{123}I-BMIPP and thallium-201 SPECT from Case 3. Normal vascular flow was ascertained by ^{201}T1 scintigraphy, while ^{123}I-BMIPP accumulation was completely absent.

Discussion

The present study, for the first time, has clarified that subjects with type I CD36 deficiency are totally defective in myocardial LCFA accumulation despite of apparently normal hepatic LCFA accumulation. Consequently, these findings suggest that CD36 might take part in the specific uptake of myocardial LCFA.

We have recently experienced total lack of myocardial LCFA accumulation in a patient with hypertrophic cardiomyopathy (HCM) (Case 1) accompanied with type I CD36 deficiency [22]. Because partially impaired myocardial LCFA accumulation in HCM was reported [18, 19], we could not definitely conclude that the complete deficiency of myocardial LCFA uptake seen in that case was due to CD36 deficiency. Furthermore, genetic analysis for CD36 was not performed in that case. In the present study, we analyzed CD36 gene in addition to flow cytometric analyses for the diagnosis of CD36 deficiency. Genomic analysis revealed that Case 1 and 2 were homozygous for a ^{478}C→T mutation. Case 3 was a compound heterozygous for the dinucleotide deletion of exon V and the single nucleotide insertion of exon X. Case 4 was also speculated to be a compound heterozygous for the dinucleotide deletion of exon V and an unknown gene abnormality.

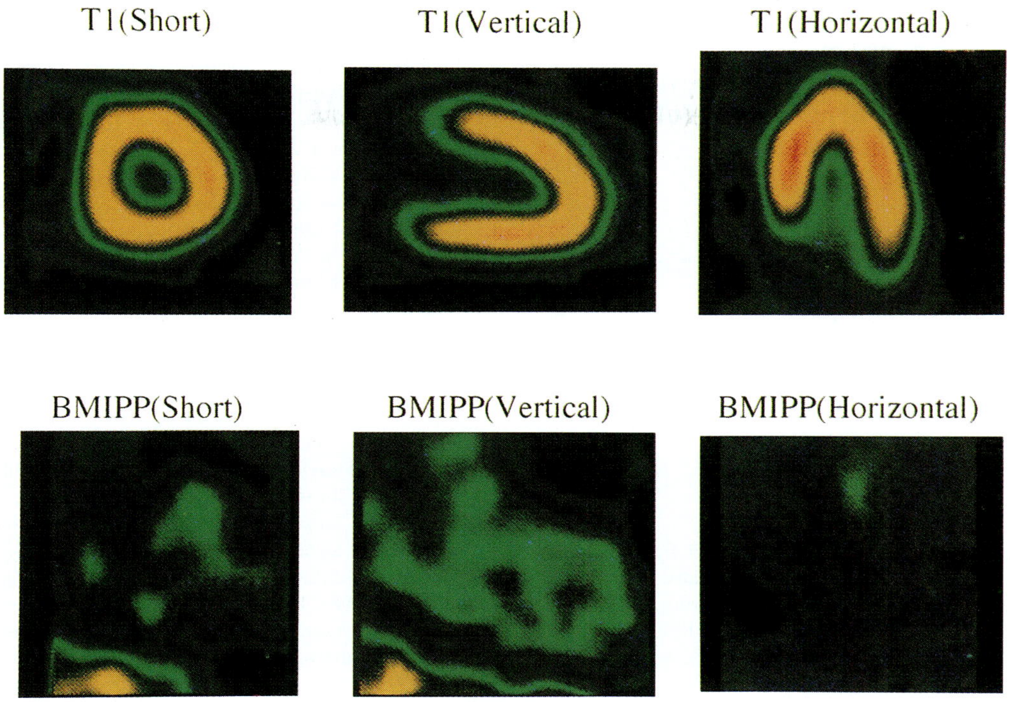

Fig. 4. Comparison of images of ^{201}Tl and ^{123}I-BMIPP SPECT. Thallium 201 (148 MBq) was injected intravenously under the same conditions as the ^{123}I-BMIPP study with an interval of at least 1 week. Left ventricular short-axis (short), vertical long-axis (vertical) and horizontal long-axis (horizontal) slices were obtained 20 min after injection in Case 3. The uptake of BMIPP (yellow color) in the liver can be seen in the bottom panel as a marker of BMIPP uptake in CD36 deficiency.

Although all subjects showed a preserved myocardial perfusion, their myocardial LCFA accumulation were completely absent and, in addition, their hepatic accumulation of LCFA were apparently normal. Taken together with these findings and experimental data (Abumrad and Tanaka), we concluded that the cause of total defect of myocardial LCFA accumulation seen in these subjects was due to abnormalities of CD36 gene.

The myocardium utilizes a large number of substrates to provide the energy required for contraction. Under normal aerobic fasting circumstances, LCFA is the preferred fuel and glucose utilization is minimal [23, 24]. In spite of the importance of LCFA as a major myocardial energy substrate, the mechanism(s) of myocardial LCFA uptake has not yet been identified. Hitherto, many debates over the processes underlying the transmembrane LCFA transport have been presented. Some investigators have claimed a mechanism of simple diffusion [25, 26]. On the other hand, others have asserted a facilitated transport system from the evidence of saturable uptake [27]. The first membrane protein involved in the uptake of LCFA was a 40 kDa plasma membrane protein of rat liver cells [28], which was subsequently demonstrated to be present in the intestine [29] and heart [30] in rats. Following, a 60-kDa protein [31], and an adipocyte long-chain fatty acid transport protein [32] were reported to be a putative membrane LCFA transporter. In addition, an 88-kDa membrane protein was enlisted as a candidate protein for LCFA uptake, in which amino acids sequence was homologous to human CD36 [2, 33, 34]. However, these were all *in vitro* experimental studies using animals. Hitherto myocardial LCFA uptake system has not been definitively identified *in vivo,* especially in human. The present study demonstrates the first evidence that CD36 molecule actually participates the myocardial LCFA uptake *in vivo*.

Metabolic and morphologic abnormalities brought about by impaired myocardial LCFA uptake are currently not known. It has been suggested that impaired LCFA metabolism is associated with cardiac hypertrophy [35, 36]. We previously reported that chemical intervention of myocardial LCFA transporter (homologous to human CD36) induced cardiac hypertrophy in rats [37] and also reported that carnitine-deficient mice show hypertrophic type of cardiomyopathy [38]. In the present study, three subjects showed normal myocardial wall thickness by echocardiography (Case 2, 3, 4), although Case 1 was HCM. We have recently reported that the prevalence of CD36 deficiency, including type II, is high in patients with HCM [22]. HCM is commonly diagnosed macroscopically by echocardiography. The microscopic features of cardiac myocytes in the subjects investigated remain unknown. In order to determine the abnormalities brought about by impaired myocardial LCFA uptake, further microscopic investigation of cardiac myocytes and an epidemiological study about the prevalence of CD36 deficiency in HCM patients will be necessary.

Acknowledgements

This work was supported by Grants-in Aid for Scientific Research from The Ministry of Education, Science, Sports and Culture in Japan (No. 07557074).

References

1. Abumrad NA, El-Maghrabi MR, Amri EZ, Lopez E, Grialdi PA: Cloning of a rat adipocyte membrane protein implicated in binding or transport of long-chain fatty acids that is induced during preadipocyte differentiation. Homology with human CD36. J Biol Chem 268: 17665–17668, 1993
2. Tanaka T, Kawamura K: Isolation of myocardial membrane long-chain fatty acid-binding protein; homology with a rat membrane protein implicated in the binding or transport of long-chain fatty acids. J Mol Cell Cardiol 27: 1613–1622, 1995
3. Greenwalt DE, Lipsky RH, Ockenhouse CF, Ikeda H, Tandon NN, Jamieson GA: Membrane glycoprotein CD36: A review of its roles in adherence, signal transduction and transfusion medicine. Blood 80: 1105–1115, 1992
4. Tandon NN, Kralisz NU, Jamieson GA: Identification of glycoprotein IV (CD36) as a primary receptor for platelet-collagen adhesion. J Biol Chem 264: 7576–7583, 1989
5. Asch AS, Barnwell J, Silverstein RL, Nachman RL: Isolation of the thrombospondin membrane receptor. J Clin Invest 79: 1054–1061, 1987
6. Endemann G, Stanton LW, Madden KS, Bryant CM, White RT, Protter AA: CD36 is a receptor for oxidized low density lipoprotein. J Biol Chem 268: 11811–11816, 1993
7. Ikeda H, Mitani T, Ohnuma M, Haga H, Ohtzuka S, Kato T, Nakase T, Sekiguchi S: A new platelet-specific antigen, Nak^a involved in the refractoriness of HLA-matched platelet transfusion. Vox Sang 57: 213–217, 1989
8. Tomiyama Y, Take H, Ikeda H, Mitani T, Furubayashi T, Mizutani H, Yamamoto N, Tandon NN, Sekiguchi S, Jamieson GA, Kurata Y, Yonezawa T, Tarui S: Identification of the platelet-specific alloantigen, Nak^a, on platelet membrane glycoprotein IV. Blood 75: 684–687, 1990.
9. Kashiwagi H, Honda S, Tomiyama Y, Mizutani H, Take H, Honda Y, Kosugi S, Kanayama Y, Kurata Y, Matsuzawa Y: A novel polymorphism in glycoprotein IV (replacement of proline-90 by serine) predominates in subjects with platelet GPIV. Thromb Haemostas 69: 481–484, 1993
10. Kashiwagi H, Tomiyama Y, Kosugi S, Shiraga M, Lipsky RH, Kanayama Y, Kurata Y, Matsuzawa Y: Identification of molecular defects in a subject with type I CD36 deficiency. Blood 83: 3545–3552, 1994
11. Kashiwagi H, Tomiyama Y, Honda S, Kosugi S, Shiraga M, Nagao N, Sekiguchi S, Kanayama Y, Kurata Y, Matsuzawa Y: Molecular basis of CD36 deficiency; evidence that a 478C-T substitution (proline90-serine) in CD36 cDNA accounts for CD36 deficiency. J Clin Invest 95: 1040–1046, 1995
12. Kashiwagi H, Tomiyama Y, Nozaki S, Honda S, Kosugi S, Shiraga M, Nakagawa T, Nagao N, Kanakura Y, Kurata Y, Matstizawa Y: A single nucleotide insertion in codon 317 of the CD36 gene leads to CD36 deficiency. Arterioscler Thromb Vasc Biol 16: 1026–1032, 1996
13. Take H, Kashiwagi H, Tomiyama Y, Honda S, Honda Y, Mizutani H, Furubayashi T, Karasuno T, Nishiura T, Kanayama Y, Kurata Y, Matsuzawa Y: Expression of GPIV and Nak' antigen on monocytes in Nak^a negative subjects whose platelets lack GPIV. Brit J Haematol 84: 387–391, 1993
14. Yamamoto N, Akamatsu N, Sakuraba H, Yamazaki H, Tanoue K: Platelet glycoprotein IV (CD36) deficiency is associated with the absence (Type I) or the presence (Type II) of glycoprotein IV on monocytes. Blood 83: 392–397, 1994
15. Knapp Mr, Kropp J: Iodine-123-labeled fatty acids for myocardial single photonemission tomography: Current status and future perspectives. Eur J Nucl Med 22: 361–381, 1995
16. Knapp Mr, Franken P, Kropp J: Cardiac SPECT with iodine-123-labeled fatty acids: Evaluation of myocardial viability with BMIPP. J Nucl Med 36: 1022–1030, 1995
17. Tateno M, Tamaki N, Yukihiro M, Kudoh T, Hattori N, Tadamura E, Nohara-R, Suzuki T, Endo K, Konishi J: Assessment of fatty acid uptake in ischemic heart disease without myocardial infarction. J Nucl Med 37: 1981–1985, 1996
18. Chen SL, Uehara T, Morozumi T, Yamagami H, Kusuoka H, Nishimura T: Myocardial metabolism of ^{123}I-BMIPP in patients with hypertrophic cardiomyopathy: Assessment by radial long-axis SPET. Nucl Med Commun 16: 336–343, 1995
19. Ohtsuki K, Sugihara H, Umamoto I, Nakamura T, Nakagawa T, Nakagawa M: Clinical evaluation of hypertrophic cardiomyopathy by myocardial scintigraphy using 123I-labelled 15-(p-iodophenyl)-3-R, S-methylpentadecanoic acid (123I-BMIPP). Nucl Med Commun 15: 441–447, 1994
20. Hwang EHH, Yamashita A, Takemori H, *et al.*: Absent myocardial I-123 BMIPP uptake in a family. Ann Nucl Med 10: 445–448, 1996
21. Kudoh T, Tamaki N, Magata Y, Konishi J, Nohara R, Iwasaki A, Ono S, Ohtake Y, Sugihara H, Kuze K, Tsujimura Y, Miyazaki T: Metabolism substrate with negative myocardial uptake of iodine-123-BMIPP. J Nucl Med 38: 548–553, 1997
22. Tanaka T, Sohmiya K, Kawamura K: Is CD36 deficiency an etiology of hereditary hypertrophic cardiomyopathy? J Mol Cell Cardiol 29: 121–127, 1997
23. Van der Vusse GJ, Glatz JFC, Starn HCG, Reneman RS: Fatty acid homeostasis in the normoxic and ischemic heart. Physiol Rev 72: 881–940, 1992
24. Rodrigues B, Cam MC, McNeffi JH: Myocardial substrate metabolism; implications for diabetic cardiomyopathy. J Mol Cell Cardiol 27: 169–179, 1995
25. DeGrella RF, Light RJ: Uptake and metabolism of fatty acids by dispersed adult rat heart myocytes. II. Inhibition by albumin and fatty acid homologues, and the effect of temperature and metabolic reagents. J Biol Chem 255: 9739–9745, 1980
26. Rose H, Hennecke T, Kammermeier H: Sarcolemmal fatty acid transfer in isolated cardiomyocytes governed by albumin/membrane-lipid partition. J Mol Cell Cardiol 22: 833–892, 1990
27. Harmon CM, Luce P, Beth AH, Abumrad NA: Labeling of adipocyte membranes by sulfo-N-succinimidyl derivates of long-chain fatty acids: Inhibition of fatty acid transporter. J Membrane Biol 121: 261–268, 1991
28. Stremmel W, Strohmeyer G, Borchard F, Kochwa S, Berk PD: Isolation and partial characterization of a fatty acid binding protein in rat liver plasma membranes. Proc Natl Acad Sci USA 82: 4–8, 1985
29. Stremmel W, Lotz G, Strohmeyer G, Berk PD: Identification, isolation, and partial characterization of a fatty acid binding protein from rat jejunal microvillous membranes. J Clin Invest 75: 1068–1076, 1985

30. Stremmel W: Fatty acid uptake by isolated rat heart myocytes represents a carrier-mediated transport system. J Clin Invest 81: 844–853, 1988
31. Fujii S, Kawaguchi H, Yasuda H: Purification of high affinity fatty acid receptors in rat myocardial sarcolemmal membranes. Lipids 22: 544–546, 1987
32. Schaffer JE, Lodish HF: Expression cloning and characterization of a novel adipocyte long chain fatty acid transport protein. Cell 79: 427–436, 1994
33. Harmon CM, Aburnrad NA: Binding of sulfosuccinimidyl fatty acids to adipocyte membrane proteins: Isolation and amino-terminal sequence of an 88-M protein implicated in transport of long-chain fatty acids. J Mem Biol 133: 43–49, 1993
34. Abumrad NA, Park JH, Park CR: Permeation of long-chain fatty acid into adipocytes. Kinetics, specificity, and evidence for involvement of a membrane protein. J Biol Chem 259: 8945–8953, 1984
35. Kurata C, Kobayashi A, Yamazaki N: Dual tracer autoradiographic study with thallium-201 and radioiodinated fatty acid in cardiomyopathic hamsters. J Nucl Med 30: 80–87, 1989
36. Takeishi Y, Chiba J, Abe S, Tonooka I, Komatani A, Tomoike H: Heterogeneous myocardial distribution of iodine-123 15–(p-iodophenyl)-3-R, S-methylpentadecanoic acid (BMIPP) in patients with hypertrophic cardiomyopathy. Eur J Nucl Med 19: 775–782, 1992
37. Kusaka Y, Tanaka T, Okamoto F, Terasaki F, Matsunaga Y, Miyazaki H, Kawamura K: Effect of sulfo-N-succinimidyl palmitate on the rat heart: Myocardial long-chain fatty acid uptake and cardiac hypertrophy. J Mol Cell Cardiol 27: 1605–1612, 1995
38. Horiuchi M, Yoshida H, Kobayashi K, Kuriwaki K, Yoshimine K, Tomomura M, Koizumi T, Nikaido H, Hayakawa J, Kuwajima, M, Saheki T: Cardiac hypertrophy in juvenile visceral steatosis (jvs) mice with systemic carnitine deficiency. FEBS Lett 326: 267–271, 1993

Molecular and Cellular Biochemistry **192:** 137–142, 1999.

Structural and functional studies on different human FABP types

J.H. Veerkamp, H.T.B. van Moerkerk, C.F.M. Prinsen, T.H. van Kuppevelt

Department of Biochemistry, University of Nijmegen, Nijmegen, The Netherlands

Abstract

Interaction of various ligands with recombinant proteins of 5 human FABP types was studied by radiochemical and fluorescence procedures. Liver, heart, intestinal and myelin FABP showed a higher affinity for oleic acid than adipocyte FABP. Intestinal and adipocyte FABP had a relatively high K_d value for arachidonic acid. Liver and intestinal FABP showed high affinity for DAUDA in contrast to the other FABP types. ANS was only well bound by liver and adipocyte FABP. Retinol was not bound by any FABP type, retinoic acid only by adipocyte FABP. Data indicate the importance of both electrostatic and hydrophobic interaction for the ligand-FABP binding. The immunological crossreactivity between six human FABP types including epidermal FABP and their respective antibodies raised in rabbit, chicken and mouse appeared to be low and may suggest heterogeneity of protein surface. (Mol Cell Biochem **192**: 137–142, 1999)

Key words: fatty acid-binding protein, myelin FABP, 11-dansylamino-undecanoic acid, 1-anilinonaphtalene-8-sulfonic acid, retinoic acid, immunological crossreactivity

Introduction

Nine types of cytoplasmic FABP have been well identified up to now [1–3]. They are 14–15 kDa proteins of 126–134 amino acids and are named after the first tissue of isolation and/or identification. Some types are limited to one tissue (intestinal, adipocyte, myelin, brain, testicular). Heart FABP occurs in many tissues, heart, skeletal, and smooth muscles, aorta, kidney, lung, placenta, and brain. Some tissues contain more FABP types, either in different cell types (brain, kidney ovary, stomach) or in the same cell (enterocyte). These tissues show spatio-temporally differential expression of different FABP types [4–6]. Figure 1 shows an alignment of the amino acid sequences for seven human FABP types, and for the recently identified rat brain FABP type [2]. They show 22–73% similarity of sequence. Although FABPs have a similar clam shell-like structure the conformation of the fatty acid within the binding cavity is different [7, 8].

Many studies with different techniques have been performed on fatty acid binding to different FABP types [9, 11], but comparative studies on more FABP types and more fatty acids with the same procedure are sparse [12, 13]. In this report we describe the binding characteristics of oleic, palmitic and arachidonic acid to recombinant preparations of human heart, liver, intestinal, adipocyte and myelin FABP. In addition, we studied their interaction with 11-dansylamino-undecanoic acid (DAUDA), 1-anilinonaphtalene-8-sulfonic acid (ANS), retinol, and retinoic acid. The former fluorescent probes have been used previously with some FABP types [13–16] in addition to 9-anthroyloxy-labeled fatty acids [13]. A systematic study of the binding of these former fluorescent ligands and of retinoids to different FABP types lacks in the literature. For a rough comparison of the external surface of the FABP we studied the reactivity of the 5 human FABP types mentioned and of human epidermal FABP with antibodies raised against the different types in rabbit, chicken and mice.

Materials and methods

Materials

Recombinant proteins of human liver and heart FABP were prepared as described previously [13, 17] Human adipocyte

Address for offprints: J.H. Veerkamp, Department of Biochemistry, University of Nijmegen, P.O. Box 9101, 6500 HB Nijmegen, The Netherlands

```
           1
H-FABP     MVDA..FLGT WKLVDSKNFD DYMKSLGVGF ATRQVASMT. .KPTTIIEKN 45
B-FABP     MVDA..FCAT WKLTDSQNFD EYMKALGVGF ATRQVGNVT. .KPTVIISQE 45
My-FABP    .SNK..FLGT WKLVSSENFD DYMKALGVGL ATRKLGNLA. .KPTVIISKK 45
A-FABP     MCDA..FVGT WKLVSSENFD DYMKEVGVGF ATRKVAGMA. .KPNMIISVN 45
E-FABP     MATVQQLEGR WRLVDSKGFD EYMKELGVGI ALRKMGAMA. .KPDCIITCD 47
I-FABP     MA....FDST WKVDRSENYD KFMEKMGVNI VKRKLAAHDN LK..LTITQE 43
L-FABP     MS....FSGK YQLQSQENFE AFMKAIGLPE EL..IQKGKD IKGVSEIVQN 43
Il-FABP    MA....FTGK FEMESEKNYD EFMKLLGISS DV..IEKARN FKIVTEVQQD 43

H-FABP     GDILTLKTHS TFKNTEIS.F KLGVEFDETT ADDRKVKSIV TLD.GGKLVH 93
B-FABP     GGKVVIRTQC TFKNTEIS.F QLGEEFEETS IDDRNCKSVI RLD.GDKLIH 93
My-FABP    GDIITIRTES TFKNTEIS.F KLGQEFEETT ADNRKTKSIV TLQ.RGSLNQ 93
A-FABP     GDVITIKSES TFKNTEIS.F ILGQEFDEVT ADDRKVKSTI TLD.GGVLVH 93
E-FABP     GKNLTIKTES TLKTTQFS.C TLGEKFEETT ADGRKTQTVC NFT.DGALVQ 95
I-FABP     GNKFTVKESS AFRNIEVV.F ELGVTFNYNL ADGTELRGTW SLE.GNKLIG 91
L-FABP     GKHFKFTITA GSKVIQNE.F TVGEECELET MTGEKVKTVV QLEGDNKLVT 92
Il-FABP    GQDFTWSQHY SGGHTMTNKF TVGKESNIQT MGGKTFKATV QME.GGKLVV 92

H-FABP     LQKWD...GQ ETTLVRELID G.KLILTLTH GTAVCTRTYE KEA        132
B-FABP     VQKWD...GK ETNCVREIKD G.KMVVTLTF GDVVAVRCYE KA.        131
My-FABP    VQRWN...GK ETTIKRKLVD G.KMVAECKM KGVVCTRIYE KV.        131
A-FABP     VQKWD...GK STTIKRKRED D.KLVVECVM KGVTSTRVYE RA.        131
E-FABP     HQEWD...GK ESTITRKLKD G.KLVVECVM NNVTCTRIYE KVE        134
I-FABP     KFKRTD.NGN ELNTVREIIG D.ELVQTYVY EGVEAKRIFK KD.        131
L-FABP     TF.....KN. .IKSVTELNG D.IITNTMTL GDIVFKRISK RI.        126
Il-FABP    NF.....PN. .YHQTSEIVG D.KLVEVSTI GGVTYERVSK RLA        127
```

Fig. 1. Alignment of the amino acid sequences of the members of the FABP family. All sequences are for human proteins, except B-FABP, which is from rat. Positions of well-conserved amino acids (identical residues present in at least 5 molecules) are shaded. H-FABP – heart FABP; B-FABP – brain FABP; My-FABP – myelin FABP; A-FABP – adipocyte FABP; E-FABP – epidermal FABP; I-FABP – intestinal FABP; L-FABP – liver FABP; II-FABP – ileal FABP.

and intestinal FABP were obtained by similar procedures from cultures of *Escherichia coli* cells provided with these cDNAs inserted into a pET expression vector. These cells were a kind gift of Dr A.M. Kleinfeld (Medical Biology Institute, La Jolla, CA, USA) and Dr J.C. Sacchettini (Albert Einstein College of Medicine, Bronx, NY, USA), respectively. Human myelin FABP cDNA in pUC19 vector (obtained from Dr Hayasaka, Yamagata University School of Medicine, Yamagata, Japan) was amplified and modified with PCR to obtain NdeI and BamHI sites, checked on sequence in pCRII vector and ligated in the pET 11a expression vector. Myelin FABP was expressed in transformed BL21(DE3) cells and isolated from supernatant after sonication, treatment with 1% protamine sulfate, acidification with acetic acid to pH 5.0 and 50% $(NH_4)_2SO_4$ precipitation. The supernatant was dialysed against 10 mM Na acetate (pH 5.0) and myelin FABP purified by DEAE Sepharose and Sephadex G50 chromatography.

[1-^{14}C]fatty acids were from Amersham International, Little Chalford, UK; DAUDA and 1,8-ANS from Molecular Probes, Junction City, OR, USA.

Binding assays

Immediately before binding assays, FABP preparations were delipidated with the Lipidex procedure [18]. All fatty acid binding assays were carried out at 37°C in a volume of 0.5 ml 10 mM Tris-I-IC1 (pH 8.0) containing 6–10 μg FABP and 0–3 μM [1-^{14}C] fatty acid. Unbound fatty acid was removed with Lipidex [18]. K_d values were determined by Scatchard analysis of binding data.

Emission spectra (400–550 nm) and fluorescence enhancement of 1 μM DAUDA, ANS, retinoic acid or retinol were measured in 1 ml 10 mM Tris-HCI (pH 8.0)/1% ethanol at FABP concentrations of 0–150 μg/ml. Excitation wavelengths used were 335, 369, 350 and 330 run for DAUDA, ANS, retinoic acid and retinol, respectively. Assays were carried out at 25°C at a Shimadzu 500 fluorimeter.

Tryptophan fluorescence of FABP preparations (30 μg protein, excitation at 285 mn, emission at 330 mn) was assayed in the presence of 0–8 μM retinoic acid or retinol in 1 ml Tris-HCl (pH 8.0)/1 % ethanol. Protein was determined with the Lowry procedure.

Immunological experiments

Antisera were raised in rabbits as described previously [19]. Cross-reactivities of anti-FABP sera were determined in ELISA with 400 ng FABP coated per well. Peroxidase-conjugated anti-rabbit IgG (γ-chain specific) was used as second step antibody, and reaction visualized by incubation with o-phenylene diamine. Cross-reactivity was defined as the reciprocal of the serum dilution required to half maximal absorbance and expressed as percentage of the value obtained with the specific antigen-antiserum combination [19]. Human epidermal FABP and its antiserum were obtained from Dr G. Siegenthaler (Laboratory Clinic Dermatology, University of Geneve, Switzerland).

Results and discussions

We realize that the Lipidex procedure results in a systematic overestimate of the equilibrium value of the free fatty acid concentration and therefore gives relatively higher dissociation constants than the use ofADIFAB [12]. For comparative purpose the former procedure gives, however, interesting data. Our 5 human FABP types show clearly distinct fatty acid-binding activities, also depending on the type of fatty acid (Table 1). Adipocyte FABP showed the highest K_d values like with the ADIFAB procedure [12]. Intestinal FABP showed also a higher K_d value for arachidonic acid than for palmitic and oleic acid. Data on liver and heart FABP agree with previous results [9, 13, 20], and show similar fatty acid-dependency as with the ADIFAB procedure [12]. Myelin FABP has a similar affinity pattern as heart FABP. This FABP type was not studied before. Only oleic acid binding was shown by gel filtration [21]. The binding affinities of murine brain FABP are similar to our data for human heart and myelin FABP, but this FABP type did not bind palmitic acid and had a very high affinity for docosahexaenoic acid [22]. The specificity of fatty acid binding by FABP types may relate to their different function. They may transfer fatty acids to specific cellular organelles for metabolism, but may also modulate differentially systems and processes that can be influenced by free fatty acids as nuclear receptors and ion channels.

DAUDA was proposed and studied as an indicator of liver FABP [13–15]. It is poorly bound by heart FABP [13]. Now we found, that also intestinal FABP binds this fluorescent ligand well (Fig. 2A) in contrast to myelin, heart and adipocyte FABP (Fig. 2B). Liver and intestinal FABP caused both a marked shift in the fluorescence emission maximum of DAUDA from 533 to 501 and 492 nm respectively, and a high fluorescence enhancement. Both FABP types display high-affinity association (K_d values of DAUDA 1.4 and 1. 1 μM, respectively). The shift (to about 520 NM) and the

Table 1. K_d values of human FABP types.

FABP type	Palmitic acid	Oleic acid	Arachidonic acid
Liver	4.02 ± 0.11	0.89 ± 0.03	0.44 ± 0.04
Heart	0.96 ± 0.05	0.43 ± 0.04	0.37 ± 0.02
Adipocyte	2.86 ± 0.59	1.56 ± 0.25	2.49 ± 0.22
Myelin	0.62 ± 0.16	0.31 ± 0.03	0.37 ± 0.02
Intestinal	0.60 ± 0.08	0.57 ± 0.04	1.49 ± 0.16

K_d values (in μM) were determined with [^{14}C] fatty acid (0–3 μM) and 0.4 μM protein. Unbound fatty acid was removed with Lipidex. Values are means ± S.D. for 3–7 independent experiments.

enhancement were much less for the other FABP types. These and previous [13, 15] data suggest that the location of the fluorescent group of DAUDA is different in the two groups of FABP types.

ANS showed previously interaction with rat and human liver FABP [23], rat intestinal FABP [24], and murine adipocyte and epidermal FABP [16]. All our human FABP types showed maximal fluorescence emission with this fluorescent probe at wave lengths of 465–475 nm, and also fluorescence enhancement at increasing protein concentrations (Fig. 3). In contrast to DAUDA, intestinal FABP showed with ANS a markedly lower response than the other FABP types. Liver and adipocyte FABP caused a relatively high increase in fluorescence with high affinity (Kd values 2.0 and 1.1 μM, respectively). Other FABP types show low-affinity binding. Our data on human adipocyte FABP are in accordance with those on murine adipocyte FABP [16]. Murine epidermal FABP showed comparable high-affinity binding and fluorescence with ANS as adipocyte FABP, but cellular retinoic acid- and retinol-binding proteins showed low-affinity binding [16].

We tested the interaction of the FABP types with retinol and retinoic acid by assay of retinoid and tryptophan fluorescence. Addition of up to 150 μg FABP to 1 μM retinol or retinoic acid caused no change of fluorescence emission wave length or fluorescence yield in contrast to cellular retinol and retinoic acid-binding protein [25, 26]. The degree of quenching of tryptophan fluorescence is related to the relative location and orientation of the ligand. Retinol (up to 8 μM) did not change the intrinsic tryptophan fluorescence of human heart, intestinal, myelin and adipocyte FABP like previously observed for intestinal [27] and adipocyte [28] FABP. Human liver FABP has no tryptophan, so the quenching assay is not possible with this FABP type. For myelin FABP retinol binding was previously suggested on base of comigration at gel filtration [21]. In contrast to wild type intestinal FABP, its Arg106Gln mutant bound retinol, although with less affinity than cellular retinol-binding protein [27].

In spite of the similar content of 2 tryptophans per molecule in all these FABP types (Fig. 1), adipocyte FABP showed a much higher intrinsic tryptophan fluorescence than heart,

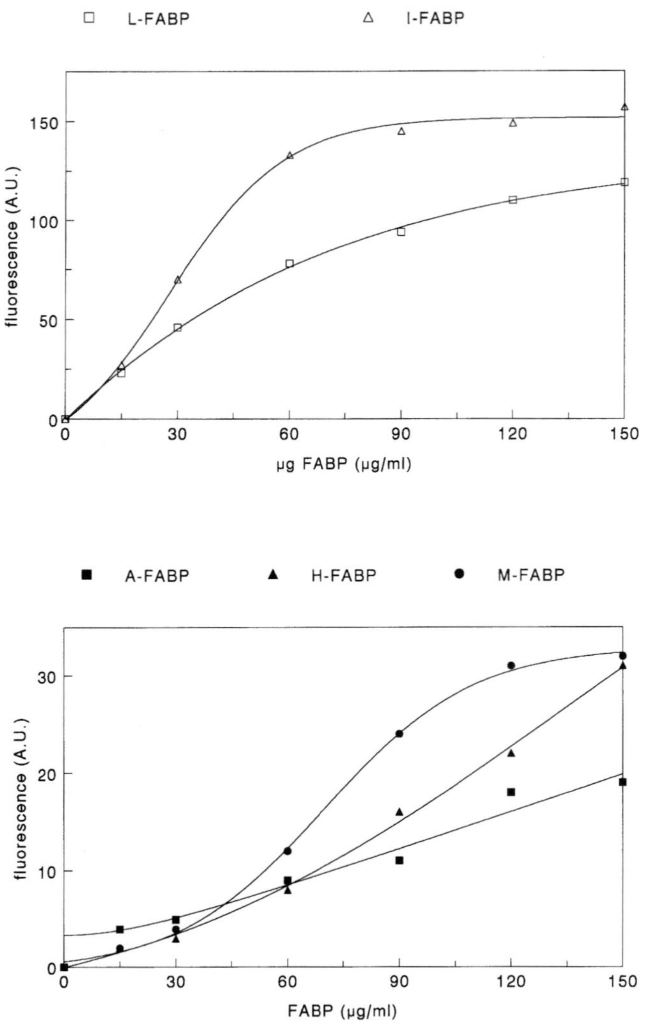

Fig. 2. Fluorescence of 5 human FABP types with 11-dansylamino-undecanoic acid (DAUDA). Different concentrations FABP were incubated with 1 μM DAUDA. Fluorescence (in arbitrary units) at maximal emission wavelength is shown as mean for 3 independent experiments. Variation coefficient was less than 5%. (A) Liver FABP (L-FABP) and intestinal FABP (I-FABP). (B) Adipocyte FABP (A-FABP), heart FABP (H-FABP) and myelin FABP (M-FABP).

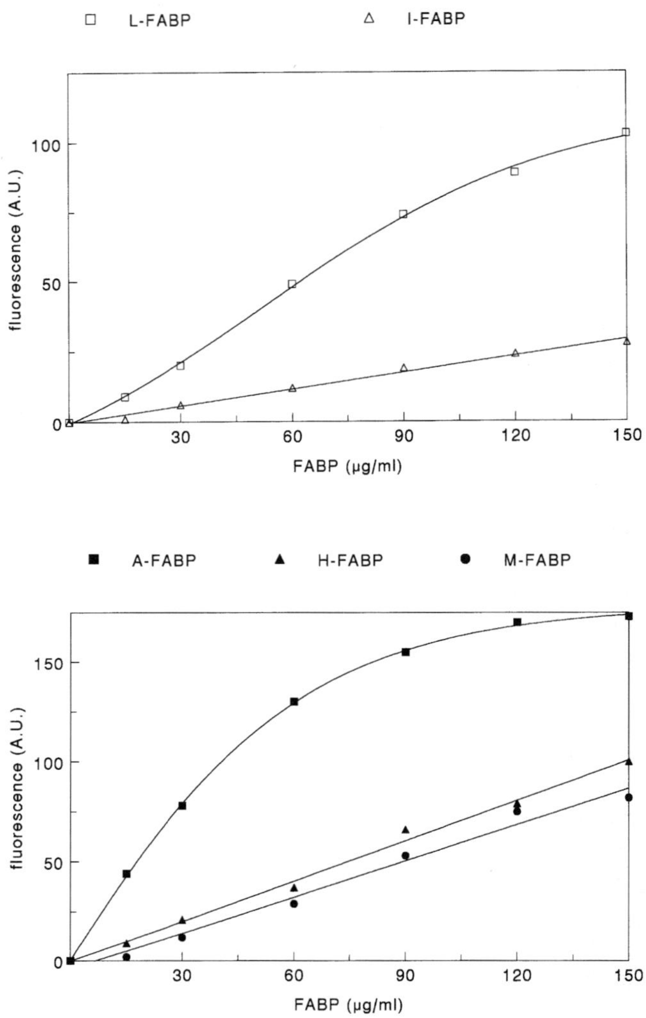

Fig. 3. Fluorescence of 5 human FABP types with 1,8-anilinonaphthalene-sulfonic acid (1,8-ANS). Different concentrations FABP were incubated with 1 μM 1,8-ANS. Fluorescence (in arbitrary units) at maximal emission wavelength is shown as mean for 3 independent experiments. Variation coefficient was less than 5%. (A) L-FABP and I-FABP. (B) A-FABP, H-FABP and M-FABP.

myelin and intestinal FABP (Fig. 4). Quenching of tryptophan fluorescence at retinoic acid addition was also most prominent, absolutely and relatively, with adipocyte FABP. Intestinal FABP did not show quenching, heart and myelin FABPs showed only minor effects. Murine adipocyte FABP showed also binding of retinoic acid [28] and a comparable decrease (about 50%) in intrinsic tryptophan fluorescence was observed with human adipocyte FABP by Baxa *et al.* [29]. Myelin FABP showed binding of retinoic acid at gel filtration [21]. Rat intestinal FABP showed also no significant interaction with retinoic acid in contrast to its Arg106G1n mutant [27]. Data indicate that FABPs generally do not bind retinoids well except adipocyte FABP, which interacts with retinoic acid. The ligand-protein electrostatic interaction, but also other properties of the binding cavity, appear to be important for the interaction of FABP with natural fatty acids, fluorescent probes and retinoids (Table 2). The functional diversity of FABP was also indicated by previous studies which showed the more general ligand-binding properties of liver FABP in contrast to heart and intestinal FABP [9–11, 13, 23, 30].

Different transfer rates and transfer mechanisms have been found for anthroyloxylabeled fatty acids from liver, heart, intestinal and adipocyte FABP to membranes [31, 32]. The collisional interaction of the latter 3 FABP types with

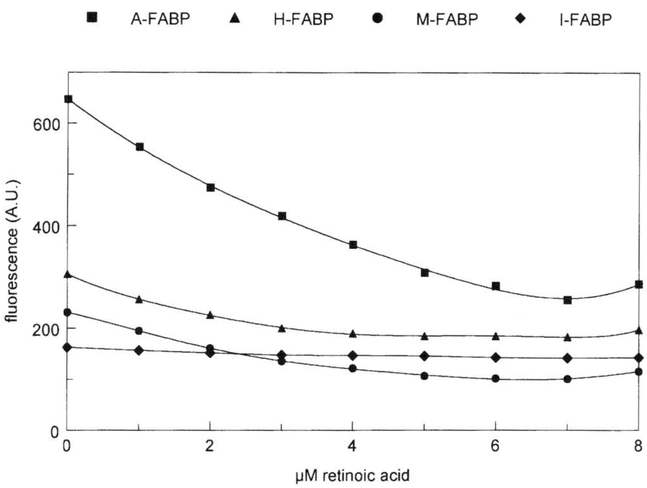

Fig. 4. Tryptophan fluorescence of human FABPs in the presence of 0–8 μM retinoic acid. Protein concentration was 30 μg/ml. Excitation was at 285 nm, emission at 330 nm. Values are means of 2 independent experiments.

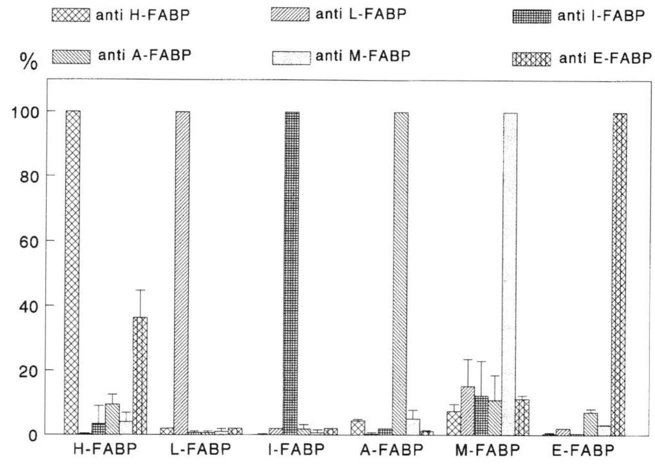

Fig. 5. Immunological reactivity of 6 human FABP types and their respective rabbit antisera. Antisera were applied in ELISA at a concentration giving 50% of the maximal response with their corresponding antigen. Wells were coated with 400 Ng FABP. Values are given as percentage of the binding of the specific antiserum to its specific antigen and are means ± S.D. of 3 independent assays.

phospholipid membranes suggested the important role of the surface charge of membrane and FABP. Blocking and mutation of certain lysine residues changed rate and mechanism of transfer with heart and adipocyte FABP [31, 33, 34]. We tested the immunological cross-reactivity as an indicator of surface properties for 6 human FABP types (Fig. 5). Liver, intestinal, adipocyte and epidermal FABP types showed only significant reactivity with their own antisera raised in rabbits. Myelin FABP showed some cross-reactivity with all other antisera, heart FABP had affinity for anti-heart and anti-epidermal FABP antisera. Absence of cross-reactivity was also previously established for FABPs from liver and heart and their antisera in the case of human, pig and rat [19]. Murine monoclonal antibodies could be prepared which showed specificity for human heart, liver or intestinal FABP. After injection of layer hens with FABPs, antibodies could be isolated from their eggs, with specificity against human liver, myelin, intestinal or adipocyte FABP. Immunological data suggests that the outer surface of FABP types is heterogenous in agreement with the primary sequence of these parts of the proteins. This specificity of antisera is also of special importance for specific quantitation of FABP types by ELISA in body fluids and immunohistochemical studies of tissue samples.

Table 2. Affinity of ligands to different FABP types

FABP type	Oleic acid	DAUDA	ANS	Retinol	Retinoic acid
Liver FABP	+ +	+ +	+ +	–	–
Intestinal FABP	+ +	+ +	±	–	–
Heart FABP	+ +	±	+	–	±
Adipocyte FABP	+	±	+ +	–	+
Myelin FABP	+ +	+	+	–	±
Epidermal FABP	+ +		+ +	–	+

Symbols: + +, high; +, low; ±, weak; –, no.

References

1. Veerkamp JH, Maatman RGHJ: Cytoplasmic fatty acid-binding proteins: Their structure and genes. Prog Lipid Res 34: 17–52, 1995
2. Bennett E, Stenvers KL, Lund PK, Popko B: Cloning and characterization of a cDNA encoding a novel fatty acid binding protein from rat brain. J Neurochem 63: 1616–1624, 1994
3. Oko R, Morales CR: A novel testicular protein, with sequence similarities to a family of lipid binding proteins, is a major component of the rat sperm perinuclear theca. Dev Biol 166: 235–245, 1994
4. Owada Y, Yoshimoto T, Kondo H: Spatio-temporally differential expression of genes for three members of fatty acid binding proteins in developing and mature rat brains. J Chem Neuroanat 12: 113–122, 1996.
5. Iseki S, Kanda T, Hitomi M, Ono T: Ontogenic appearance of three fatty acid binding proteins in the rat stomach. Anat Rec 229: 51–60, 1991
6. Iseki S, Amano O, Fujii H, Kanda T, Ono T: Immunohistochemical localization of two types of fatty acid-binding proteins in rat ovaries during postnatal development and in immature rat ovaries treated with gonadotropins. Anat Rec 241: 235–243, 1995
7. Banaszak L, Winter N, Xu Z, Bernlohr DA, Cowan S, Jones TA: Lipid-binding proteins: A family of fatty acid and retinoid transport proteins. Adv Protein Chem 45: 89–151, 1994
8. Thompson J, Winter N, Terway D, Bratt J, Banaszak L: The crystal structure of the liver fatty acid-binding protein. A complex with two bound oleates. J Biol Chem 272: 7140–7150, 1997
9. Paulussen RJA, Veerkamp JH: Intracellular fatty acid-binding proteins, characteristics and function. In: H.J. Hilderson (ed). Subcellular Biochemistry, Vol 16. Plenum Press, New York, 1990, pp 175–226

10. Rolf B, Oudenampsen-Krüger E, Börchers T, Faergeman NJ, Knudsen J, Lezius A, Spener F: Analysis of the ligand binding properties of recombinant bovine liver-type fatty acid binding protein. Biochim Biophys Acta 1259: 245–253, 1995
11. Glatz JFC, Van der Vusse GJ: Cellular fatty acid-binding proteins: Their function and physiological significance. Prog Lipid Res 35: 243–282, 1996
12. Richieri GV, Ogata RT, Kleinfeld AM: Equilibrium constants for the binding of fatty acids with fatty acid-binding proteins from adipocyte, intestine, heart, and liver measured with the fluorescent probe ADIFAB. J Biol Chem 269: 23918–23930, 1994
13. Maatman RGHJ, Van Moerkerk HTB, Nooren IMA, Van Zoelen EJJ, Veerkamp JH: Expression of human liver fatty acid-binding protein in *Escherichia coli* and comparative analysis of its binding characteristics with muscle fatty acid-binding protein. Biochim Biophys Acta 1214: 1–10, 1994
14. Wilkinson TCI, Wilton DC: Studies on fatty acid-binding proteins. The detection and quantification of the protein from rat liver by using a fluorescent fatty acid analogue. Biochem J 238: 419–424, 1986
15. Peeters RA, In 't Groen MAPM, De Moel MP, Van Moerkerk HTB, Veerkamp JH: The binding affinity of fatty acid-binding proteins from human, pig and rat liver for different fluorescent fatty acids and other ligands. Int J Biochem 21: 407–418, 1989
16. Kane CD, Bernlohr DA: A simple assay for intracellular lipid-binding proteins using displacement of 1-anilinonaphthalene 8-sulfonic acid. Anal Biochem. 233: 197–204, 1996
17. Peeters RA, Ena JM, Veerkamp JH: Expression in *Escherichia coli* and characterization of the fatty-acid-binding protein from human muscle. Biochem J 278: 361–364, 1991
18. Glatz JFC, Baerwaldt CCF, Veerkamp JH, Kempen WM: Diurnal variation of cytosolic fatty acid-binding protein content and of palmitate oxidation in rat liver and heart. J Biol Chem 259: 4295–4300, 1984
19. Paulussen RJA, Van Moerkerk HTB, Veerkamp JH: Immunochemical quantitation of fatty acid-binding proteins. Tissue distribution of liver and heart FABP types in human and porcine tissues. Int J Biochem 22: 393–398, 1990
20. Paulussen RJA, Van der Logt CPE, Veerkamp M: Characterization and binding properties of fatty acid-binding proteins from human, pig, and rat heart. Arch Biochem Biophys 264: 533–545, 1988
21. Uyemura K, Yoshimura K, Suzuki M, Kitamura K: Lipid binding activities of the P2 protein in peripheral nerve myelin. Neurochem Res 9: 1509–1514, 1984
22. Xu LZ Sanchez R, Sali A, Heintz N: Ligand specificity of brain lipid-binding protein. J Biol Chem 271: 24711–24719, 1996
23. Takikawa H, Kaplowitz N: Binding of bile acids, oleic acid, and organic anions by rat and human hepatic Z protein. Arch Biochem Biophys 251: 385–392, 1986
24. Kirk WR, Kurian E, Prendergast FG: Characterization of the sources of protein ligand affinity: 1-sulfonato-8-(1′)anilinonaphthalene binding to intestinal fatty acid binding protein. Biophys J 70: 69–83, 1996
25. Okuno M, Kato M, Moriwaki H, Kanai M, Muto Y: Purification and partial characterization of cellular retinoic acid-binding protein from human placenta. Biochim Biophys Acta 923: 116–124, 1987
26. Malpeli G, Stoppini M, Zapponi MC, Folli C, Berni R: Interactions with retinol and retinoids of bovine cellular retinol-binding protein. Eur J Biochem 229: 486–493, 1995
27. Jakoby MG, Miller KR, Toner JJ, Bauman A, Cheng L, Li E, Cistola DP: Ligand-protein electrostatic interactions govern the specificity of retinol- and fatty acid-binding proteins. Biochemistry 32: 872–878, 1993
28. Matarese V, Bernlohr DA: Purification of murine adipocyte lipid-binding protein. Characterization as a fatty acid-and retinoic acid-binding protein. J Biol Chem 263: 14544–14551, 1988
29. Baxa CA, Sha RS, Buelt MK, Smith AJ, Matarese V, Chinander LL, Boundy KL, Bernlohr DA: Human adipocyte lipid-binding protein: purification of the protein and cloning of its complementary DNA. Biochemistry 28: 8683–8690, 1989
30. Storch J: Diversity of fatty acid-binding protein structure and function: Studies with fluorescent ligands. Mol Cell Biochem 123: 45–53, 1993
31. Storch J, Herr FM, Hsu KT, Kim HK, Liou HL, Smith ER: The role of membranes and intracellular binding proteins in cytoplasmic transport of hydrophobic molecules: Fatty acid-binding proteins. Comp Biochem Physiol 115B: 333–339, 1996
32. Hsu K-T, Storch J: Fatty acid transfer from liver and intestinal fatty acid-binding proteins to membranes occurs by different mechanisms. J Biol Chem 271: 13317–13323, 1996
33. Herr FM, Matarese V, Bernlohr DA, Storch J: Surface lysine residues modulate the collisional transfer of fatty acid from adipocyte fatty acid binding protein to membranes. Biochemistry 34: 11840–11845, 1995
34. Herr FM, Aronson J, Storch J: Role of portal region lysine residues in electrostatic interactions between heart fatty acid binding protein and phospholipid membranes. Biochemistry 35: 1296–1303, 1996

Molecular and Cellular Biochemistry **192**: 143–156, 1999.

The third leg: Molecular dynamics simulations of lipid binding proteins

Thomas B. Woolf and Michael Tychko
Johns Hopkins University, Department of Physiology, Baltimore, MD, USA

Abstract

Molecular dynamics computer simulations can provide a third leg which balances the contributions of both structural biology and binding studies performed on the lipid binding protein family. In this context, these calculations help to establish a dialogue between all three communities, by relating experimental observables with details of structure. Working towards this connection is important, since experience has shown the difficulty of inferring thermodynamic properties from a single static conformation. The challenge is exemplified by ongoing attempts to interpret the impact of mutagenesis on structure and function (i.e. binding). A detailed atomic-level understanding of this system could be achieved with the support of all three legs, paving the way towards rational design of proteins with novel specificities. This paper provides an outline of the connections possible between experiment and theory concerning lipid binding proteins. (Mol Cell Biochem **192**: 143–156, 1999)

Key words: molecular dynamics, LBP, FABP, structure-function. protein-lipid interactions, rational drug design

Introduction

Computer calculations can provide a necessary third component to the understanding of lipid binding proteins (LBPs). This component provides a route to connect detailed structural information with binding and cellular function measurements. The approach is described in this contribution. The importance of simulation techniques lies in their ability to rationalize binding measurements at the atomic level using ensembles generated from experimental structures. Because the methods of computer calculations and their application to LBPs may well be new to many readers, the first part of the article is primarily a review of the concepts and methods of molecular dynamics (MD) computer calculations. The second part of the chapter describes calculations on three intracellular LBPs (iLBPs). The third, and final section, describes the expected goals of these calculations and their role in connecting information from structure and function to provide the balance of a third leg.

Molecular dynamics computer calculations

The methods of MD computer calculations have been developed over many years of effort by the chemical physics and biophysics communities. The approach has been summarized in two classic texts: *Computer Simulation of Liquids* by Allen and Tildesley [1] and *Proteins: A theoretical perspective of dynamics, structure, and thermodynamics* by Brooks *et al.* [2]. The description to follow is necessarily much briefer, being meant as an introduction. More details should be pursued from the books.

The concept of MD dates from the early history of computers and the realization that thermodynamic quantities related to the liquid state could be calculated with a sufficiently large statistical mechanical ensemble of conformations. The earliest calculations used vdW spheres to describe deviations from ideal gas behavior. This led to increasingly sophisticated models of the liquid environment. The development of the methods led to the realization that proteins could be studied with the same tools. That is, the thermodynamic quantities related to the room temperature fluctuations of protein motion could be calculated on a computer. The first paper to show that protein motions could be calculated was from McCammon *et al.* [3]. The history of the application of MD computer calculations to proteins dates to this point.

Practical concerns - boundary conditions

Biomolecular systems simulated with MD typically contain between 5000 and 20000 atoms, if the solvent is represented

Address for offrpints: T.B. Woolf, John Hopkins University, Department of Physiology, 725 N. Wolfe Street, Baltimore, MD 21205, USA

explicitly. Atomic bonds are represented as springs; additional harmonic-like terms apply to angles and dihedrals. These energetic terms are referred to as bonded interactions, and can generally be calculated rapidly. By contrast, the calculation of non-bonded (electrostatic and vdW) interactions is very time consuming: while the number of bonds, angles and dihedrals scales roughly linearly with the number of atoms in the system, the nonbonded terms are in principle calculated between all pairs of atoms. The time to compute them scales roughly as the number of atoms squared, and for large systems this process dominates total CPU usage.

The rapid increase in computational effort required with increasing system size places simulators in a quandary. On one hand, they would like to simulate as large a system as possible, to better model the bulk solution behavior measured experimentally; on the other, an ever increasing price must be paid even for incremental increases in system size. A good portion of the art form lies in striking an appropriate balance that generates reasonable answers to the questions at hand. Various methods have been developed to deal with this dilemma, and some are clearly more approximate than others. A discussion of three rather commonly used techniques follows: The 'water-droplet' model, periodic boundary conditions with cutoffs, and particle mesh Ewald summation.

As the name suggests, the 'water-droplet' model is used to solvate the system of interest with a large spherical shell of waters. A small potential is then applied so that solvent atoms near the periphery experience small restraining forces that prevent their exit from the simulation. The goal is to moderately perturb those waters near the boundary with a minimal effect on the inside of the system. This approach is related to the stochastic boundary potentials first introduced by Brooks and Karplus [4], and allows effective solvation with fewer waters than are required for full periodic boundary conditions. It is expected that the method will be especially effective for ligand interactions within a protein cavity far from the surface. The motions of peripheral residues might be influenced by the effective surface term, so their behavior should be treated with slightly less confidence.

The second method is implemented by surrounding the system with images of itself, such that an atom on the edge of the simulation box 'sees' atoms on the opposite side of the box, rather than a vacuum. Non-bonded interactions are then truncated for atom pairs separated by more than a certain distance, usually with some form of smoothing function to reduce artifacts [5]. This technique allows the simulation of what is effectively an infinite system (in the sense of having no edges) reasonably quickly. However, recent work has shown that cutoff schemes can distort the simulation, especially when there are formally charged molecules, as in the case of DNA [6 ,7].

Ewald summations avoid truncation artifacts by treating the simulation box as the unit cell of a crystal, and using a mixed direct and fourier sum to calculate the total energy. Despite their accuracy, conventional Ewald summation methods were considered too computationally expensive to be useful. However, the recently developed particle-mesh implementation [8] is as fast as cutoffs for most systems, and appears to be emerging as the method of choice for future work.

Software and parameters

All MD software packages require a set of parameters that accounts for the different atom types whose interactions are modeled by the potential function. These parameters are initially obtained from experimental and quantum mechanical calculations performed on small molecules which are chemically representative of macromolecular functional groups. The choice of parameters to describe the bond, angle, dihedral, electrostatic, and vdW interactions between atoms is nontrivial. Several distinct sets of parameters are available (CHARMM, GROMOS, AMBER, etc). Although these sets differ in their descriptions of molecules, each is designed as a self-consistent whole to produce realistic molecular behavior within the context of a particular package. It should be emphasized, that it is not possible to 'mix-and-match' different sets to tune a simulation towards a specific result. Rather, the careful simulator uses the parameter set appropriate to the system and software package and analyzes the results.

Once the dynamics package, parameters, and system size have been chosen, someone familiar only with menu-driven packages might think that running the actual simulation would be quite simple. Unfortunately, this is not the case. Although in principle simple dynamics will eventually equilibrate the system from any initial state, in practice minimization and restrained dynamics are used to do so more efficiently. Various properties of the system (temperature, pressure, energy. overall structure, etc.) must be monitored, and only after the system has equilibrated can its behavior be analyzed.

Relative free energy calculations

Unfortunately, calculating free energies of binding directly via MD is not feasible, because binding events occur on a relatively long time scale. However, some techniques, commonly referred to as alchemical methods, have been developed to estimate the difference in binding free energy caused by mutation [9]. These methods treat the mutation as a perturbation of the native protein. All of these methods attempt to gradually change the initial system (native protein) to a new system (mutant) over the course of the simulation. However, the process of defining the two systems, and the path between them, can be difficult, and the simulations require large amounts of computer time. The development of new methods is currently underway in several groups (e.g. Aqvist, Gilson, Brooks, Woolf). It is hoped that these methods will be practical enough to allow large numbers of ligands or mutants to be screened.

A thought experiment might help to illustrate the types of calculations that MD and free energy perturbation can provide. It has been argued [10, 11] that the 'bent' conformation of the fatty acid ligand in many iLBP x-ray structures is energetically unfavorable relative to that found in solution. The premise is that an unfavored conformation in solution is also unfavored within the iLBP protein cavity. From this starting point, the argument is made that the relative binding affinity of different ligands could be calculated by the degree of change in their solution preferred conformation to their conformation within the protein cavity.

This argument is incorrect, because it considers only the internal energy of the fatty acid, and neglects the environment: Ligand interactions, both favorable and unfavorable, and bound (protein) versus free (water). The protein and water both alter the energy surface for the ligand, in different ways. Conversely, the presence of the fatty acid can alter the preferred conformations of the water and the protein. Any realistic estimate of binding free energy must take all of this into account.

Another way of describing the situation is that a computer calculation can be imagined that integrates the mean forces required to move the ligand through a set of all possible conformations. As the ligand, e.g. moves from an extended chain form to a 'bent' form, the mean force required for the motion is determined. The result would be that within the protein cavity, the extended form is much less favorable than the 'bent' form. The protein cavity has changed the energy surface for ligand motion. In summation, an estimate of the free energy change in a ligand from solution to protein requires a calculation of all conformations in each setting and the response of the environment in each setting.

Calculations on three iLBPs

Three iLBPs have been simulated using MD methods: Adipocyte fatty acid binding protein (A-FABP), human muscle fatty acid binding protein (M-FABP), and intestinal fatty acid binding protein (I-FABP). All three share the β-clam structure which is characteristic of iLBPs, despite their sequence heterogeneity (A-FABP and M-FABP are 64% identical, while either is about 30% identical to I-FABP). It is worthwhile to point out that A and M-FABPs have identical residues in the region of the cavity immediately near the ligand's head group, while I-FABP is different (Fig. 1). Table 1 shows the pdb identifier and the total simulation length for each of the systems. The two

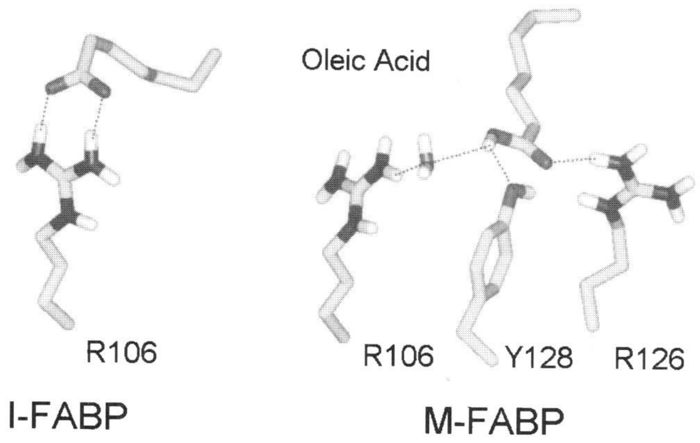

Fig. 1. Water and protein structure around the oleic acid headgroup in the M and I-FABPs. The structures are from points during the molecular dynamics simulations.

5 nsec simulations of I-FABP (apo and binding oleic acid) are among the longest of current MD simulations, which are more commonly 0.25–1 nsec in length.

The method for constructing these simulations is now briefly described. For more detail on the system construction and analysis, see our submitted and upcoming publications [12–14]. The starting point in each case was the original pdb x-ray coordinate set. Hydrogens were then added with attention to their local environment. The new structure was then relaxed into the CHARMM potential function. This is needed since refinement methods often use a different function than MD. Thus, the local minima of the refined structure is not expected to be at a minima in the dynamics program. The relaxed structure was then solvated using a 'water-droplet' model for the external environment. This provides a solvation shell for the protein and is less computationally demanding than periodic boundary conditions. Waters, in the overlay sphere, located within 2.6 Å of any heavy atom in the original structure were deleted. This provided a reasonable initial estimate of the water surrounding the system. The total simulation size was on the order of 8,000

Table 1.

FABP-type	PDB code	MD trajectory
A-FABP : stearate	1lif	1 ns
A-FABP: oleate	1lid	1 ns
M-FABP: stearate	1hmt	1 ns
M-FABP: oleate	1hms	1 ns
M-FABP: elaidate	1hmr	1 ns
I-FABP: palmitate	2ifb	2 ns
I-FABP: oleate	1sac (Sacchettini)	5 ns
I-FABP: myristate	1icm	2 ns
I-FABP apo form	1ifc	5 ns

atoms (70% solvent). Equilibration then proceeded through minimization and relaxation dynamics. The production trajectory was generated by integrating Newton's equations of motion with a 2 fsec time step (conformations were saved every 50 fsec). The resulting dataset for all nine simulations occupies approximately 38 giga-bytes of disk space.

Proton placement on the lipid headgroup

An interesting example case of the use of MD to suggest differences in structure/ function is seen with the hydrogen bond networks surrounding the lipid headgroups (Fig. 1). The authors of the x-ray crystal structures have commented on the rich hydrogen bonding networks that exist around the headgroup regions [15–20]. In particular, several of the FABPs have 'bridging' waters between protein and ligand, in addition to direct proteindigand hydrogen bonds.

Measurements of the pK of the ligand in I-FABP [21] have suggested that the headgroup is charged similarly to free fatty acids in solution. Although the analogous experiments have not yet been performed on either A or M-FABP, two independent sets of calculations lead to the conclusion that lipids bound to these two proteins tend to be protonated. First, Poisson-Boltzmann calculations using Delphi/Qnifft [22–23] suggest that the ligand's pK is shifted roughly 5 units upon binding. These calculations did not consider all potential proton placements, and a fuller set of calculations is planned with all likely sidechain protonation states evaluated as well as the possibility of H_3O^+ for the bridging water. Similar calculations on lipids bound to I-FABP suggest no such change, consistent with experiment. Moreover, MD calculations performed with charged lipid bound to M-FABP were grossly unstable, diverging dramatically from the crystal structure within the first 50 psec of dynamics. This behavior further reinforces the Poisson-Boltzmann results. By contrast, otherwise identical simulations using neutral lipids were stable over 1 nsec of dynamics, with no evidence of the instability seen in the charged simulations. Moreover, simulations of charged lipids bound to I-FABP were also stable, over 5 nsec of dynamics.

Finally, this finding is testable by the same types of NMR experiments that measure no change in the pK of the fatty acid headgroup in the I-FABP system. It is planned, in collaboration with the lab of David Cistola, that NMR measurements will be made to determine more about the A and M-FABP systems.

L72A and L78A: Examples of alchemical mutations

Detailed relative free energy calculations of changes in binding affinity upon mutation were computed for two sites in the I-FABP system. This system and the sites were chosen for comparison with the binding affinity and mutagenesis data being collected by the Kleinfeld group [24–27]. In particular, the experimental work suggests ΔΔG's of –1.5 kcal/mol and 0.3 kcal/mol for (L72A and L78A, respectively) in affinity for palmitate [27]. The calculations are briefly described here. Further detail will be found in a later publication [28].

During the course of alchemical free energy calculations, some atoms are slowly changed from one chemical type to another. This is typically accomplished by breaking the transformation into several stages, typically called windows, and performing dynamics for each window. Although this transformation is physically unrealistic (at least there have been no such results published recently), the overall change in free energy along this path should still be correct, since free energy is a state function, and hence independent of path. However, the system must sample a reasonable subset of phase space in each window, in order to create an equilibrium pathway between states and converge numerically. Once the windows have been calculated, they can be summed to reconstruct the free energy curve for the transformation.

In these simulations, leucine residues were transformed into alanine. The backbone atoms were left untouched, while the β-methylene group was transformed into a methyl, and the remaining atoms removed from the simulation. Each window was equilibrated for 10 psec, and data collected for at least 25 psec. Each window was evaluated to determine if sufficient sampling had taken place, and additional dynamics was run if necessary, since incomplete sampling would lead to systematic errors. This is a problem with all free energy methods, and is not unique to the iLBP system [9, 29]. The transformations were performed on both the apo and holo (palmitate) forms, and the differences between the free energy changes were used to assess the overall change in affinity. Each calculation required approximately two months on a dedicated workstation (SGI R4400).

Figure 3 shows the four relative free energy calculations along the reaction path. The curves show reasonable behavior, with no large shifts along the reaction coordinate or obvious discontinuities. The windows with the largest set of deviations are those at the ends, as is typical of this method. This is emphasized in Fig. 4, where four windows are shown for each of the two alchemical mutations. The 0.9 lambda window (nearly complete transition from Leu to Ala) had the largest fluctuations in each case, which did not converge by 100 psec. This implies that further sampling and possibly additional sampling at 0.95 lambda may be necessary to improve the error at these points.

The results are disappointing considering the amount of computer time invested. The L72A mutation was predicted to change binding affinity by –1.8 kcal/mol, which is quite reasonable, but the effect of the L78A mutation was overestimated as 3.5 kcal/mol. The most likely cause of the large error is

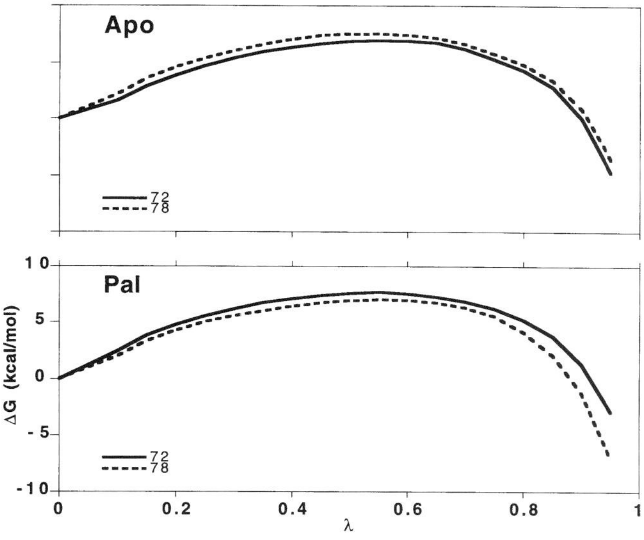

Fig. 2. ΔΔG along the alchemical reaction coordinate for changes of L72A and L78A in I-FABP. Both the holo (palmitate) and apo changes are illustrated.

insufficient sampling in some windows. In particular, the points in lambda near the final alanine do not seem fully converged.

It is clear from these results that these methods are too expensive to be used to consistently predict the behavior of the large number of mutations that are being examined by Kleinfeld's lab. More rapid methods for computing relative binding affinity (even qualitatively) will have to be derived for this purpose. The development of methods of this type is being pursued in our group as well as in other theoretical groups.

Linear response estimations of relative binding affinity

One possible qualitative estimate of relative free energy changes has been suggested based on the idea of linear response [30–34]. The concept is similar to electron transfer (Marcus) theory, in suggesting that the free energy curves that describe changes in the system are roughly parabolic in form. With this assumption, it becomes possible to relate a single trajectory to regions of phase space that have not been explored (e.g. properties of a mutant protein based on a trajectory of the native form). That is, if a single trajectory has converged to provide an estimate of the local free energy, then extrapolation from this initial point can be made with some confidence if the shape for perturbation is along a parabola or similar defined curve. The use of this method for biomolecules has been pioneered by Aqvist [30–32].

The linear response analysis to the free energy uses the trajectory to provide a baseline reference for perturbation. For each conformation from the trajectory, the mutated residue is substituted for the native (in this case alanine for leucine), and the change in overall electrostatic and vdW energy for the system is evaluated. These changes are averaged over the trajectory, and the free energy of binding is estimated from:

$$\Delta G_{bind} = 1/2 \left\langle \Delta V_{electrostatic} \right\rangle + \alpha \left\langle \Delta V_{vdW} \right\rangle$$

The electrostatic term is analogous to Marcus theory, and derivable from the Born equation. The vdW term, although not theoretically rigorous, is included because of the importance of vdW interactions in biomolecular systems. This method has been applied directly to the iLBPs [28] and does not provide excellent agreement with experiment, even when alpha is treated as a fittable parameter. However, some form of linear response may be able to give qualitative insight into the $\Delta\Delta G$ of transfer.

Interaction energy analysis

Molecular dynamics simulations can provide information unavailable by other means. For example, the interaction energy between atoms can be directly computed. Although not equivalent to the rigorous thermodynamic analyses described above, strong interaction energies may suggest the importance of particular atoms and residues in binding. An example of interaction analysis is shown in Fig. 5. Further analysis and discussion is found in two submitted papers and one paper still in preparation [12–14]. This analysis shows that the interaction energy between the ligand and protein is dominated by alkane chain vdW interactions in A and M-FABPs, and electrostatic headgroup interactions in I-FABP. This emphasizes the importance of the ligand protonation state in determining the motions and structures near the headgroup. Strong electrostatic protein:lipid interactions have been postulated in CRBPs by analogy to I-FABP [35].

Further breakdown of the interaction energy on the residue level provides insights into those residues with the strongest interaction with the ligand. This is shown in Fig. 6 for A and M-FABPs. The result is intriguing in suggesting energetic differences between similar structures. In one case R126 of M-FABP had the strongest headgroup interactions, while R106 had very minor interactions. In A-FABP the trend was reversed, as R106 had the most significant interactions with the stearic acid ligand.

Covariance of motions in protein and ligand

The coupling of atomic motions in a biomolecular simulation can provide insights into the 'essential dynamics' of the system [36]. The primary motional patterns, or modes, can be determined, and used to define the motions which may

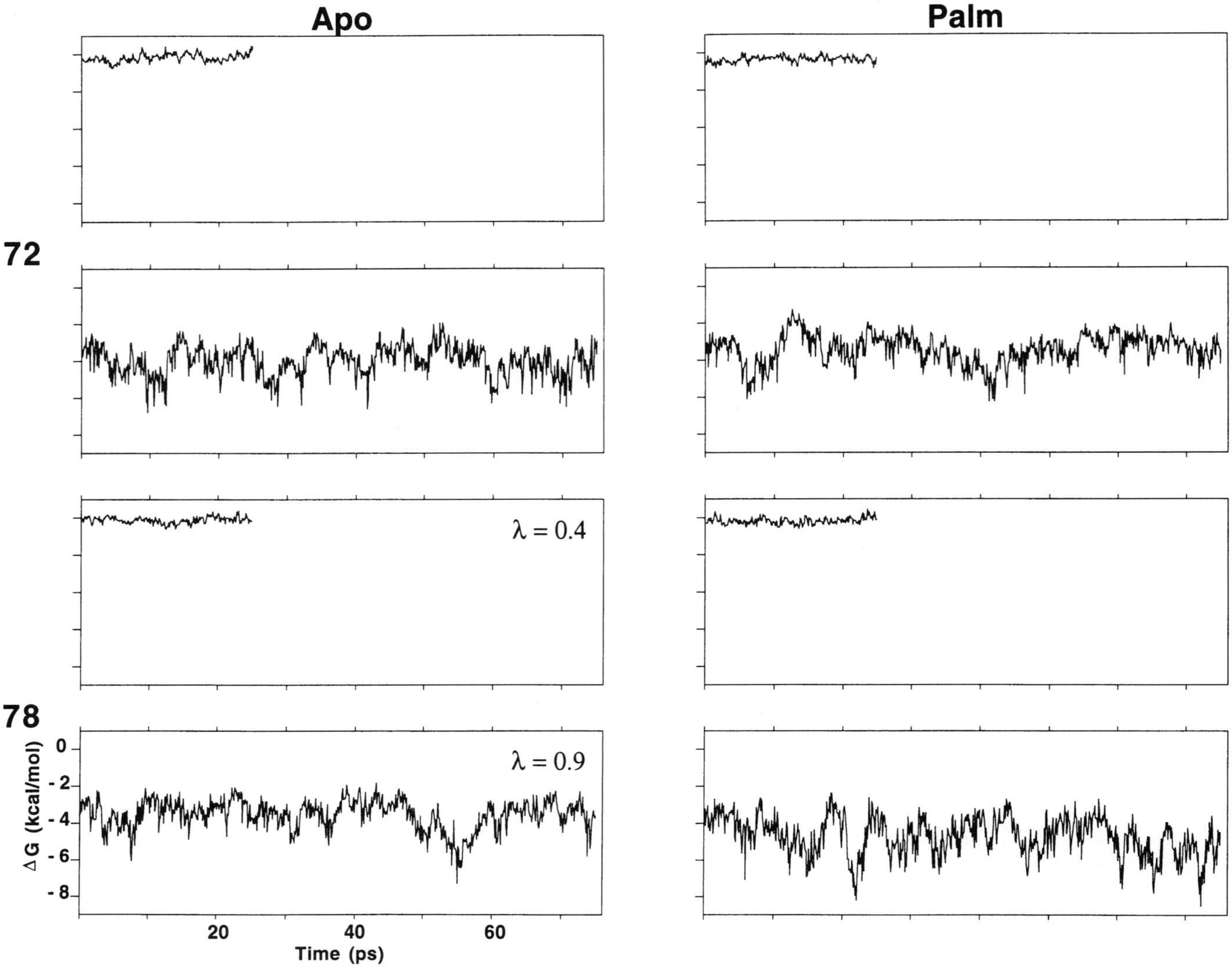

Fig. 3. ΔΔG convergence in the reaction coordinate lambda-windows for the L72A and L78A alchemical changes. Both apo and holo are shown. The 0.4 and 0.9 lambda windows are shown in four cases. Note the much larger fluctuations in the 0.9 lambda windows relative to the midpoint 0.4 window.

be most important for function. This is done by calculating the zero-time covariance averaged over the trajectory. For each conformation in the trajectory, the average displacement of an atom relative to another is computed. The average of this quantity is the zero-time covariance. When many pairs are examined, patterns emerge which describe the global motions of the system. Regions which are relatively uncoupled show small covariances, while regions which are strongly coupled have covariances near one. It is also possible for motions to be anticorrelated, yielding negative covariances. In summary, this analysis reveals groups of atoms which tend to move in concert with one another.

Covariances were determined between the ligand heavy atoms and the protein, as well as between the ligand and some interior solvent molecules. Figure 7 shows the covariance matrices for A and M-FABPs with stearic acid bound. The patterns of covariances were strikingly different in these simulations. For example, the M-FABP was much more tightly coupled to the stearic acid than the A-FABP. Furthermore, in M-FABP the pattern of coupling alternates several times from correlated to anti-correlated along the length of the alkane chain. A different pattern emerged for A-FABP, which had smooth correlations running along the alkane chain, with a slight break at the headgroup. Furthermore, the stearic acid in A-FABP has much stronger interactions with the water than with the protein. This suggests a different mechanism of selectivity could be involved between the two systems. The water could play a larger role in binding for the A-FABP system than for the M-FABP.

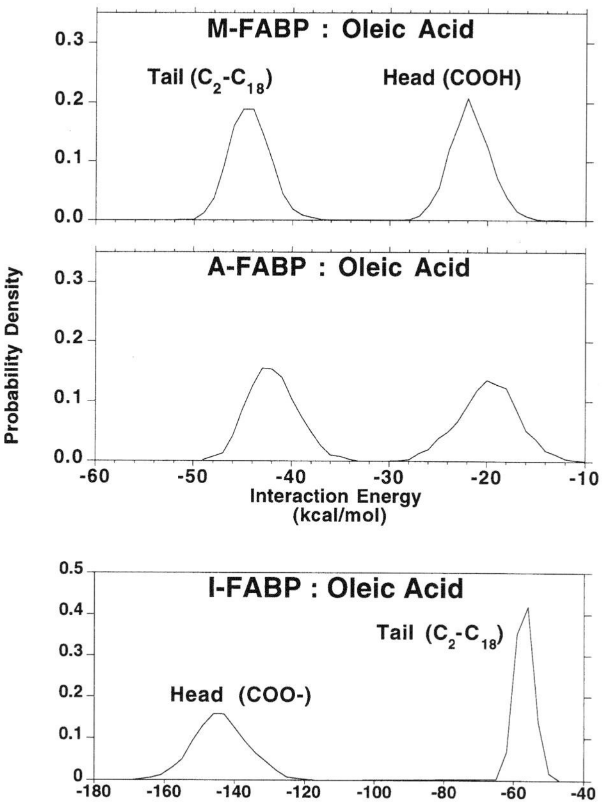

Fig. 4. The head and tail interaction energy distributions between the ligand and the protein for three FABPs. Note that the scale is larger for I-FABP.

Importance scale

It was recently suggested that interaction energy and covariance analyses could be combined to predict the importance of specific residues in ligand binding [12, 13]. The basic premise is simple: those residues which interact favorably and move in concert with the ligand are likely to be the ones influencing its binding. This is a qualitative technique only, since there is no attempt to connect it with more complex calculations. However, it is relatively easy to apply and may provide insights unavailable from other analyses applied to MD trajectories.

Figure 8 presents the application of this concept to all five MD simulations that have been performed on A and M-FABPs. The data describing interaction energies and covariances was normalized and summed for either variant of the protein, as well as for all five runs. The grand total plot reveals residues that are presumably important for general binding function, while the individual totals for A and M-FABPs (2 and 3 simulations respectively) highlight certain sites that may be crucial for particular variant's specificity. For example, R106 scores very highly in A-FABP, but negligibly in M-FABP.

Internal volume calculations: Problems and pitfalls

Several software programs are in use for visualization of molecular structure and calculation of molecular surfaces and volumes. The programs use different algorithms and define the volumes and surfaces in different ways. This creates problems with using the software to define the volume of the cavity found in iLBPs. One main problem is that the programs are optimized for finding relatively small enclosed volumes. The determination of a closed volume with an opening to the outside (e.g. at the D-E gap and the location of the two α-helices) presents a problem to the software. When the gaps are closed, either with a large probe choice or with a set of pseudo-atom plugs, the volume estimates can still vary widely from one program to another.

An illustration of this problem is shown in Fig. 9, which presents results for A, M, and I-FABP x-ray structures. The programs VOIDOO and GRASP were used to calculate the internal volumes. With default values for probe size and grid spacing, the calculated internal volumes (from the same structure) varied from about 1000-1100 Å^3 (GRASP) to 200–400 Å^3 (VOIDOO). To further emphasize the types of numbers that can come from the programs, the grid size was varied from 0.4–1.0 in VOIDOO. The variation in calculated internal volumes was extreme.

The determination of changes in internal volume throughout the course of a MD trajectory can be imagined, but the reality is that the artifacts observed in the previous paragraphs make it difficult for a reasonable determination of volume during the trajectory time-course. This is not to argue that the numbers are uninteresting, only that the definition of the internal volume carries with it a degree of arbitrariness that has seldom been commented on within the LBP community. Thus, even changes in volume as a function of simulation time would not necessarily give insight into function, since the size and even direction of volume changes can vary with the options used to estimate the internal volume of the protein.

Water dynamics

The behavior of the internal water during the course of the MD trajectory is a subject of interest on several levels. First, the water behavior may be related to the effective pK values for the ligand and the amino acid sidechains within the protein cavity. Second, the dynamic nature of the water exchange could be related to functional binding affinity. Third, the network of hydrogen bonds provided by the water may be important for structural stability of the protein and ligand interactions within the cavity. For these reasons the water structure was examined in some detail. A few highlights of this analysis are presented in this section.

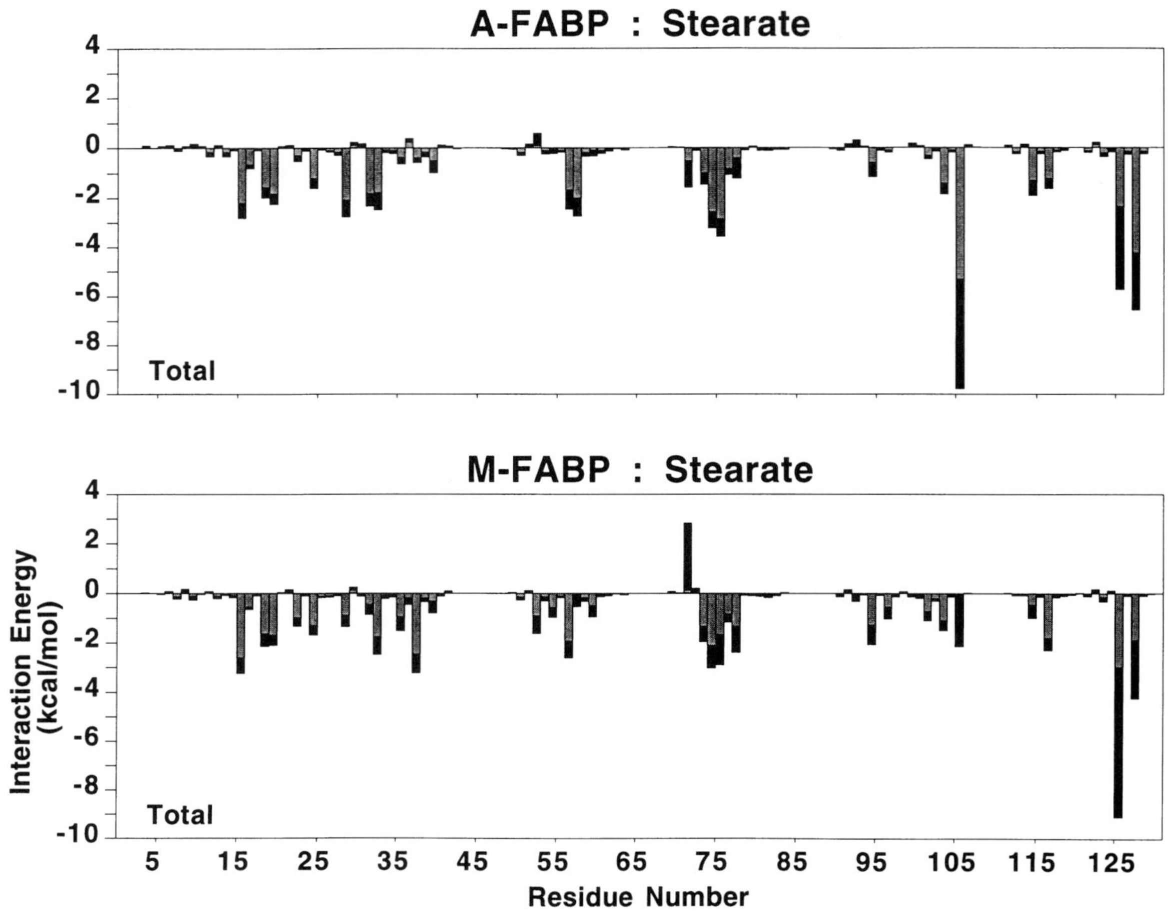

Fig. 5. Individual amino acid interaction energies with oleic acid bound by A and M-FABP. The lighter bar shows the average while the darker tip is the positive RMS deviation from the average.

The forty waters which spent the most time near the ligand were determined in each simulation. These are listed in Tables 2 and 3 for stearic acid bound to A and M-FABP. The effective diffusion constant was estimated from the slope of the mean square displacement plotted against time. This quantity may not have converged on the time scale of the simulation, but it does give an indication that several populations of water molecule behavior are present within the system. Some water molecules moved about as quickly as those outside the protein cavity, while others had less than 1/100th of that mobility.

Figure 10 illustrates the headgroup region of the I-FABP: oleic acid system, and several of the neighboring water molecules. Two time-series are presented which serve as qualitative examinations of the hydrogen bonding between ligand carboxylate and solvent. Though the indicated waters were relatively confined to this region, their hydrogen bonding with the ligand was still transient. This further emphasizes the types of additional insight MD can provide once a crystal structure is available.

Connections of theory and experiment in the LBPs

Recent mutagenesis experiments in the Kleinfeld lab have shown entropy/ enthalpy compensation is a common result of alanine mutations in I-FABP [27]. No one amino acid appears to dominate binding to the extent that changing it to alanine completely destroys function, implying that favorable binding is achieved through many relatively small protein: lipid interactions. However, analysis is complicated by the fact that a mutation can have global effects on protein stability and dynamics which indirectly influence binding in addition to the more intuitive local effects within the binding cavity. For example, a mutation which destabilizes the native, ligand-binding state could decrease overall binding affinity without disrupting the interactions stabilizing binding. Such a mutation would not need to be within the binding pocket, and could not be deduced by visually examining the crystal structure.

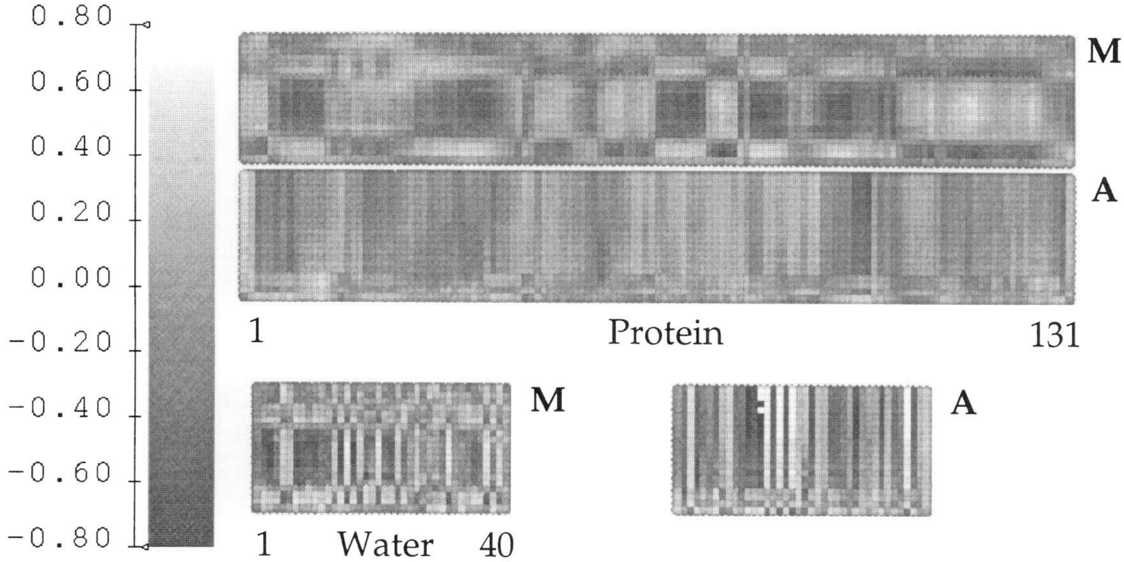

Fig. 6. Zero-time covariance plot of the correlated motion of protein and water with heavy atoms of stearic acid. The top two panels show the amino acid covariance with the ligand (vertical scale). The bottom two panels show the 40 most common nearby waters and their zero-time covariance with the ligand. The scale runs from a dark anti-correlated motion to a light positively correlated motion. An intermediate gray is zero average correlation.

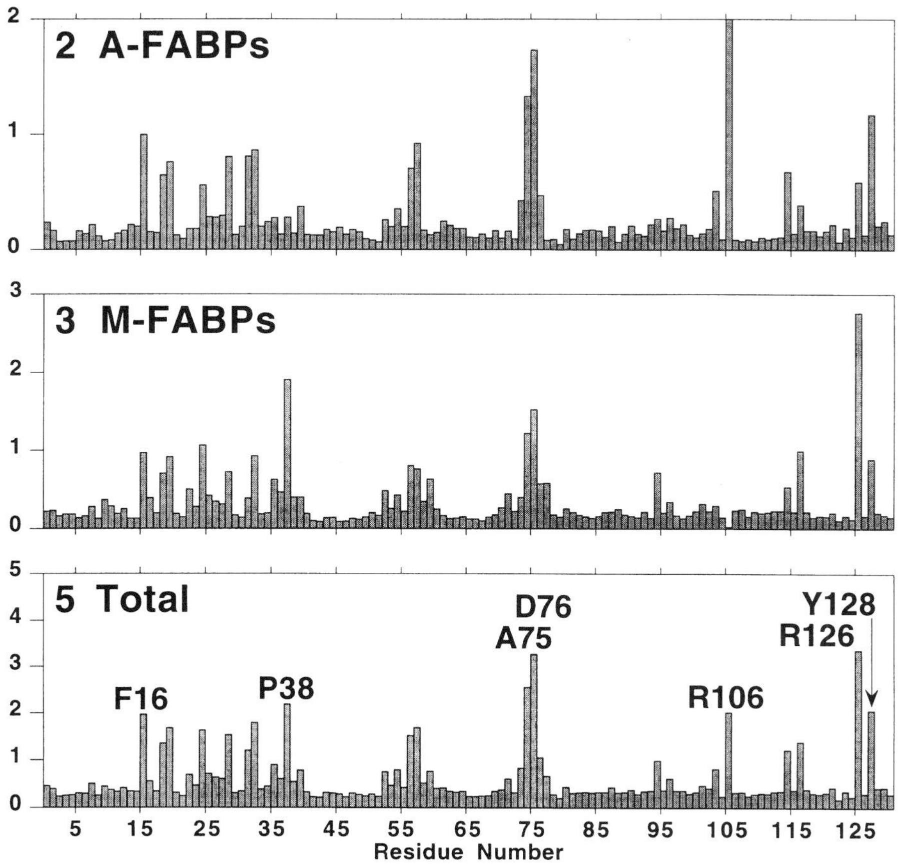

Fig. 7. The sum of normalized importance scale values for A and M-FABP simulations. The top panel is the sum of stearic and oleic acid simulations for A-FABP. The middle panel is the sum of stearic, oleic, and elaidic acid simulations for M-FABP. The bottom panel is the sum of all five simulations. The residues with high values are suggested to have a greater relative importance in the binding.

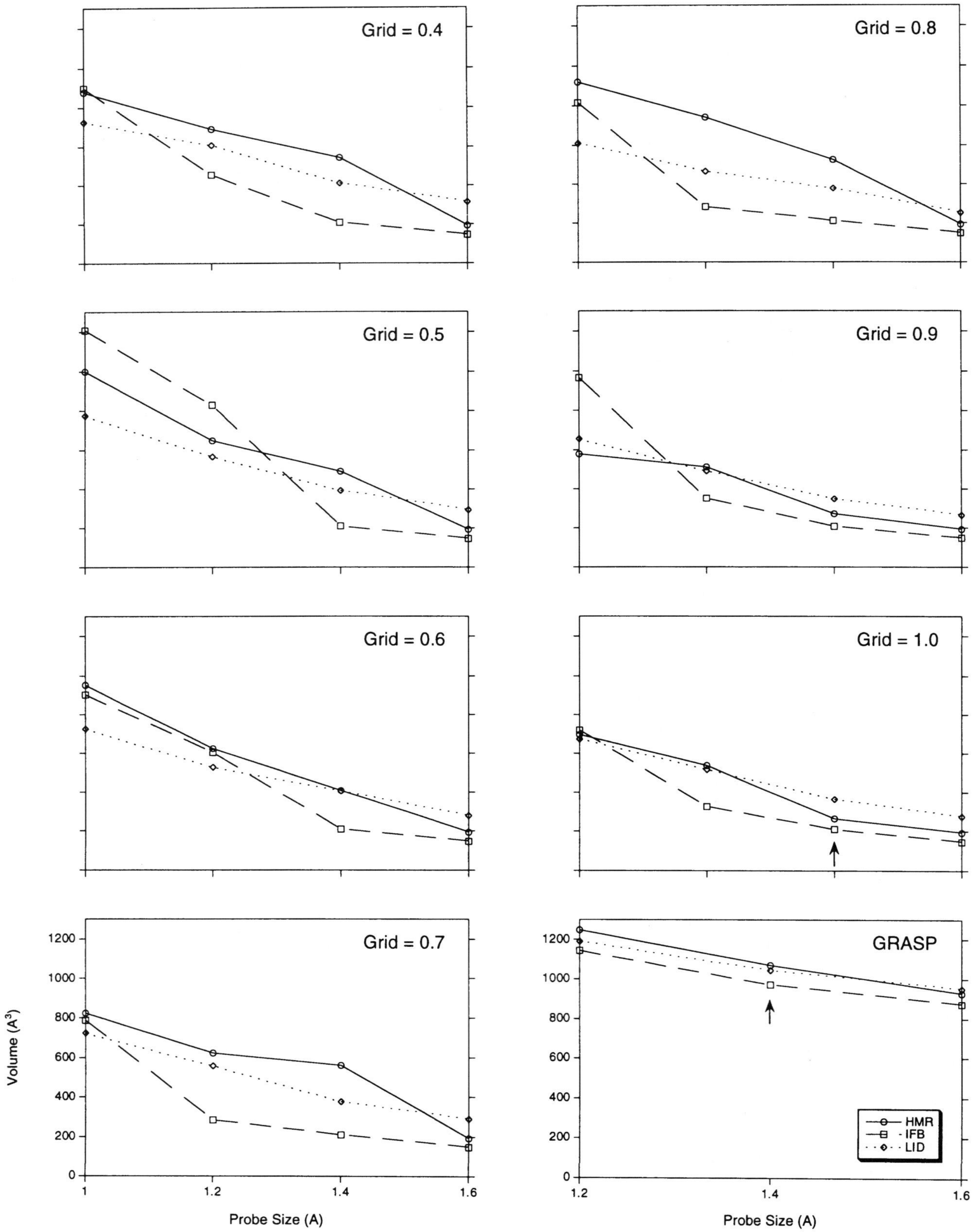

Fig. 8. The calculated internal cavity volume for three FABPs depends on the program (GRASP or VOIDOO) and parameters used. The grid spacing in VOIDOO was varied from 0.4–1.0 A with a range of probe sizes. Arrows indicate each program's default values.

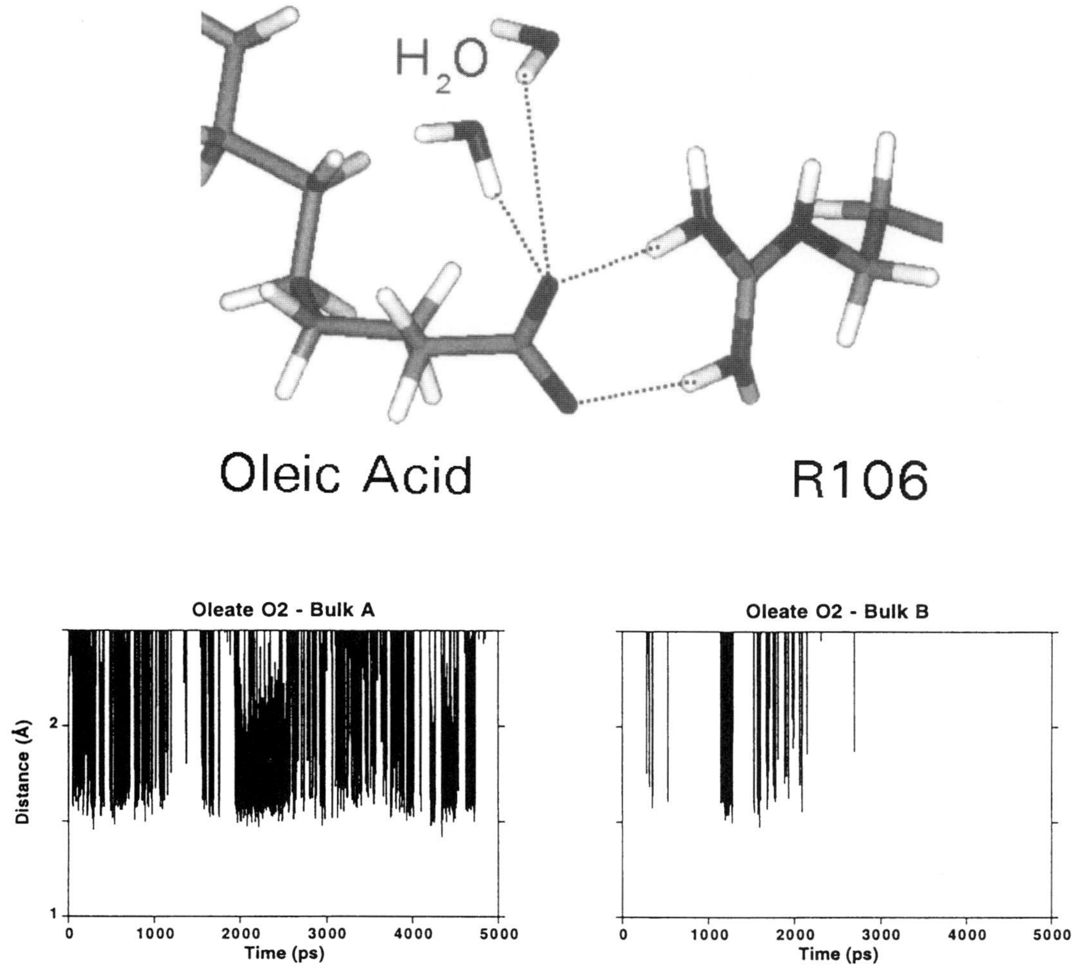

Fig. 9. Time-series for two waters interacting with the oleic acid headgroup in the I-FABP are shown. The dynamic nature of hydrogen bonding exchange is shown over the course of 5 nsec.

Understanding the thermodynamic interplay within this system is challenging. Molecular dynamics will, in time, be able to provide insight into the mechanisms. At this point, these simulations are for the most part unable to predict these effects well. However, there are some promising areas currently being developed.

There are several difficulties involved in directly calculating changes in binding affinity with mutation or changing ligands. The largest of these is the limitation available CPU power places on sampling. Moreover, free energy methods are relatively young and are still under active development. Similarly, the parameters and functions used to describe the molecular energy surface are constantly being refined. As computer speed continues to improve and as methods continue to be developed for free energy calculations, the ability to connect simulated results with experiments will grow. The improvements seen over of the last 10 years support this view (see [9] for a review).

The ability to accurately predict the effect of atomic level changes on the free energy of binding would enable the design of proteins that bind specific ligands with high affinity. Measuring the structural and binding behavior of predicted sequences would then provide valuable information both about the LBP system and the prediction process. Such experiments could be used to refine the prediction schemes while directly expanding the empirical knowledge base. This sort of iterative approach has been used successfully for rational drug design [37], and there is no apparent reason why it can not be applied successfully to the LBPs. Indeed, the existence of a large number of LBPs with a wide range of binding behavior is evidence that the basic β-clam structure can be tuned to specifically bind different ligands. Single amino acid mutations in I-FABP and CRBP-II have demonstrated that modifications to internal amino acids can change binding preference [35].

It would be desirable to calculate not just the thermodynamics of binding but the kinetics as well. The on and off

Table 2. Nearest Water Neighbors to the Stearic Acid in A-FABP

Contacts (#/trj)	Estimated Diffusion Constant (um²/msec)	Most Frequent Near Neighbors
9530	0.05	O1 C 1 O2 C2 C3 C6
8493	0.02	O2 C1 C3 C2 O1 C4
4300	0.13	C5 C3 C6 O2 C8 Cl C2 C4
4001	0.07	O2 Cl O1 C2 C3 C5
3961	0.07	C7 C6 C5 C4 C8 O1 C9 C2
3815	0.13	C6 C5 O2 C7 C4 C8 O1 Cl
3731	0.17	C6 C8 C7 C4 C13 O1 O2 Cl
3388	0.18	02 C6 C3 C8 Cl C4 01 C5
3119	0.10	C6 C7 C8 C5 C4 C13 C3 O2
2709	0.13	C4 O2 C3 C13 C6 C11 C15
2160	0.07	O2 C1 O1 C3 C2 C4 C6 C8
1518	0.50	O2 C11 O1 C13 C6 C3 C1 C8
1386	0.47	C14 C13 C15 C10 C16 C8 C12
1278	0.38	C13 C2 O1 C3 C1 O2 C15
1164	0.52	C3 C5 O2 C11 C1 C8 C4 C13
1015	0.08	C5 C6 C4 O1 O2 C1 C3 C7
980	0.55	C8 C5 C6 C11 C10 C4 C7
972	0.08	C13 C3 C6 C11 C8 C4 C14
879	0.37	C13 C14 C15 C3 C16 O2 C11
705	0.47	C13 C15 C14 C17 C11 C16 C12
699	0.52	C17 C18 C16 C15 C14 C12
597	0.35	C11 C10 C6 C12 C13 C8 O2
499	1.10	C17 C18 C16 C15 C12
416	0.007	C2 C5 O1 C3 C4 O2 C1 C6
388	0.70	C14 C15 C13 C16 C12 C18 C17
377	0.60	C18C17 C16 C15 C14 C13 C12
371	0.80	C15 C16 C17 C13 C14 C18 C12
364	0.70	C18C16 C17 C13 C12 C15 C14
342	0.30	CIO CII C12 C14 C8C13 C9
341	0.50	C16 C14 C18 C17 C15 C12
269	0.90	C18 C17 C16 C13 C15 C14 C12
252	0.57	C17 C18 C16 C15
242	0.80	C16 C18 C14 C17 C15 C13 C12
230	0.72	C17 C16 C18 C15 C14 C13
223	0.75	C 17 C18 C14 C16 C12 C13
194	0.78	C18 C17 C16
175	0.07	C5 C4 C6 C3
174	0.62	C15 C17 C13 C18 C16 C14
165	0.67	C15 C13 C14 C16 C17 C18
164	0.53	C18C17 C16 C15

Table 3: Nearest Water Neighbors to the Stearic Acid in M-FABP

Contacts (#/tri)	Estimated Diffusion Constant (um²/msec)	Most Frequent Near Neighbors
6495	0.07	O1 Cl O2 C2
6067	0.28	C8 C14 C9 C12 C10 C13 C6
5655	0.03	C2 O2 C1 C3 O1 C4 C5
5623	0.15	O2 C11 O1 C16 C18 C17
5149	0.05	O2 O1 Cl C6 C14 C4 C10 C13
4904	0.02	O2 C1 C4 O1 C6
4514	0.47	O1 Cl O2 C18 C17
3510	0.07	C4 C6 C5 C8 O2 C7 C1
3474	0.10	C3 C2 C17 O1 C15 C18 C1
3235	0.08	C2 O2 C1 O1 C3 C4
2964	0.20	O2 O1 C1 C4 C2 C18 C17
2387	0.50	C8 C6 C10 C9 C12 C7 C4 O2
2148	0.12	O1 O2 C2 C4 C1 C3 C5 C18
1941	0.08	C6 02 C8 C9 C7 C4 C2 Cl
1890	0.60	C9 C7 C8 C1O C6 C11 C12
1370	0.52	O2 O1 C1 C2 C4
949	0.12	O1 C4 C3 C2 C1 C5 C6 O2
853	0.68	C8 C10 C9 C12 C6 C7 C11
827	0.75	C12 C13 C14 C11 C10 C15 C9
793	0.60	C9 C10 C11 C8 C12 C7 C13
625	0.70	C9 C10 C8 C11 C12
487	0.57	C17 C18 C15 C16 O1 C13
421	0.92	C11 C9 C10 C12 C13 C8
418	0.68	C10 C9 C11 C8 C12 C13 C14
300	0.77	C8 C10 C9 C7 C6
253	0.65	C11 C10 C13 C9 C12
239	0.80	C9 C11 C10 C12 C8
226	0.68	C11 C9 C12 C8 C10
220	0.67	C15 C14 C13 C16
209	1.0	C11 C12 C10 C9
177	1.2	C7 C6 C8 C5 C4
174	1.0	C10 C12 C11 C9 C8
164	1.2	C9 C8 C7 C11 C10
161	0.75	C11 C12 C10 C9
132	0.017	O2 C2 C4 C1 O1 C3 C5
129	0.75	C9 C11 C10 C8 C12
109	0.0067	C3 C4 C2 O1 O2 C5 C1
97	0.42	C12 C11
96	0.45	C11 C12 C13 C10 C9
90	1.1	C12 C11 C10

rates have been estimated by the Kleinfeld group [26]. These values cannot be calculated directly from conventional MD simulations since an event must occur a statistically significant number of times before its rate can be analyzed, and the typical lifetimes of both the bound and free states are significantly longer than the longest feasible trajectories. This does not mean that MD can't be used to comment on these phenomena.

Umbrella sampling techniques can be used to force the system to cross large kinetic barriers and reconstruct the statistical behavior of the unperturbed system. The reaction coordinate could be loosely defined as the ligand entering or leaving the binding cavity, and in principle a series of calculations could calculate the free energy change upon insertion or removal of the ligand from the protein. The kinetic constants would then be estimated from the free energy curve, for example using Kramer's theory. This basic approach has been used previously to estimate the isomerization kinetics in simple alkanes and to predict the ion conduction rate through model ion channels [38, 39]. Application of this technique to LBPs is complicated by the difficulty involved in determining the appropriate binding path, or reaction coordinate. Moreover, since binding a ligand can significantly alter the global dynamics of a protein, thorough (and expensive) sampling is necessary along the binding path. This problem would be especially acute at the estimates of the rate limiting step for entry and exit.

It may be possible to gain insights into ligand entry/exit mechanisms by analyzing the principle components of motion from a long simulation. The method of quasi-harmonic dynamics projects the motions from a trajectory onto modes of mean displacement. Instead of describing the motions in terms of the (x,y,z) changes for each atom, principle component modes use coupled motions of large numbers of atoms as drawn from the trajectory. In this sense, they are analogous to the zero-time covariances described above. These modes are similar to those obtained from a normal mode calculation on a single structure, except that they can include much larger anharmonic motions not present in a harmonic approximation from a single state. Certain long time scale modes may reveal some slow transitions in protein structure which permit ligand entry/exit. These modes are currently being calculated for the four longer trajectories of FABPs available to date.

Another connection point with experiment is provided by consideration of the types of interactions that might occur between iLBPs and lipid bilayers which are related to transport of the ligand *in vivo*. These interactions have been proposed to have an electrostatic effect from experimental work [40, 41]. The calculation of such effects could proceed on two levels. The first, and simplest level, is the use of Poisson-Boltzmann. calculations based on protein and a model of the lipid bilayer. Such calculations have been used for MARCKS and similar proteins/peptides by the McLaughlin and Honig groups [42]. The advantage of such calculations is that they are reasonably rapid and converge to a single numerical value. The disadvantage is that they are limited by the quality of the single conformations used and the geometries used for docking. That is, if electrostatics alone is dominant, then the models may give good results. If conformational changes occur with docking, then more detailed models may be needed. More detailed models for docking with membrane can be imagined, but are expensive from a computational standpoint. In particular, the simulation of a protein:lipid interaction with a reasonable explicit water environment could be on the order of 30–50,000 atoms. While this can be imagined, it is not a realistic current simulation. Thus, a first step towards the simulation of electrostatic interactions within protein/bilayer systems is probably best attempted using the Poisson-Boltzmann approach.

Conclusion

Molecular dynamics computer calculations can provide atomic level insights into the connection between structure and function. This is particularly exciting for the lipid binding proteins because of the large number of high quality x-ray structures already available [15], and the large quantity of binding affinity data [24–27]. Current MD calculations are limited by computer speed and algorithm development, but are already capable of elucidating the dynamics of LBPs on the nanosecond time scale. Moreover, relative free energy calculations promise a growing ability to calculate directly measurable quantities. This will provide a mechanism for detailed comparison of molecular models of protein structure with measurement of binding function. By enhancing the communication between these two basic fields, MD may become the third leg supporting and balancing our knowledge of LBPs.

Acknowledgments

The Maui High Performance Computer Center (MHPCC) and the Whitaker Computer Center in the Department of Biomedical Engineering are thanked for their grants of computer time in the calculations. Funding for the research was provided by the American Heart Association through a grant-in-aid and by the Department of Physiology and the Bard Foundation. Helpful comments on the manuscript were provided by Alan Grossfield.

References

1. Allen MP, Tildesley DJ: Computer Simulation of Liquids. Clarendon Press, Oxford, 1987
2. Brooks CLI, Karplus M, Pettitt BM: Proteins: A theoretical perspecitive of dynamics, structure, and thermodynamics. John Wiley and Sons, New York, 1988
3. McCammon JA, Gelin BR, Karplus M: Dynamics of folded proteins. Nature 267: 585–590, 1977
4. Brunger AT, Huber R, Karplus M: Trypsinogen-Trypsin transition: A molecular dynamics study of induced conformational change in the activation domain. Biochem 26: 5153–5162, 1987
5. Steinbach PJ, Brooks BR: New spherical-cutoff methods for long-range forces in macromolecular simulation. J Comp Chem 15: 667–683, 1994
6. Auffinger P, Louise-May S, Westhof E: Multiple molecular dynamics simulations of the anticodon loop of $tRNA^{Asp}$ in aqueous solution with counterions. J Am Chem Soc 117: 6720–6726, 1995
7. Auffinger P, Louise-May S, Westhof E: Molecular dynamics simulations of the anticodon hairpin of $tRNA^{Asp}$: Structuring effects of C-H--O hydrogen bonds and of long-range hydration forces. J Am Chem Soc 118: 1181–1189, 1996
8. Darden T, York D, Pedersen L: Particle mesh Ewald: an N log(N) method for Ewald sums in large systems. J Chem Phys 98: 10089–10092, 1993
9. Kollman P: Free energy calculations: Applications to chemical and biochemical phenomena. Chem Rev 93: 2395–2417, 1993
10. Young ACM, Scapin G, Kromminga A, Patel S13, Veerkamp JH, Sacchettini JC: Structural studies on human muscle fatty acid binding protein at 1.4 Å resolution: Binding interactions with three C18 fatty acids. Structure 2: 523–534, 1994
11. The 3rd International Conference on Lipid-Binding Proteins. Minneapolis, Minnesota. 1997
12. Woolf TB: Simulations of fatty acid binding proteins suggest sites important for function I: stearic acid. Biophys J 74: 681–693, 1998

13. Woolf TB, Tychko M: Simulations of fatty acid binding proteins II: Sites for discrimination of monounsaturated ligands. Biophys J 74: 694–707, 1998
14. Woolf TB, Grossfield A, Tychko M: Differences between apo and three holo intestinal lipid binding proteins seen by molecular dynamics computer calculations. Submitted (1998)
15. Banaszak L, Winter N, Xu Z, Bernlohr DA, Cowan S, Jones TA: Lipid-binding proteins: A family of fatty acid and retinoid transport proteins. Adv Protein Chem 45: 89–151, 1994
16. Eads J, Sacchettini JC, Kromminga A, Cordon JI: *Escherichia coli*-derived rat intestinal fatty acid binding protein with bound myristate at 1.5 Å resolution and I-FABP Arg106--->Gln with bound oleate at 1.74 Å resolution. J Biol Chem 268: 26375–26785, 1993
17. LaLonde JM, Bernlohr DA, Banaszak Q: X-ray crystallographic structures of adipocyte lipid-binding protein complexed with palmitate and hexadecanesulfonic acid. Properties of cavity binding sites. Biochemistry 33: 4885–4895, 1994
18. Sacchettini JC, Gordon JI, Banaszak LJ: Crystal structure of rat intestinal fatty-acid-binding protein: Refinement and analysis of the *Escherichia coli*-derived protein with bound palmitate. J Mol Biol 208: 327–3391, 1989
19. Sacchettini JC, Gordon JI: Rat intestinal fatty acid binding protein. A model system for analyzing the forces that can bind fatty acids to proteins. J Biol Chem 268: 18399–18402, 1993
20. Scapin G, Gordon JI, Sacchettini JC: Refinement of the structure of recombinant rat intestinal fatty acid-binding apoprotein at 1.2 Å resolution. J Biol Chem 267: 4253–4269, 1992
21. Cistola DP, Sacchettini JC, Banaszak LJ, Walsh MT, Gordon JI: Fatty acid interactions with rat intestinal and liver fatty acid-binding proteins expressed in *Escherichia coli*. J Biol Chem 264: 2700–2710, 1989
22. Antosiewicz J, McCammon JA, Cilson MK: Prediction of pH-dependent properties of proteins. J Mol Biol 238: 415–436, 1994
23. Sharp KA, Honig B: Electrostatic interactions in macromolecules: Theory and applications. Annu Rev Biophys Chem 19: 301–332, 1990
24. Richieri CV, Ogata RT, Kleinfeld AM: Equilibrium constants for the binding of fatty acids with fatty acid-binding proteins from adipocyte, intenstine, heart, and liver measured with the fluorescent probe ADIFAB. J Biol Chem 269: 23918–23930, 1994
25. Richieri CV, Ogata RT, Kleinfeld AM: Thermodynamics of fatty acid binding to fatty acid-binding proteins and fatty acid partition between water and membranes measured using the fluorescent probe ADIFAB. J Biol Chem 270: 15076–15084, 1995
26. Richieri CV, Ogata RT, Kleinfeld AM: Kinetics of fatty acid interactions with fatty acid binding proteins from adipocyte, heart, and intestine. J Biol Chem 271: 11291–11300, 1996
27. Richieri CV, Low PJ, Ogata RT, Kleinfeld AM: Mutants of rat intestinal fatty acid binding protein illustrate the critical role played by enthalphy-entropy compensation in ligand binding. J Biol Chem 272: 16737–16740, 1997
28. Woolf TB, Tychko M: In preparation
29. Lau FTK, Karplus M: Molecular recognition in proteins: Simulation analysis of substrate binding by a TyrosyPtRNA synthetase mutant. J Mol Biol 236: 1049–1066, 1994
30. Aqvist J, Medina C, Samuelsson JE: A new method for predicting binding affinity in computer-aided drug design. Protein Eng 7: 385–391, 1994
31. Aqvist J, Mowbray SL: Sugar recognition by a glucose/galactose receptor. J Biol Chem 270: 9978–9981, 1995
32. Aqvist J, Hansson T: On the validity of electrostatic linear response in polar solvents. J Phys Chem 100: 9512–9521, 1996
33. King C, Barford RA: Calculation of electrostatic free energy differences with a time-saving approximate method. J Phys Chem 97: 8798–8802, 1993
34. Levy RM, Belhadj M, Kitchen DB: Gaussian fluctuation formula for electrostatic free-energy changes in solution. J Chem Phys 95: 3627–3633, 1991
35. Jakoby MG, Miller KR, Toner JJ, Bauman A, Cheng L, Li E, Cistola DR: Ligand-protein electrostatic interactions govern the specificity of retinol- and fatty acid-binding proteins. Biochemistry 32: 872–878, 1993
36. Amadei A, Linssen BM, Berendsen HJC: Essential dynamics of proteins. Proteins 17: 412–425, 1993
37. Kuntz ID, Meng EC, Schoichet BK: Structure-Based Molecular Design-Acc Chem Res 27: 117–123, 1994
38. Roux B, Karplus M: Ion transport in a model gramicidin channel. Structure and Thermodynamics. Biophys J 59: 961–981, 1991
39. Pastor RW, Venable RM: Molecular and stochastic dynamics simulation of lipid membranes. Computer Simulation of Biomolecular Systems Vol 2. Leiden, 1993
40. Herr FM, Matarese V, Bernlohr DA, Storch J: Surface lysine residues modulate the collisional transfer of fatty acid from adipocyte fatty acid binding protein to membranes. Biochemistry 34: 11840–11845, 1995
41. Herr FM, Aronson J, Storch J: Role of portal region lysine residues in electrostatic interactions between heart fatty acid binding protein and phospholipid membranes. Biochemistry 35: 1296–1303, 1996
42. Ben-Tal N, Honig B, Peitzach R, Denisov G, McLaughlin S: Binding of small basic peptides to membranes containing acidic lipids: Theoretical models and experimental results. Biophys J 71: 561–575, 1996

Molecular and Cellular Biochemistry **192**: 157–161, 1999.

Lipid-transfer proteins from plants: Structure and binding properties

Françoise Guerbette, Michèle Grosbois, Alain Jolliot-Croquin, Jean-Claude Kader and Alain Zachowski

Laboratoire de Physiologie Cellulaire et Moléculaire, Unit CNRS 2135, Université Pierre et Marie Curie, Paris, France

Abstract

Plant cells contain lipid-transfer proteins (LTPs) able to transfer phospholipids between membranes *in vitro*. Plant LTPs share in common structural and functional features. Recent structural studies carried out by NMR and X-ray crystallography on an LTP isolated from maize seeds have showed that this protein involves four helices packed against a C-terminal region and stabilized by four disulfide bridges. A most striking feature of this structure is the existence of an internal hydrophobic cavity running through the whole molecule and able to accomodate acyl chains. It was thus of interest to study the ability of maize LTP to bind hydrophobic ligands such as acyl chains or lysophosphatidylcholine and to determine the effect of this binding on phospholipid transfer. The binding abilities of maize LTP, presented in this paper, are discussed and compared to those of lipid-binding proteins from animal tissues. (Mol Cell Biochem **192**: 157–161, 1999)

Key words: lipid, transfer, binding, proteins, plants, maize

Abbrevations: LTP – lipid-transfer proteins; FABP – fatty acid binding protein; ALBP – adipocyte lipid-binding protein; SCP-2 – sterol carrier protein - 2; PC – phosphatidylcholine; lyso-PC – lysophosphatidylcholine

Introduction

Lipid transfer proteins (LTPs) are defined by their ability to transfer phospholipids between membranes *in vitro*. They have been characterized in a large variety of organisms including mammalian tissues, plants, yeast, fungi and some bacteria. It has been suggested - but not yet well demonstrated - that these proteins participate in extracellular flux of lipids, essential for the biogenesis and renewal of endomembranes [1].

The situation is more complicated for plant LTPs which are still in search of an *in vivo* role. It is now clear that these proteins are present in numerous plants and share in common structural and functional properties : small size (about 9 kDa), high isoelectric point (9–10), presence of eight cysteine residues, among 90–93 amino acids, located in conserved positions and forming four disulfide bridges [2, 3]. Structural studies by NMR and X-ray crystallography have been carried out on plant LTPs. These proteins comprise a single compact domain with four alpha-helices and a long C-terminal region. They have a tunnel-like hydrophobic cavity able to accomodate fatty acids [4, 5]. A complex between maize LTP and palmitic acid has been indeed recently crystallized [4]. This recent structural model of ns-LTPs gives a basis for understanding the function of LTPs in relation to the transfer and binding of molecules containing acyl chains (phospholipids, fatty acids, acyl CoA or other acyl group containing molecules).

As it was found that various cDNAs coding for LTPs contained a signal peptide suggesting a possible secretion of these proteins, a role in cell wall edification, and more precisely biosynthesis of cutin, has been proposed [6–8]. In this perpective it was of interest to study the interaction of LTPs with other ligands than phospholipids originally used to define this class of proteins, such as fatty acids which are necessary substrates for the formation of cutin. LTP isolated from maize seedlings is a good model for such studies. This

Author for offprints: J.-C. Kader, Laboratoire de Physiologie Cellulaire et Moléculaire, Unit CNRS 2135, Université Pierre et Marie Curie , Tour 53, 4 place Jussieu, 75252 Paris cedex 05 (France)

paper presents novel informations on the binding of acyl chains and of other ligands such as lysophosphatidylcholine on maize LTP and on the effect of this binding on the phospholipid transfer process.

Materials and methods

Materials

Maize seeds (*Zea mays* L., cv. Mona) were a gift from Pioneer France-Maïs.

Protein preparation

LTP was prepared from maize seeds after 3 days of germination on moist vermiculite in the dark at 30°C. Steps of purification were as defined previously [9]. Purified LTP used was the isoform fully characterized [10].

Lyso PC binding assay

Radioactive lysophospholipids were obtained by action of phospholipase A1 on phosphatidylcholine biosynthetically radiolabelled by aging potato slices in the presence of (^{3}H) methyl choline as precursor [9]. Seventy-three nmoles of (^{3}H) lyso-PC were dried under nitrogen then 1.5 ml of pure LTP fraction (100 nmoles) was added in two steps, separated by a strong agitation in the presence of small glass beads. Volume was adjusted to 2 ml with elution buffer. After incubation under agitation at 30°C during 30 min, the sample was loaded onto a Sephadex G-100 column to separate the LTP/lyso-PC complex from free LTP and lysoderivative. Fractions of 4 ml were collected and 0.5 ml aliquots subjected to liquid scintillation counting. LTP elution was followed by ELISA with specific polyclonal IgGs directed against the maize LTP, purified by affinity chromatography, as described elsewhere [11].

Acyl-CoA binding assay

The binding of (^{14}C) oleoyl-CoA to LTP was estimated after separation of LTP/ ligand complex by FPLC on a Mono S column. For binding assay, pure LTP (39 nmoles) was incubated with 1.8 nmole of (^{14}C) oleoyl-CoA, in a final volume of 0.7 ml, at 0°C during 30 min. Elution from Mono S column was performed with a NaCl gradient from 0–250 mM and 1 ml fractions were collected. Radioactivity was measured on 0.5 ml aliquots.

PC transfer activity of LTP complexes

LTP complexes were formed with either lyso-PC or oleoyl-CoA under the conditions described above. PC transfer activity catalyzed by the different complexes was assayed as previously reported [12] by measuring the transfer of (^{3}H)phosphatidylcholine from sonicated liposomes to maize mitochondria.

Results

Formation of a lyso-PC/LTP complex

After incubation of LTP with lyso-PC, the protein eluted from Sephadex G-100 column as two peaks (Fig. 1). Radioactivity associated to lyso-PC was detected in the second peak indicating that the LTP complexed with a lysoderivative was more retained than free LTP. Fifty-eight percents of the initial radioactivity of lyso-PC was found associated to LTP, indicative of a n apparent association constant (K_a) of ca. 50 nM^{-1}.

Transfer activity of lyso-PC/LTP complexes

PC transfer activity was impaired by a preincubation of LTP with lyso-PC (Fig. 2). Increasing amounts of lysoderivatives

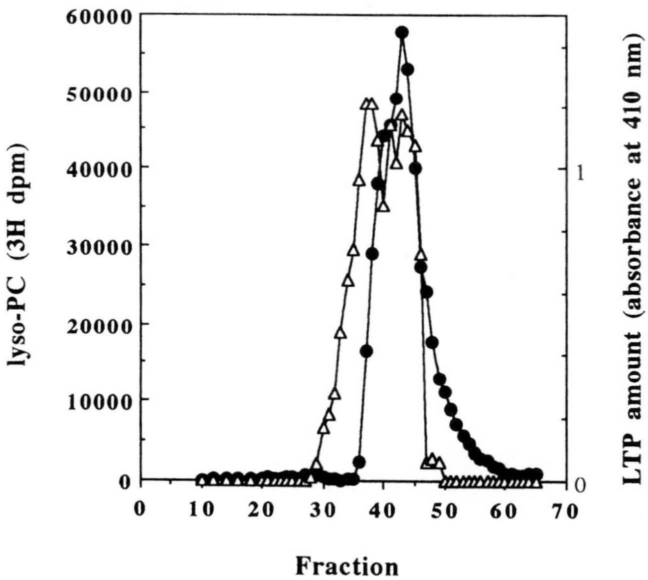

Fig. 1. Sephadex G-100 chromatography of lysoPC-LTP complex. Pure maize LTP (100 nmoles) was incubated for 30 min at 30°C with ^{3}H-lysoPC (73 nmoles) then the different components were separated by gel chromatography. LysoPC was followed by its radioactivity (●). LTP was assayed using an ELISA test with affinity purified specific IgG (Δ).

inhibited the capability of LTP to transfer PC from liposomes to mitochondria. The dose-response was sensitive to the molecular category of lyso-PC. As can be seen, the transfer catalyzed by 50 nmoles LTP was 50% inhibited by approximately 0.6 nmoles lyso-palmitoyl-PC or 0.8 nmoles lyso-oleoyl-PC.

Formation of an oleoyl-CoA/LTP complex

The LTP/oleoyl-CoA complex was separated from the free forms of both components by FPLC on a Mono S column (Fig. 3). It was found to be associated with 33% of the initial radioactivity of oleoyl-CoA present in the assay.

Transfer activity of oleoyl-CoA/LTP complexes

Binding of oleoyl-CoA on LTP before PC-transfer assay diminished the transfer efficiency. However, oleoyl-CoA appeared to be less potent than lyso-PC in this role, as the transfer associated with 100 nmoles of LTP was only decreased by 25% after incubation with 50 nmoles of oleoyl-CoA (Fig. 4).

Discussion

Maize LTP was characterized by its *in vitro* capability to transfer phospholipids between two membranes [10]. Here

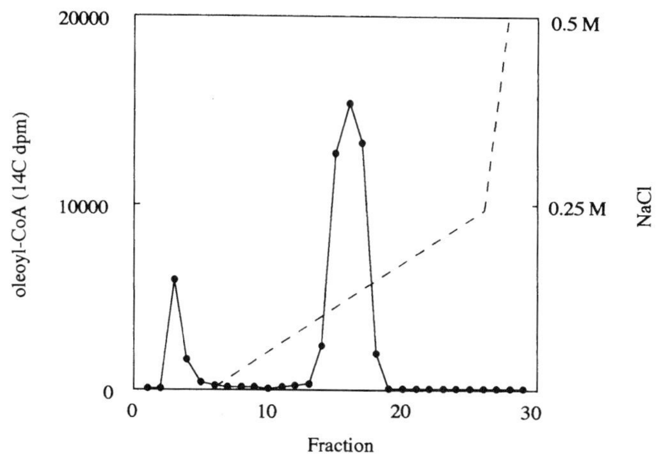

Fig. 3. FPLC chromatography on mono-S column of oleoyl-CoA/LTP complex. Maize LTP (39 nmoles) was incubated for 30 min at 0°C with ^{14}C-oleoyl-CoA then deposited on a mono-S column. The complex was then eluted by a NaCl gradient. Oleoyl-CoA was followed by its radioactivity (●).

we showed that it can also bind acyl-CoA esters and lyso derivatives. This property has been already described in the case of LTPs such as the ones from carrot [8], spinach [13], rapeseed [14] or castor bean [15].

The fact that the complexes could be separated from both the unbound ligand and the free protein was indicative of two properties. First, the complex life has to be long enough to allow the separation which indicates a relatively high

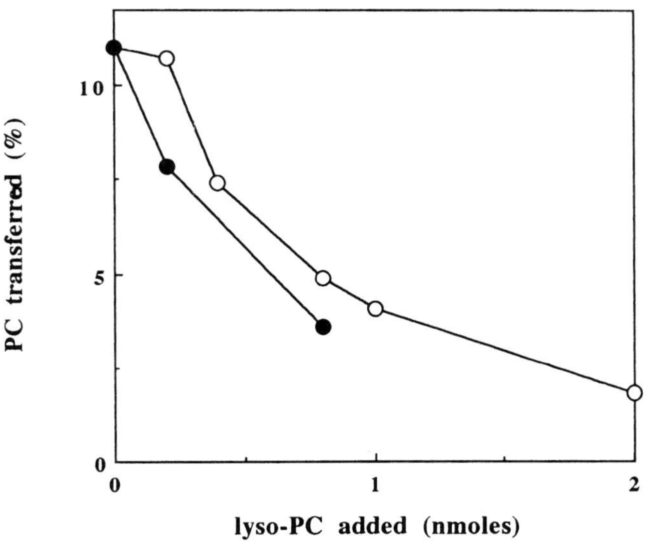

Fig. 2. Effect of lysoPC binding on PC transfer catalyzed by LTP. A complex of LTP and either oleoyl-lysoPC (○) or palmitoyl-lysoPC (●) was formed before the transfer assay. LTP amount was 50 nmoles and lysoPC amount was as indicated in abscissa. PC transfer was measured between sonicated small unilamellar vesicles and maize mitochondria as described previously [12].

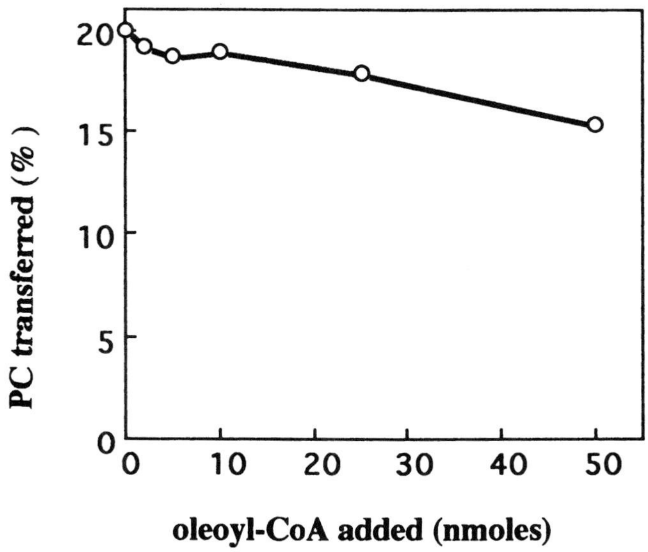

Fig. 4. Effect of oleoyl-CoA binding on PC transfer activity of maize LTP. A complex between LTP (100 nmoles) and the indicated amount of acyl-CoA was formed before PC transfer activity was assayed as indicated in Material and methods and in legend of Fig. 2.

binding constant. Secondly, since the complex was eluted from the chromatographic support after the free LTP, their structure should be different. This suggests that binding of the acyl chain should induce a change in conformation of the protein. The association between maize LTP and acyl-CoA esters or lyso derivatives appeared to be tighter than between the protein and a phospholipid, based on previous studies with PC showing a weak binding [12] and on the inhibitory effect of lyso-PC or oleoyl-CoA binding towards the phospholipid transfer rate. However, lysoderivatives were more potent inhibitors than acyl-CoA. In the case of LTP isolated from rape seedlings, oleoyl-CoA could completely abolish the PC transfer towards the mitochondrial membrane [14]. It is thus interesting to note that the relative affinities for the different ligands vary according to the source of LTP.

This acyl binding ability of maize LTP is in agreement with the structural studies made by NMR or X-ray crystallography on maize LTP indicating the presence of an hydrophobic cavity able to accomodate one or two acyl chains [4–5]. Recent NMR studies performed on barley LTP have led to similar conclusion: The structure of the complex between barley LTP and palmitoyl-CoA revealed a hydrophobic binding site that can expand to accomodate both large and smal ligands [16].

It is of interest to note that several lipid-binding proteins have been characterized in mammalian cells. Fatty-acid binding proteins constitute a family of low molecular mass proteins that are suggested to function as intracellular transporters of fatty acids [17]. In contrast with plant LTPs, FABPs exhibit a predominantly beta-sheet structure [18]. Several structural studies have been carried out on the various FABP categories. For example, crystallographic studies have recently established that the fatty acid binding site of adipocyte lipid-binding protein (ALBP) is within an internalized water-filled cavity [18]. Rat liver FABP (L-FABP) isoforms I and II have two fatty acids or fatty acyl CoA binding sites [19]. This is coherent with the crystal structure of the recombinant form of rat liver fatty acid-binding protein complexed with oleic acid; two fatty acid molecules were showed to bind within a central cavity [20].

Other types of mammalian proteins which are able, like plant LTPs, to interact both with phospholipids and fatty acids, have been found. This is the case with sterol carrier protein-2 (SCP-2) which has ben showed to bind fluorescent fatty acyl-CoAs at a single site with high affinity [21]. It remains to establish the relations between FABPs, SCP-2 and acyl-CoA-binding proteins (ACBP) [22]. In addition, it is to be noted that plant cells contain ACBP-like proteins which have been characterized in rape seeds [23]. It is not known if maize seeds contain similar proteins.

Future studies on maize LTP will involve recombinant approaches to help progress in the knowledge of the binding process of lipids on LTP and of the interactions of LTP with membranes. At a longer term, LTP gene antisense or over-expression strategy will be followed in order to reduce or to increase LTP amount in maize tissues and thus to study the consequences on lipid metabolism and physiology of the transgenic plants.

References

1. Wirtz KWA: Phospholipid transfer proteins. Ann Rev Biochem 60: 73–99, 1991
2. Kader JC: Lipid-transfer proteins in plants. Annu Rev Plant Physiol Plant Mol Biol 47: 627–654, 1996
3. Kader JC: Lipid-transfer proteins: A puzzling family of plant proteins. Trends Plant Sci 2: 66–70, 1997
4. Shin DH, Lee JY, Hwang KY, Kim KK, Su SW: High-resolution crystal structure of the non-specific lipid-transfer protein from maize seedlings. Structure 3: 189–199, 1995
5. Gomar J, Petit MC, Sodano P, Sy D, Marion D, Kader JC, Vovelle F, Ptak M: Solution structure and lipid binding of a nonspecific lipid transfer protein extracted from maize seeds. Protein Science 5: 565–577, 1996
6. Sterk P, Booij H, Scheleekens GA, Van Kammen A, de Vries SC: Cell-specific expression of the carrot EP2 lipid transfer protein gene. Plant Cell 3: 907–921, 1991
7. Pyee J, Yu H, Kolattukudy PE: Identification of a lipid transfer protein as the major protein in the surface wax of broccoli (Brassica oleracea) leaves. Arch Biochem Biophys 311: 460–468, 1994
8. Hendriks T, Meijer EA, Thoma S, Kader JC, De Vries SC: The carrot extracellular lipid transfer protein EP2: In: G. Coruzzi, P. Puigdomenech, NATO ASI, (eds). Quantitative aspects with respect to its putative role in cutin synthesis. Plant Molecular Biology, Springer Verlag, Berlin, 1994, pp 85–94
9. Douady D, Guerbette F, Kader JC: Purification of phospholipid transfer protein from maize seeds using a two-step chromatographic procedure. Physiol Vég 23: 373–380, 1984
10. Sossountzov L, Ruiz-Avial L, Vignols F, Jolliot A, Arondel V, Tchang F, Grosbois M, Guerbette F, Miginiac E, Delseny M, Puigdomenech P, Kader JC: Spatial and temporal expression of a maize lipid tranfer protein gene. Plant Cell 3: 923–933, 1991
11. Grosbois M, Guerbette F, Kader JC: Changes in level and activity of phospholipid transfer protein during maturation and germination of maize seeds. Plant Physiol 90: 1560–1564, 1989
12. Grosbois M, Guerbette F, Jolliot F, Quintin F, Kader JC: Control of maize lipid transfer protein activity by oxido-reducing conditions. Biochim Biophys Acta 1170: 197–203, 1993
13. Rickers J, Spener F, Kader JC: A phospholipid transfer protein that binds fatty acids. FEBS Lett 180: 29–32, 1985
14. Ostergaard J, Vergnolle C, Schoentgen F, Kader JC: Acyl-binding/lipid-transfer proteins from rape seedlings, a novel category of proteins interacting with lipids. Biochim Biophys Acta 1170: 109–117, 1993
15. Tsuboi S, Osafune T, Tsugeki R, Nishimura M, Yamada M: Nonspecific lipid transfer protein in castor bean cotyledon cells: Subcellular localization and a possible role in lipid metabolism. J Biochem (Tokyo) 111: 500–508, 1992
16. Lerche MH, Kragelund BB, Bech LM, Poulsen FM: Barley lipid-transfer protein complexed with palmitoyl CoA: The structure reveals a hydrophobic site that can expand to fit both large and small lipid-like ligands. Structure 5: 291–306, 1997
17. Gericke A, Smith ER, Moore DJ, Mendelsohn R, Storch J: Adipocyte fatty acid-binding protein: interaction with phospholipid membranes and thermal stability studied by FTIR spectroscopy. Biochemistry 36: 8311–8317, 1997

18. Ory J, Kane CD, Simpson MA, Banaszak LJ, Bernlohr DA: Biochemical and crystallographic analyses of a portal mutant of the adipocyte lipid-binding protein. J Biol Chem 272: 9793–9801, 1997
19. Frolov A, Cho TH, Murphy EJ, Schroeder F: Isoforms of rat liver fatty acid binding protein differ in structure and affinity for fatty acids and fatty acyl CoAs. Biochemistry 36: 6545–6555, 1997
20. Thompson J, Winter N, Terwey D, Bratt J, Banaszak L: The crystal structure of the liver fatty acid-binding protein. A complex with two bound oleates. J Biol Chem 272: 7140–7150, 1997
21. Frolov A, Cho TH, Billheimer JT, Schroeder F: Sterol carrier protein-2, a new fatty acyl coenzyme A-binding protein. J Biol Chem 271: 31878–31884, 1996
22. Kerkhoff C, Beuck M, Threige-Rasmussen J, Spener F, Knudsen J, Schmitz G: Acyl-CoA binding protein (ACBP) regulates acyl-CoA: Cholesterol acyltransferase (ACAT) in human mononuclear phagocytes. Biochim Biophys Acta 1346: 163–172, 1997
23. Hills MJ, Dann R, Lydiate D, Sharep A: Molecular cloning of a cDNA from *Brassica napus L.* homologous of acyl-CoA-binding protein. Plant Mol Biol 25: 917–920, 1994

Molecular and Cellular Biochemistry **192:** 163–165, 1999.

Index to Volume 192

Developments in Molecular and Cellular Biochemistry

Series Editor: Naranjan S. Dhalla, Ph.D., M.D. (Hon.), FACC

1. V.A. Najjar (ed.): *Biological Effects of Glutamic Acid and Its Derivatives.* 1981 ISBN 90-6193-841-4
2. V.A. Najjar (ed.): *Immunologically Active Peptides.* 1981 ISBN 90-6193-842-2
3. V.A. Najjar (ed.): *Enzyme Induction and Modulation.* 1983 ISBN 0-89838-583-0
4. V.A. Najjar and L. Lorand (eds.): *Transglutaminase.* 1984 ISBN 0-89838-593-8
5. G.J. van der Vusse (ed.): *Lipid Metabolism in Normoxic and Ischemic Heart.* 1989 ISBN 0-7923-0479-9
6. J.F.C. Glatz and G.J. van der Vusse (eds.): *Cellular Fatty Acid-Binding Proteins.* 1990 ISBN 0-7923-0896-4
7. H.E. Morgan (ed.): *Molecular Mechanisms of Cellular Growth.* 1991 ISBN 0-7923-1183-3
8. G.J. van der Vusse and H. Stam (eds.): *Lipid Metabolism in the Healthy and Diseased Heart.* 1992 ISBN 0-7923-1850-1
9. Y. Yazaki and S. Mochizuki (eds.): *Cellular Function and Metabolism.* 1993 ISBN 0-7923-2158-8
10. J.F.C. Glatz and G.J. van der Vusse (eds.): *Cellular Fatty-Acid-Binding Proteins, II.* 1993 ISBN 0-7923-2395-5
11. R.L. Khandelwal and J.H. Wang (eds.): *Reversible Protein Phosphorylation in Cell Regulation.* 1993 ISBN 0-7923-2637-7
12. J. Moss and P. Zahradka (eds.): *ADP-Ribosylation: Metabolic Effects and Regulatory Functions.* 1994 ISBN 0-7923-2951-1
13. V.A. Saks and R. Ventura-Clapier (eds.): *Cellular Bioenergetics: Role of Coupled Creatine Kinases.* 1994 ISBN 0-7923-2952-X
14. J. Slezák and A. Ziegelhöffer (eds.): *Cellular Interactions in Cardiac Pathophysiology.* 1995 ISBN 0-7923-3573-2
15. J.A. Barnes, H.G. Coore, A.H. Mohammed and R.K. Sharma (eds.): *Signal Transduction Mechanisms.* 1995 ISBN 0-7923-3663-1
16. A.K. Srivastava and J.-L. Chiasson (eds.): *Vanadium Compounds: Biochemical and Therapeutic Applications.* 1995 ISBN 0-7923-3763-8
17. J.M.J. Lamers and P.D. Verdouw (eds.): *Biochemistry of Signal Transduction in Myocardium.* 1996 ISBN 0-7923-4067-1
18. E.-G. Krause and R. Vetter (eds.): *Biochemical Mechanisms in Heart Function.* 1996 ISBN 0-7923-4118-X
19. R. Vetter and E.-G. Krause (eds.): *Biochemical Regulation of Myocardium.* 1996 ISBN 0-7923-4259-3
20. G.N. Pierce and W.C. Claycomb (eds.): *Novel Methods in Molecular and Cellular Biochemistry of Muscle.* 1997 ISBN 0-7923-4387-5
21. F.N. Gellerich and S. Zierz (eds.): *Detection of Mitochondrial Diseases.* 1997 ISBN 0-7923-9925-0
22. P.K. Singal, V. Panagia and G.N. Pierce (eds.): *The Cellular Basis of Cardiovascular Function in Health and Disease.* 1997 ISBN 0-7923-9974-9
23. S. Abdel-aleem and J.E. Lowe (eds.): *Cardiac Metabolism in Health and Disease.* 1998 ISBN 0-7923-8104-1
24. A.K. Srivastava and B. Posner (eds.): *Insulin Action.* 1998 ISBN 0-7923-8113-0

Developments in Molecular and Cellular Biochemistry

25. V.A. Saks, R. Ventura-Clapier, X. Leverve, A. Rossi and M. Rigoulet (eds.): *Bioenergetics of the Cell: Quantitative Aspects*. 1998 ISBN 0-7923-8118-1
26. G.N. Pierce, H. Rupp, T. Izumi and A. Grynberg (eds.): *Molecular and Cellular Effects of Nutrition on Disease Processes*. 1998 ISBN 0-7923-8171-8
27. K. Ahmed, E. Chambaz and O.G. Issinger (eds.): *Molecular and Cellular View of Protein Kinase CK2*. 1999 ISBN 0-7923-8208-0
28. M.V. Cohen, J.M. Downey, R.J. Gelpi and J. Slezák (eds.): *Myocardial Ischemia and Reperfusion*. 1998 ISBN 0-7923-8173-4
29. D.A. Bernlohr and L. Banaszak (eds.): *Lipid Binding Proteins within Molecular and Cellular Biochemistry*. 1999 ISBN 0-7923-8223-4
30. R. Alvarez-Gonzalez (ed.): *ADP-Ribosylation Reactions: From Bacterial Pathogenesis to Cancer*. 1999 ISBN 0-7923-8235-8
31. S. Imai and M. Endo (eds.): *Muscle Physiology and Biochemistry*. 1999 ISBN 0-7923-8264-1

KLUWER ACADEMIC PUBLISHERS – DORDRECHT / BOSTON / LONDON